ADVANCES IN CHEMICAL PHYSICS

VOLUME XCVI

EDITORIAL BOARD

Advances in CHEMICAL PHYSICS

Edited by

I. PRIGOGINE

Center for Studies in Statistical Mechanics and Complex Systems
The University of Texas
Austin, Texas
and
International Solvay Institutes
Université Libre de Bruxelles
Brussels, Belgium

and

STUART A. RICE

Department of Chemistry
and
The James Franck Institute
The University of Chicago
Chicago, Illinois

VOLUME XCVI

AN INTERSCIENCE® PUBLICATION
JOHN WILEY & SONS, INC.
NEW YORK • CHICHESTER • BRISBANE • TORONTO • SINGAPORE

This text is printed on acid-free paper.

An Interscience® Publication

Library of Congress Catalog Number: 58-9935

ISBN 0-471-15652-3

Printed in the United States of America

10 9 8 7 6 5 4 3 2 1

CONTRIBUTORS TO VOLUME XCVI

ROBERT J. GORDON, Department of Chemistry, University of Illinois at Chicago, Chicago, Illinois

FRANK GROSSMAN, Universität Freiburg, Fakultät für Physik, Freiburg, Germany

GREGORY E. HALL, Department of Chemistry, Brookhaven National Laboratory, Upton, New York

MARSHA I. LESTER, Department of Chemistry, University of Pennsylvania, Philadelphia, Pennsylvania

VINCENT MCKOY, A.A. Noyes Laboratory of Chemical Physics, California Institute of Technology, Pasadena, California

MIGUEL ANGEL SEPÚLVEDA, Institute for Fundamental Chemistry, Kyoto, Japan

CARL WINSTEAD, A.A. Noyes Laboratory of Chemical Physics, California Institute of Technology, Pasadena, California

INTRODUCTION

Few of us can any longer keep up with the flood of scientific literature, even in specialized subfields. Any attempt to do more and be broadly educated with respect to a large domain of science has the appearance of tilting at windmills. Yet the synthesis of ideas drawn from different subjects into new, powerful, general concepts is as valuable as ever, and the desire to remain educated persists in all scientists. This series, *Advances in Chemical Physics*, is devoted to helping the reader obtain general information about a wide variety of topics in chemical physics, a field that we interpret very broadly. Our intent is to have experts present comprehensive analyses of subjects of interest and to encourage the expression of individual points of view. We hope that this approach to the presentation of an overview of a subject will both stimulate new research and serve as a personalized learning text for beginners in a field.

I. PRIGOGINE
STUART A. RICE

CONTENTS

APPLICATIONS OF DOPPLER SPECTROSCOPY TO PHOTOFRAGMENTATION

ROBERT J. GORDON

Department of Chemistry, University of Illinois at Chicago, Chicago, IL 60607

GREGORY E. HALL

Department of Chemistry, Brookhaven National Laboratory, Upton, NY 11973

CONTENTS

I. INTRODUCTION

Doppler absorption spectroscopy is a powerful tool for studying the elementary reaction dynamics of atoms and molecules. If a particle moving

Advances in Chemical Physics, Volume XCVI, Edited by I. Prigogine and Stuart A. Rice.
ISBN 0-471-15652-3

with velocity **v** absorbs a photon, the absorption frequency is shifted by an amount $\Delta\nu$,

$$\Delta\nu/\nu_0 = w/c, \tag{1.1}$$

where ν_0 is the absorption frequency of the stationary particle, w is the component of **v** along the line of sight, and c is the speed of light. When the frequency resolution of the measurement is sufficient, a mere glance at the absorption profile will give the maximum speed of the particle. From a more careful analysis of the shape of the Doppler profile (or of several profiles taken under selected conditions), it is possible to deduce the entire *distribution* of speeds and recoil angles of the particle. For molecules with preferential alignment of their rotation axes, Doppler spectroscopy is also sensitive to this distribution of angular momentum directions. Many problems in the study of gas phase reactions can be addressed by such a technique, and the application of Doppler spectroscopy to molecular reaction dynamics is now a mature subject. A number of previous review articles treat different aspects of the field [1–4].

The goals of the present chapter are twofold. The first is purely pedagogic. We present here an elementary derivation of many of the key classical and semiclassical results needed by the experimentalist for the interpretation of atomic and molecular Doppler spectra. Much of this material is scattered in the literature, and we felt that a unified treatment would be useful to newcomers to the field. Second, we review a number of selected topics, such as multiphoton effects and anisotropy in rotationally resolved predissociation, which may not have been covered adequately in the review literature. The remainder of this section presents an overview of Doppler spectroscopy as a probe of photofragment anisotropy, translational energy distributions, and vector correlations in photofragmentation and photoinduced bimolecular reactions. Sections II–V deal with the Doppler profiles of atomic fragments produced by single-photon molecular photodissociation. In Section II, the angular distribution is derived for an atom recoiling with a fixed kinetic energy. The Doppler profile measured along an arbitrary laboratory angle is derived in Section III. In Section IV, rotation of the parent molecule is included, still assuming a constant kinetic energy of the fragment. In Section V, a distribution of recoil speeds is included, and methods for inverting the Doppler profile to obtain the speed distribution function are discussed. In Section VI, the previous results are generalized to deal with absorption of more than one photon. In Section VII, the earlier results are further generalized to treat the important case of molecular photofragments, where correlation between the velocity and rotational angular momentum vectors is present.

It may at first seem surprising that it is possible to extract a three-dimensional distribution function from a one-dimensional projection of the velocity. Nevertheless, research over the past 30 years has produced an array of subtle techniques that allow such information to be extracted from Doppler profiles. Zare and Herschbach [5] were the first to show that the angular distribution of atomic fragments produced in the photodissociation of a diatomic molecule may be deduced from the Doppler line shape of the fragment. This result stems from the fact that the probability that the parent molecule absorbs a photon is proportional to $\cos^2 \theta$, where θ is the angle between the electric vector **E** of the photolysis source and the transition dipole moment $\boldsymbol{\mu}$ of the molecule. For a diatomic molecule, $\boldsymbol{\mu}$ must point either along the molecular axis or perpendicular to it. In the first case, the fragment atoms have a $\cos^2 \theta$ angular distribution, and if the atoms are probed in the same direction as **E,** the Doppler profile will have a *maximum* at its center. In the second case, the fragments have a $\sin^2 \theta$ distribution and the Doppler profile has a *minimum* at the center.

Since the fragments of a diatomic molecule have fixed speeds (determined by the dissociation energy of the molecule, the frequency of the absorbed photon, and the atomic energy levels populated), the angular distribution may be determined uniquely from the shape of the Doppler profile. These angular distributions of photoproducts were first verified in the classic "photolysis mapping" experiment of Solomon [6], in which recoiling atoms from Br_2 and I_2 reacted on the surface of a bulb coated with a Tellurium film. Shortly after that, Busch and Wilson [7] and Diesen et al. [8] established the field on a quantitative footing by using a mass spectrometer to detect fragment atoms in a molecular beam machine. A new generation of experiments taking advantage of Doppler spectroscopy to probe molecular photofragmentation began when Welge [9] used Lyman-α Doppler spectroscopy to probe the H photofragments from single-photon dissociation of HI.

If the parent molecule contains more than two atoms, the problem is complicated by the fact that some of the available energy may be deposited into internal degrees of freedom of the molecular fragments, and the Doppler profile will depend on both the speed and the angular distributions of the detected particle. The two distribution functions may be recovered from the Doppler profile if their effects can be somehow uncoupled from one another. One way of accomplishing this is to measure the Doppler profile at a "magic angle," at which the angular contribution for an atomic fragment (or more generally, for any fragment lacking rotational polarization) drops out [10]. The speed distribution can then be determined uniquely from the derivative of the Doppler line shape [11].

The resolution of the speed distribution function can be dramatically enhanced by aligning the photolysis and detection sources coaxially and introducing a delay between the pump and probe pulses. The effect of this delay is to allow particles not traveling along the line of sight to fly out of the field of view, making this Doppler measurement resemble a one-dimensional core through the three-dimensional velocity distribution. Wittig and coworkers [12] first applied this technique to the photodissociation of H_2S and determined the SH fragment vibrational state populations from the H atom Doppler profiles.

Much of the early history of Doppler spectroscopy of photofragments dealt with measuring atomic velocity distributions. As experimental attention turned to *molecular* fragments, it soon became clear that the Doppler spectra depend not only on the fragment speed and angular distributions, but also on rotational polarization of the fragment and its correlation with recoil velocity [13–17]. It had long been known that the determination of state populations by optical spectroscopy required corrections of the measured intensities for rotational alignment effects [18–20]. Considering the rotational alignment as an explicit function of the Doppler shift leads to a quantitative treatment of the vector correlations between the parent molecule transition dipole moment, $\boldsymbol{\mu}$, the fragment velocity, $\mathbf{v}$, and the fragment rotational angular momentum, $\mathbf{J}$. These molecular vector correlations can be determined by comparing the Doppler profiles in several experimental geometries for transitions belonging to different rotational branches probing the same product state. From such measurements it is possible to obtain information about the excited state symmetry and about exit channel torques and geometries that cannot be readily gained by any other method. Several reviews have focussed on the Doppler spectroscopy of vector correlations in photodissociation [1].

Applications to bimolecular reactions have taken a parallel track. It was first shown by Kinsey [11] how a set of Doppler profiles of product molecules in the interaction region of a crossed molecular beam apparatus could be analyzed to yield differential cross sections. Early experiments of this type were performed in the late 1970s and early 1980s [21, 22]. A renewed interest in Doppler analysis of crossed beam reactions has occurred more recently [23–25]. Doppler methods in double-resonance experiments have also been creatively employed to study collisional processes in molecules with Doppler-selected velocities, notably by MIT groups [26, 27] and by McCaffery and coworkers [28]. Velocity-dependent inelastic cross sections, both total and differential, can be inferred from a close analysis of the polarized double-resonance experiments with two high-resolution continuous lasers, or a single high-resolution laser and polarization-sensitive dispersed fluorescence detection.

The success of Doppler techniques to characterize correlated velocity and angular momentum distributions of state-selected photoproducts led several groups to consider the new information that might be obtained from such pump-probe experiments by merely waiting for a collision. Several variations of such a ''stretched'' pump-probe experiment are possible. The sequence of inelastic collisions in a buffer gas that reduces the initial photofragment velocity distribution to a thermal, isotropic one was investigated by recording the evolution of Doppler profiles at increasing collision number for light [29] and heavy [10] atomic photofragments. A more cleanly isolated single inelastic collision process can instead be studied by probing the Doppler profiles of the initially cold target molecules, following excitation to a previously unpopulated state by collision with a fast and anisotropic photofragment. Early experiments of this type were performed for inelastic scattering of H + CO_2 [30] and H + CO [31], using photolytically produced H atoms. It was recognized at this time that the Doppler profiles of the excited target molecules were a sensitive function of the differential cross sections, even in a bulb experiment. Since the laboratory-frame momentum transfer is a direct measure of the scattering angle for these nearly elastic collisions, one can see that small-angle scattering leaves the targets with nearly unchanged laboratory velocities, while backward scattering will impart a maximum velocity parallel to that of the photofragment projectile. Sideways scattering will impart an intermediate laboratory velocity, having an anisotropy that is of opposite sign to that of the photofragment. Interpretation of the laboratory velocity distribution, as probed by Doppler spectroscopy, can thus give a low-resolution view of the differential cross section, based on pump-probe measurements in a bulb. These kinematic arguments have been formalized by Shafer et al. [32] for arbitrary collision energetics and mass combinations.

Reactive variations on this theme were also developed, following the scheme:

$$AX + h\nu \longrightarrow A + X$$

$$A + BC \longrightarrow AB + C$$

with Doppler probing of reaction product AB or C [33–46]. The field made a major advance beyond simple considerations of velocity distributions as workers in the laboratories of Hancock [36, 37, 44] and Simons [36, 38, 40–42] generalized the bipolar moment formalism of Dixon [15] to include the rotational polarization effects in the Doppler spectroscopy of photoinitiated bimolecular reaction products. This field has developed rapidly in recent years and is treated in excellent recent reviews by Orr-Ewing and Zare [2, 3] and by Brouard and Simons [4].

Many features of Doppler spectroscopy are shared by other techniques that probe the correlated velocity and angular momentum distribution of a state-selected photofragment or reaction product. Resonance-enhanced multiphoton ionization, followed by a time-of-flight (REMPI-TOF) analysis can be performed, with deliberately degraded mass resolution, to map the fragment velocity component along the flight axis onto the flight time [47–51]. Collecting the entire sample of ions on a large planar detector integrates over the transverse velocity components, in a way entirely analogous to conventional Doppler spectroscopy. With a detector smaller than the transverse spread of the ion cloud, the TOF signals approach a one-dimensional core through the velocity distribution, resembling the velocity-aligned Doppler spectroscopy (VADS) technique instead of the conventional one-dimensional projection [49]. Recent work in Zare's lab has shown that two-color ionization with beams orthogonal to each other and to the flight direction allows independent Doppler selection of **x** and **y** velocity components, followed by TOF analysis of the **z** velocity [45]. The effect is similar to that of a flight tube with an optically controlled aperture, whose size that can be adjusted with the laser bandwidths and which can be positioned on two axes by independent tuning of the two lasers. Finally, we note that ion imaging methods, as pioneered by Chandler and Houston [52], can be considered as two-dimensional analogues of Doppler spectroscopy that record the projection of the three-dimensional velocity distribution onto a plane instead of a line and are complete with analogous rotational polarization effects. The full three-dimensional distribution can be obtained by a hybrid technique introduced by Kinugawa and Arikawa [53], who used planar imaging to obtain the two transverse components of the velocity distribution and Doppler spectroscopy for the component along the flight direction. In a related technique, Ni et al. [54] aligned the probe laser parallel to the photolysis laser, with a fixed delay between the two pulses. With this technique the transverse components of the velocity distribution are obtained by translating the probe beam relative to the photolysis beam, while the third component is again obtained by Doppler spectroscopy. The translated-beam technique was used by Sanov et al. [55] to measure the transverse velocity of NO produced in the photolysis of NO_2, with sufficient resolution to distinguish between different $O(^3P_J)$ states.

II. ANGULAR DISTRIBUTION OF PHOTOFRAGMENTS[1]

Consider a molecule having a transition dipole moment $\boldsymbol{\mu}$ pointing along the body-fixed $\hat{\mathbf{z}}$-axis. Suppose that this molecule is irradiated by linearly polarized light with an electric field **E** pointing along the space-fixed $\hat{\mathbf{Z}}$-axis. When the molecule dissociates, a fragment A recoils along direction $\hat{\mathbf{a}}$ in the body-fixed frame. For simplicity, we will assume initially that only one

[1] This section follows closely the derivation in Application 6 of ref. [56].

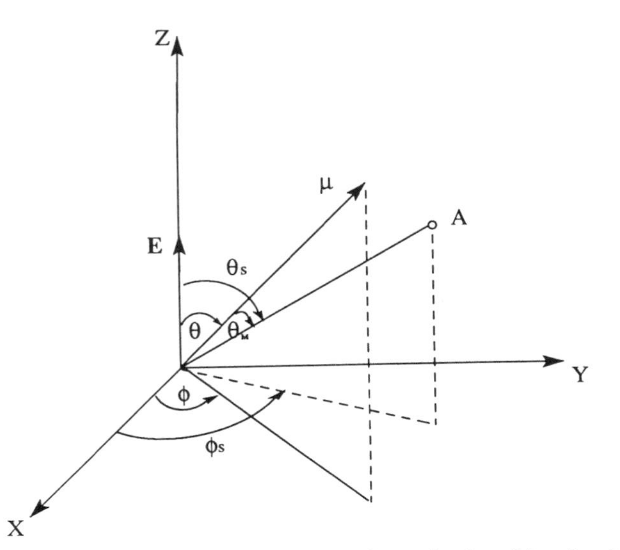

Figure 1. Laboratory-fixed and body-fixed coordinates for describing the photodissociation of a molecule. The electric field vector **E** points along the laboratory $\hat{\mathbf{Z}}$-axis, while the transition dipole moment **μ** points along the body-fixed $\hat{\mathbf{z}}$-axis. The fragment, A, has a velocity described by laboratory angles θ_s and ϕ_s.

photon is absorbed and that the detected fragment is an atom. We define the following sets of angles (see Fig. 1):

θ, ϕ, χ are Euler angles transforming $\hat{\mathbf{Z}}$ into $\hat{\mathbf{z}}$.

θ_S, ϕ_S, and χ_s are Euler angles transforming $\hat{\mathbf{a}}$ into **E**, defining the recoil direction in the space-fixed frame.

θ_M and ϕ_M are the angles between $\hat{\mathbf{a}}$ and **μ**, defining the recoil direction in the molecular frame.

The value of θ_M is determined by the symmetries of the initial and final states. For parallel transitions $\theta_M = 0$, while for perpendicular transitions $\theta_M = \pi/2$. More generally, the probability of recoil at an arbitrary value of θ_M is given by

$$f_{a,\mu}(\theta_M, \phi_M) = f_M(\theta_M) = \sum_l (2l + 1)C_l P_l(\cos\theta_M) \tag{2.1}$$

where the coefficients are given by the projections

$$C_l = \tfrac{1}{2}\int_{-1}^{1} P_l(\cos\theta_M)\, f_M(\theta_M)\, d\cos\theta_M \tag{2.2}$$

For a dipole transition, the probability of a molecule absorbing a photon is proportional to $\cos^2\theta$,

$$f_{E,\mu}(\theta) = \frac{1}{4\pi}[P_0 + 2P_2(\cos\theta)] \tag{2.3}$$

The molecules are initially oriented isotropically in space, and we assume for now that the dissociation is instantaneous. The probability of the fragment recoiling at laboratory angle θ_S is given by an integral of $f_{E,\mu}$ times f_M over all θ and ϕ,

$$f_{E,a}(\theta_S) = \frac{1}{4\pi}\int_0^{\pi}\int_0^{2\pi} f_{E,\mu}(\theta)\, f_M(\theta_M)\sin\theta\, d\theta\, d\phi \tag{2.4}$$

Using the spherical harmonic addition theorem [56], we may replace θ_M in Eq. (2.1) by θ, ϕ, θ_s, and ϕ_s to get the expression

$$f_M = \frac{1}{2}\sum_{l=0}^{\infty}(2l+1)C_l\frac{4\pi}{2l+1}\sum_{m=-l}^{l} Y^*_{l,m}(\theta,\phi)Y_{l,m}(\theta_S,\phi_S) \tag{2.5}$$

By substituting Eqs. (2.3) and (2.5) into Eq. (2.4) and integrating over ϕ, we project out the $m = 0$, $l = 0, 2$ terms. Finally, integrating over θ, we obtain

$$f_{E,a}(\theta_S) = \tfrac{1}{2}[C_0P_0 + 2C_2P_2(\cos\theta_S)] \tag{2.6}$$

Setting $C_0 = \sigma/2\pi$, where σ is the total cross section, we obtain for $\sigma = 1$

$$f_{E,a}(\theta_S) = \frac{1}{4\pi}[P_0 + \beta P_2(\cos\theta_S)] \tag{2.7}$$

where the anisotropy parameter β is given by

$$\beta = 4\pi C_2 = 4\pi\int_{-1}^{1} P_2(\cos\theta_M)\, f_M(\theta_M)d\cos\theta_M \tag{2.8}$$

In the special case of axial recoil at an angle θ_M^0,

$$f_M(\theta_M) = \frac{1}{2\pi}\delta(\cos\theta_M - \cos\theta_M^0) \tag{2.9}$$

and

$$\beta = 2P_2(\cos \theta_M^0) \tag{2.10}$$

For a parallel transition, $\beta = 2$ and $f_{E,a}(\theta_s) \propto \cos^2 \theta_s$, whereas for a perpendicular transition, $\beta = -1$ and $f_{E,a}(\theta_s) \propto \sin^2 \theta_s$.

III. DOPPLER PROFILE AT AN ARBITRARY DETECTION ANGLE

Equation (2.7) gives the angular distribution of a fragment A recoiling at an angle θ_S with respect to the polarization direction $\hat{\mathbf{Z}}$ of the photolysis laser. This is directly related to the Doppler profile for the probe beam propagating along $\hat{\mathbf{z}}$. For Doppler analysis along an arbitrary direction $\hat{\mathbf{k}}$, we need to re-express Eq. (2.7) in terms of a new set of angles, related to the direction of the probe beam (see Fig. 2):

θ_v, ϕ_v are the angles between the probe direction $\hat{\mathbf{k}}$ and the recoil direction $\hat{\mathbf{a}}$.

θ_a is the angle between $\hat{\mathbf{Z}}$ and $\hat{\mathbf{k}}$ (i.e., between the pump polarization and the Doppler analysis beam propagation directions).

Invoking the spherical harmonic addition theorem a second time to replace θ_S by θ_a and θ_v, we obtain from Eq. (2.7)

$$f_{a,k}(\theta_v, \phi_v) = \frac{1}{4\pi}\left[P_0 + \frac{4\pi}{5}\beta \sum_{m-2}^{m} Y_{2,m}^*(\theta_a, \phi_a)\, Y_{2,m}(\theta_v, \phi_v)\right] \tag{3.1}$$

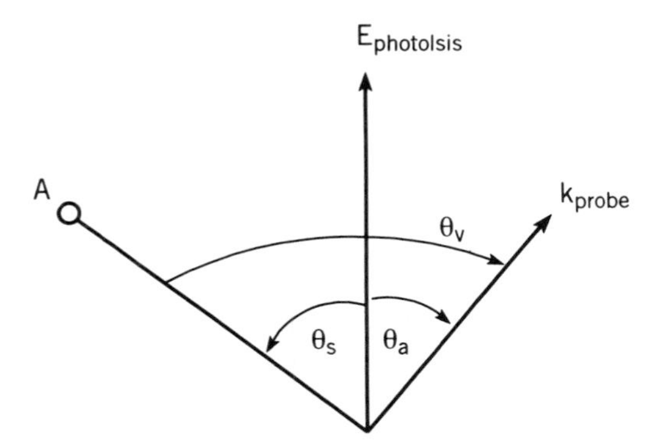

Figure 2. Transformation from the angular coordinates of the velocity of fragment A to the direction of the probe laser, $\hat{\mathbf{k}}$.

Suppose the fragment recoils along direction $\hat{\mathbf{a}}$ with velocity $\mathbf{v}$. Let w be the component of $\mathbf{v}$ along the probe direction $\hat{\mathbf{k}}$, such that

$$w = v \cos \theta_v \tag{3.2}$$

We multiply Eq. (3.1) by $\delta(w - v \cos \theta)$ and integrate it over ϕ_v and θ_v. The integral over ϕ_v projects out the $m = 0$ term, while the second integral replaces $\cos \theta_v$ by w/v, where $-1 \leq w/v \leq 1$. The result is the number density of fragments moving along the line of sight with speed w (apart from a normalization constant),

$$\begin{aligned} D(w) &= \frac{1}{4\pi} \int_0^{2\pi} \int_0^{\pi} f_{a,k}(\theta_v, \phi_v)\, \delta(w - v \cos \theta_v) \sin \theta_v \, d\theta_v \, d\phi_v \\ &= \frac{1}{2v} [1 + \beta_{\text{eff}} P_2(w/v)] \end{aligned} \tag{3.3}$$

where

$$\beta_{\text{eff}} = \beta P_2(\cos \theta_a) \tag{3.4}$$

and β is given by Eq. (2.10). Equation (3.3) is the Doppler profile, where the frequency shift, $\Delta\nu = \nu - \nu_0$, is related to w by Eq. (1.1). If the fragment atoms have a distribution of speeds, $v^2 f(v)$, the Doppler profile is obtained by averaging Eq. (3.3) over the distribution function, giving

$$D(w) = \tfrac{1}{2} \int_{|w|}^{\infty} [1 + \beta_{\text{eff}} P_2(w/v)]\, f(v)\, v\, dv \tag{3.5}$$

Equation (3.5) has the property of transforming a Boltzmann speed distribution, that is, a Gaussian function $f(v)$, into a Gaussian $D(w)$, whenever $\beta_{\text{eff}} = 0$.

IV. ROTATION OF THE PARENT MOLECULE

Rotation of the parent molecule reduces the anisotropy by averaging the recoil direction over a range of angles. Even if the molecule dissociates instantaneously, β will still be affected because the rotational velocity of the parent molecule adds vectorially to the axial velocity of the recoiling fragment. If the excited molecule lives for an appreciable fraction of a rotational period before dissociating, the anisotropy will be further decreased.

The angular distribution of the photofragments of a diatomic molecule was derived classically by Busch and Wilson [7], semiclassically by Jonah

[57], and quantum mechanically by Mukamel and Jortner [58]. Following the treatment of Busch and Wilson, we consider an angular velocity of the molecule $\boldsymbol{\omega}$ and a rotational velocity of the fragment $\mathbf{v}_r$.

$$\mathbf{v}_r = \boldsymbol{\omega} \times \mathbf{r} \tag{4.1}$$

where $\mathbf{r}$ is the radial vector connecting the center of mass of the parent molecule to the fragment A. The recoil velocity $\mathbf{v}$ is the vector sum of the axial recoil velocity $\mathbf{v}_0$ and the orthogonal rotational velocity $\mathbf{v}_r$,

$$\mathbf{v} = \mathbf{v}_0 + \mathbf{v}_r \tag{4.2}$$

The resultant $\mathbf{v}$ lies on a cone of half angle α, where

$$\tan \alpha = \mathbf{v}_r/\mathbf{v}_0 \tag{4.3}$$

as shown in Fig. 3. If we assume that $\mathbf{v}$ is distributed uniformly on this cone, then the effect of rotation on the Doppler profile is readily calculated. We define the following sets of angles:

θ_M and ϕ_M define the angle between $\mathbf{v}$ and $\boldsymbol{\mu}$.
θ_M^0 is the angle between $\mathbf{v}_0$ and $\boldsymbol{\mu}$.
α and ϕ_α define the angle between $\mathbf{v}$ and $\mathbf{v}_0$.

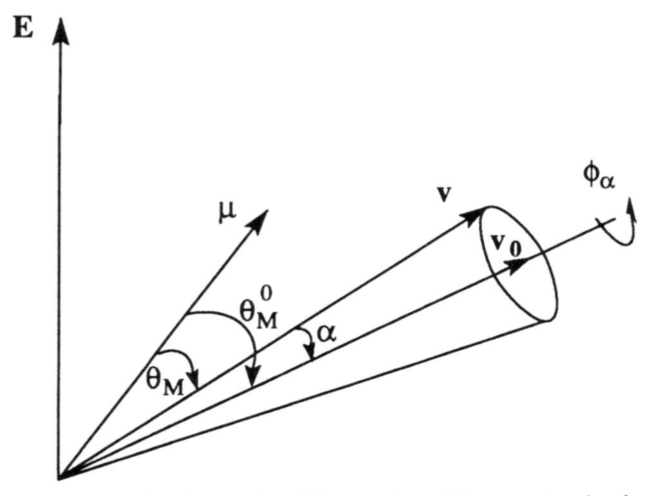

Figure 3. Coordinates describing rotation of the parent molecule.

For a particular value of ϕ_α, Eq. (2.10) gives

$$\beta = 2P_2(\cos \theta_M) \tag{4.4}$$

By invoking once again the spherical harmonic addition theorem, where θ_M is the dihedral angle between (α, ϕ_α) and $(\theta_M^0, \phi_M = 0)$, we obtain

$$\beta = \frac{8\pi}{5} \sum_{m=-2}^{m} Y^*_{2,m}(\theta_M^0 0)\, Y_{2,m}(\alpha, \phi_\alpha) \tag{4.5}$$

and averaging over ϕ_α we get

$$\beta(\theta_M^0, \alpha) = 2P_2(\cos \theta_M^0) P_2(\cos \alpha) \tag{4.6}$$

Consider next what happens if the molecule lives for a time t before dissociating. In this case the fragment is deflected by an angle $\alpha' = \alpha + \omega t$. For unimolecular decay, the probability of the molecule surviving for a time between t and $t + dt$ is given by

$$p(t)\, dt = \frac{1}{2\tau} e^{-t/\tau}\, dt \tag{4.7}$$

where τ is the average lifetime of the molecule. The time-averaged anisotropy parameter is therefore given by

$$\begin{aligned} \beta(\theta_M^0, \alpha, \omega\tau) &= 2P_2(\cos \theta_M^0) \int_0^\infty P_2[\cos(\alpha + \omega\tau)] p(t)\, dt \\ &= 2P_2(\cos \theta_M^0)\, g(\alpha, \omega\tau) \end{aligned} \tag{4.8}$$

and the effective anisotropy parameter is defined as [59]

$$\beta_{\text{eff}} = 2P_2(\cos \theta_a)\, P_2(\cos \theta_M^0)\, g(\alpha, \omega\tau) \tag{4.9}$$

where the integral in Eq. (4.8) gives

$$g(\alpha, \omega\tau) = \frac{P_2(\cos \alpha) + (\omega\tau)^2 - 3\omega\tau \sin \alpha \cos \alpha}{1 + 4(\omega\tau)^2} \tag{4.10}$$

In the limit of small α, g has the value

$$g = \frac{1 + (\omega\tau)^2}{1 + 4(\omega\tau)^2} \tag{4.11}$$

which ranges from $\frac{1}{4}$ for $\omega\tau >> 1$, to 1 for $\omega\tau = 0$. In the long lifetime limit, $\beta = \frac{1}{2}$ for a parallel transition and $-\frac{1}{4}$ for a perpendicular transition.

The assumption that **v** is distributed uniformly on a cone is equivalent to neglecting the correlation between **μ** and the rotational angular momentum **J**. This treatment will be valid of the molecule dissociates (or is predissociated) rapidly enough that the rotational structure of the upper state is unresolved, or if the photolysis source is broad enough to excite all allowed rotational levels. Yang and Bersohn [60] extended the classical treatment to polyatomic molecules, using thermally averaged rotational autocorrelation functions. Extreme prolate and oblate symmetric tops behave in the same way as diatomic molecules, but $g(\tau \rightarrow \infty)$ only varies between 0.25 and 0.20 within these limits. Mukamel and Jortner [58] showed that the quantum mechanical result is identical to the classical result in two limits. Direct dissociation of a $^1\Sigma$–$^1\Sigma$ transition was shown to give the classical $\cos^2 \theta_s$ distribution derived in Section II. In the limit of broad-band excitation and slow predissociation, the fourfold reduction in anisotropy also results. The case of rotationally resolved predissociation was not considered by Mukamel and Jortner [58], and gives qualitatively different, nonclassical behavior. It is the coherent preparation of multiple J states in the predissociating molecule that generates the time evolution of molecular frame orientation with a classical correspondence. The slow and fast dissociation limits are distinguished by the relative magnitudes of the energy spacing between successive J levels of the dissociating state and by the lifetime-broadened widths of those states. When the lifetime width is greater than the rotational spacing, Mukamel and Jortner's result reduces to the classical limit for direct dissociation ($\cos^2 \theta$ for parallel transitions and $\sin^2 \theta$ for perpendicular transitions); on the other hand, if the lifetime width is narrower than the rotational spacing, their theory gives the slow dissociation result.

This classical model must fail for predissociation with rotational resolution. Consider the relative orientations of **μ**, **J**, and the molecular axis shown in Fig. 4. In the high-J limit of a Q-branch transition, for example, the molecule rotates in a plane perpendicular to **μ** (i.e., **μ** is parallel to **J**), and the anisotropy of the fragments is independent of τ (contrary to the assumption of the classical model). P- and R-branch transitions produce molecules rotating in a plane containing **μ**, and the $\tau \rightarrow 0$ and $\tau \rightarrow \infty$ limits differ. We should emphasize that if the rotational branches can be resolved, the $\tau \rightarrow \infty$ limit has necessarily been closely approached, and no further information about the predissociation lifetime is contained in the observed anisotropy parameter. The persistence of the anisotropy for a long-lived state is illustrated by the dissociation of the quasibound $B^1\Pi_u$ state of Na_2 [61].

A semiclassical treatment of the anisotropy in the long lifetime, narrow-band excitation limit is as follows [62]: The axial recoil limit (i.e., $v_0 >> v_r$) is equivalent to identifying the laboratory-fixed recoil direction with the

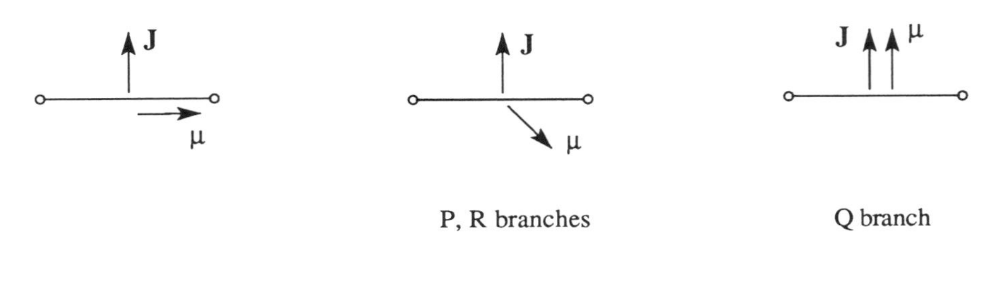

|| Transition ⊥ Transition

Figure 4. Effect of parent molecule rotational alignment on the anisotropy of the Doppler profile. For a parallel transition **μ**, is perpendicular to **J**, and rotation of the parent molecule diminishes the spatial anisotropy of the fragments. For P- and R-branch transitions in the high-**J** limit, **μ** is again perpendicular to **J**, and rotation in the plane containing **μ** also diminishes the anisotropy. In contrast, in the Q-branch transitions in the high-**J** limit, the plane of rotation is perpendicular to **μ**, and the anisotropy does not change with time.

body-fixed direction of the bond axis at the time of dissociation [58]. In a state of well-defined angular momentum, the probability of such body-fixed molecular orientations is precisely the squared rotational wave function, in its orientation representation [57]. For a symmetric top undergoing a transition from a state with initial quantum numbers J'', K'', and M'' to a final state with quantum numbers J', K', M' (or for a diatomic molecule with Λ equal to K″ and K′), the probability of recoiling at angle θ_s is given by

$$f_{E,a}(\theta_S) = \frac{2J' + 1}{8\pi} \sum_{M'',M'} P(J', K', M'; J'', K'', M'')\, |D^{J'}_{M',K'}(\theta_s, \phi_s, \chi_s)|^2 \tag{4.12}$$

In this expression, $P(J', K', M'; J'', K'', M'')$ is the transition probability and $[(2J' + 1)/8\pi]^{1/2}\, D^{J'}_{M',K'}$ is the rotational wave function of the discrete upper level. For absorption of a single photon, Eq. (4.12) is readily shown [62] to become

$$f_{E,a}(\theta_S) \propto \langle J''K''1q|J'K'\rangle^2 \sum_{M'',M'} \langle J''M''10|J'M'\rangle^2 |D^{J'}_{M',K'}(\theta_s, \phi_{s'}\, \chi_s)|^2 \tag{4.13}$$

where $(2J' + 1)\, \langle J''K''1q|J'K'\rangle^2$ is the Hönl–London factor, and $q = -1$, 0, or 1. This expression is evaluated to give

$$f_{E,a}(\theta_s) \propto \langle J''K''1\ 1|JK'' + 1\rangle^2 \, |D^1_{10}|^2 + \langle J''K''1\ 0|JK''\rangle^2 \, |D^1_{00}|^2$$
$$+ \langle J''K''1 - 1|JK - 1''\rangle^2 \, |D^1_{-10}|^2$$
$$\propto 1 + \beta \, P_2(\cos\theta_s) \tag{4.14}$$

For a P-branch transition the anisotropy parameter β has the value

$$\beta = \frac{J''(J'' - 1) - 3\,K^2}{J''(2J'' + 1)} \tag{4.15}$$

while for a Q-branch transition it has the value

$$\beta = -\frac{J''(J'' + 1) - 3K^2}{J''(J'' + 1)} \tag{4.16}$$

and for an R-branch transition it has the value

$$\beta = \frac{(J'' + 1)\,(J'' + 2) - 3K^2}{(J'' + 1)\,(2J'' + 1)} \tag{4.17}$$

In the limit of very large J'', $\beta = \frac{1}{2}$ for P- and R-branches of both parallel and perpendicular transitions, and $\beta = -1$ for a Q-branch transition. If β were written as a rotationally depolarized anisotropy parameter, as in Eq. (4.8), the $\tau \rightarrow \infty$ limit of g would appear to be $\frac{1}{4}$, $-\frac{1}{2}$, or 1, depending on the transition and branch excited, rather than $\frac{1}{4}$, independent of the type of excitation.

A more complex polyatomic example, H_2CO, has been studied by Moore and coworkers [63]. They used a similar formulation to explain the H_2 fragment anisotropy, which depends on the excited asymmetric top state of the parent molecule and the rotational transition used to access that state, but clearly has no relation to the classical $g(\tau \rightarrow \infty)$ factor. The frequent observation of isotropic fragmentation, commonly attributed to "rotational averaging," actually requires a better explanation than the $\tau \rightarrow \infty$ limit, as we have seen here.

In molecular systems with more than a single potential energy surface leading to the same final states, interference effects can further influence both the velocity anisotropy and the polarization of the fragments. While they are beyond the scope of the present chapter, we alert the reader to possible deviations from the simple classical and semiclassical results collected here, and cite some representative papers [64–69].

V. DETERMINATION OF SPEED DISTRIBUTION FUNCTIONS

A. Magic Angle Doppler Spectroscopy

The kinetic energy distribution of the recoiling fragments in a photodissociation reaction is useful for determining the dissociation mechanism. For example, if the internal state of one of the fragments is known, the internal energy distribution of the other fragment can be calculated from the distribution of relative kinetic energies. We wish to obtain this distribution function by inverting the Doppler profile of the state-selected fragment. An inherent difficulty associated with Doppler spectroscopy is that it provides only the projection of **v** along the line of sight. A method is therefore required to reconstruct the speed distribution function $f(v)$ from the observed distribution of w. From Eq. (3.5) it is apparent that if β were zero, $f(v)$ could be determined directly by differentiating $D(w)$ with respect to w, that is,

$$f(v) = -\frac{2}{v}\frac{\partial D(w)}{\partial w}\bigg|_{|w|=v} \tag{5.1}$$

One way to eliminate the term containing β is to measure the Doppler profile twice, once with $\hat{\mathbf{k}}$ parallel to **E** to obtain $D_{\parallel}$, and once with $\hat{\mathbf{k}}$ perpendicular to **E** to obtain $D_{\perp}$ [70]. In the linear combination $\frac{1}{2}D_{\parallel} + D_{\perp}$, the β term then drops out. A better technique, known as magic angle Doppler spectroscopy (MADS), is to measure a single profile at the magic angle $\theta_a = \cos^{-1}(3^{-1/2}) = 54.74°$, which guarantees that $\beta_{\text{eff}} = 0$. This method has been used by Cline et al. [10] to determine the speed distribution of $I(^2P_{1/2})$ produced by the photodissociation of n-C_3F_7I, and by Huang et al. [71] to measure $Cl(^2P_{1/2})$ and $Cl(^2P_{3/2})$ obtained from C_2H_3Cl and $C_2H_2Cl_2$.

Both of these methods are limited by the requirements of a very high signal-to-noise ratio, which is needed to calculate the derivative of the Doppler profile. This limitation is illustrated by the following numerical example [71]. Consider the model distribution function,

$$f(v) = Ce^{-(v-v_0)^2/\sigma^2} \tag{5.2}$$

where v_0 is the most probable speed, σ is the width of the distribution, and C is a normalization constant. Plots of $D(w)$ for various values of v_0/σ are shown in Fig. 5(a). We see that for $v_0/\sigma < 1$ the shape of $D(w)$ is not very sensitive to v_0, since both v_0 and σ contribute to the width of the profile. With noisy data it would be difficult to determine from the shape of the Doppler profile whether $f(v)$ peaks at zero speed.

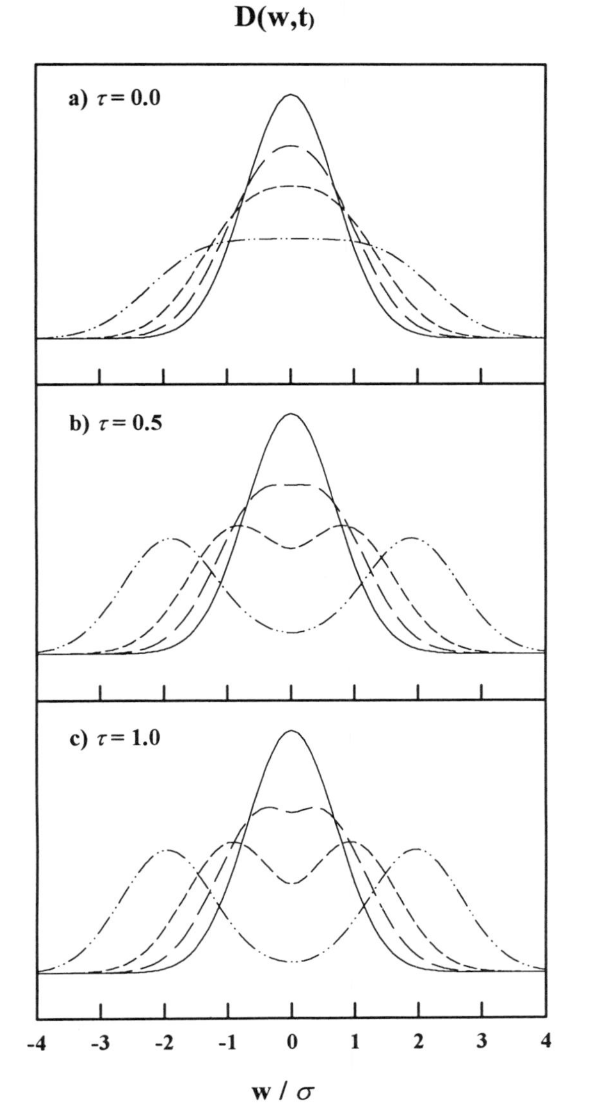

Figure 5. Velocity-aligned Doppler profiles for the model distribution function, Eq. (5.2), with different values of the reduced most probable speed v_0/σ. (*a*) Zero delay between the pump and probe. (*b*, *c*) Reduced delays of $\tau = 0.5$ and 1.0, respectively, where $\tau = \alpha t/\rho$, α is the average thermal speed of the parent molecule, and ρ is the RMS average Gaussian radius of the two laser beams.

Useful information may be extracted from the weighted integral of the magic angle Doppler profile, which is much less sensitive to noise than its derivative. In particular, it is possible to obtain the nth moment of the distribution function [72],

$$\langle v^n \rangle = \frac{4\pi}{N} \int_0^\infty f(v) v^{n+2}\, dv \tag{5.3}$$

where N is a normalization constant,

$$N = 4\pi \int_0^\infty v^2 f(v)\, dv \tag{5.4}$$

Substituting Eq. (5.1) into Eqs. (5.3) and (5.4) and integrating by parts, we obtain

$$\langle v^n \rangle = \frac{(n+1) \displaystyle\int_0^\infty D(v) v^n\, dv}{\displaystyle\int_0^\infty D(v)\, dv} \tag{5.5}$$

Also, from the tails of the Doppler profile it is possible to determine the maximum speed of the recoiling fragment,

$$\Delta \nu_{\max}/\nu_0 = \nu_{\max}/c \tag{5.6}$$

Further discussion of magic angle Doppler spectroscopy of molecular photofragments is presented in Section VII below, in the context of rotational polarization effects.

B. Velocity Aligned Doppler Spectroscopy

The requirement of a high signal-to-noise ratio for inverting a MADS profile may be alleviated by aligning the two laser beams coaxially ($\theta_a = \pi/2$) and introducing a delay between the pump and probe lasers. This method, known as velocity-aligned Doppler spectroscopy (VADS), was introduced by Wittig and coworkers [12] and further refined by Dixon et al. [73]. Coaxial alignment of the pump and probe laser beams has the useful property that fragments not moving along the line of sight fly out of the field of view and are not detected by the probe beam. This spatial discrimination allows the detector to distinguish between slow fragments moving in direction $\hat{\mathbf{k}}$ and fast fragments moving at some other angle but with the same velocity component along $\hat{\mathbf{k}}$. By this method the shape of $f(v)$ is more easily revealed, as shown in Figs. 5(*b* and *c*). A limitation of the VADS method is that the extraction of $f(v)$ from the VADS Doppler profiles depends on β and its possible velocity dependence. In the following analysis $f(v)$ may be obtained directly from the VADS profile only in the case of isotropic fragmentation.

Let $\mathbf{v}$ be the recoil velocity of the fragment, $\mathbf{c}$ be the thermal velocity of the parent molecule, and $\mathbf{u} = \mathbf{v} + \mathbf{c}$ be the resultant laboratory velocity. Let $\hat{\mathbf{z}}$ be the direction of the laser beams, and let $p_x(c_x)p_y(c_y)p_z(c_z)$ be the velocity distribution of the parent molecule. Ignoring, for the moment, motion of the parent molecule in the $\hat{\mathbf{z}}$-direction, we may write

$$D(\Delta\nu, t) = \frac{1}{4\pi}\int\int\int\int\int p(v)\,[1 - \tfrac{1}{2}\,\beta_{\text{eff}}P_2(v_z/v)]$$
$$\times\ p_x(c_x)p_y(c_y)\ \Phi(u_x, u_y)\delta(w - v_z)\ d^3v\ dc_x\ dc_y \qquad (5.7)$$

where $P_2(\cos\theta_a) = -\frac{1}{2}$ for the coaxial configuration, $\Phi(u_x, u_y)$ is the probability that the fragment is still in the field of view after delay t, and $\Delta\nu$ is related to w by Eq. (1.1). The value of Φ clearly depends on the spatial profiles of the laser beams. We will assume that both beams have Gaussian profiles, with the radius $\sigma = a$ for the photolysis laser and $\sigma = b$ for the probe. For example, for the photolysis laser,

$$I_{\text{ph}}(x, y) = \frac{1}{\pi a^2}\exp[-(x^2 + y^2)/a^2] \qquad (5.8)$$

In general the radial parameters a and b vary with axial distance z. It may happen, however, that the detector views only a small volume where the laser profiles are fairly uniform (e.g., if the fragment is ionized by the probe laser, and a collimating aperture samples only the focal region). In this case it is easy to show that

$$\Phi(u_x, u_y) = \frac{1}{\pi\rho^2}\exp[-(u^2 - w^2)t^2/\rho^2] \qquad (5.9)$$

where $\rho^2 = a^2 + b^2$.

We assume next that the parent molecules have a Boltzmann velocity distribution with most probable speed α. For the x-component of the velocity,

$$p_x(c_x) = \frac{1}{\sqrt{\pi}\alpha}\exp(-c_x^2/\alpha^2) \qquad (5.10)$$

with similar expressions for the y- and z-components. Integrating $\Phi(u_x, u_y)p_x(c_x)p_y(c_y)$ in Eq. (5.7) over c_x and c_y yields the delay function,

$$F(v, v_z, t) = \frac{\rho^2}{\rho^2 + \alpha^2 t^2} \exp[-(v^2 - v_z^2)t^2/(\rho^2 + \alpha^2 t^2)] \tag{5.11}$$

and integrating over v_z gives the VADS profile,

$$D(\Delta\nu, t) = \tfrac{1}{2} \int_{|w|}^{\infty} [1 - \tfrac{1}{2}\beta P_2(w/v)]\, F(v, w, t) f(v)\, v\, dv \tag{5.12}$$

Finally, it is necessary to average $D(\Delta\nu, t)$ over the distribution of c_z (i.e., over the thermal motion of the parent molecule along the line of sight). At this point it is convenient first to convolute $p_z(c_z)$ with the resolution function of the apparatus. Denoting the convolution of these two functions by $p(c_z)$, we get for the observed Doppler profile

$$D_c(\Delta\nu, t) = \int_{-\infty}^{\infty} D(w - c_z, t)\, p(c_z)\, dc_z \tag{5.13}$$

To extract $f(v)$ from experimental data, it is necessary to correct for the apparatus function, $p(c_z)$. A convenient way of doing this is to fit the experimental profile D_c with a series of harmonic oscillator basis functions. If $p(c_z)$ is a Gaussian function, each term in the series may be deconvoluted analytically to obtain $D(w, t)$. Setting $\beta = 0$ and differentiating with respect to w, we may invert Eq. (5.12) to obtain $f(v)$,

$$f(v) = -\frac{2}{w} \frac{\rho^2 + \alpha^2 t^2}{\rho^2} \frac{dD\,(w, t)}{dw}\bigg|_{|w|=v} + \frac{4t^2}{\rho^2} D(w, t)\bigg|_{|w|=v} \tag{5.14}$$

An example of the use VADS to determine $f(v)$ is shown in Fig. 6. In this example HCl was produced by photodissociating vinyl chloride at 193 nm [74]. A minimum in the Doppler profile after a delay of 1400 ns indicates that the speed distribution peaks near 10^3 cm/s.[2]

Although the probe frequency is scanned with the delay time fixed for the usual applications of VADS, useful information may also be obtained by varying t at a fixed laser frequency. For example, at the center of the

[2]HCl was detected by 2 + 1 resonance-enhanced multiphoton ionization, via the $F\,^1\Delta$ intermediate state. Measurements with parallel and perpendicular polarization of the probe laser did not reveal any rotational polarization of the fragment.

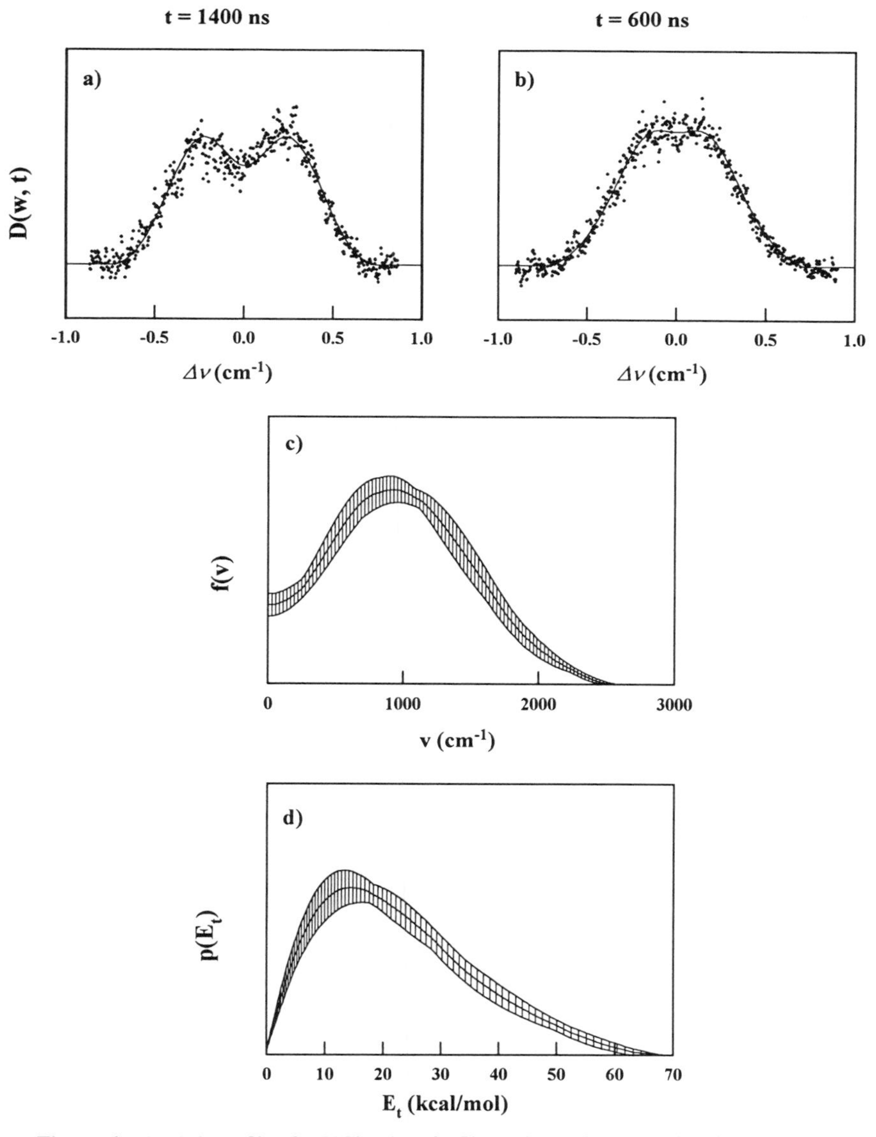

Figure 6. VADS profiles for HCl ($v' = 0$, $J' = 13$) produced by the photodissociation of CH_2CHCl at 193 nm. (*a*, *b*) Frequency scans obtained with delay times of 1400 and 600 ns, respectively. The curves are least-squares fits using harmonic oscillator basis functions. (*c*, *d*) show respectively the speed and kinetic energy distributions functions obtained with Eq. (5.14).

Doppler profile Eq. (4.7) becomes

$$D(0, t) = \frac{1 + \beta_{\text{eff}}/4}{2\pi\rho^2} \int_0^\infty e^{-v^2t^2/\rho^2} f(v)\, v\, dv \tag{5.15}$$

This result is valid for any value of β. For a Gaussian speed distribution function,

$$f(v) = v^n e^{-v^2/v_0^2} \tag{5.16}$$

the decay curve has the form

$$D(0, t) = \frac{\text{const}}{(\rho^2 + v_0^2 t^2)^{1+n/2}} \tag{5.17}$$

An example for which this type of measurement was useful is the photodissociation of vinyl chloride to produce hydrogen molecules [72]:

$$CH_2CHCl + h\nu(193 \text{ nm}) \longrightarrow H_2 + HCCCl \tag{5.18}$$

The hydrogen molecules were detected by resonance-enhanced multiphoton ionization.[3] By varying the intensity of the photolysis laser, we knew that H_2 was produced by a single 193-nm photon. It is possible, however, that the primary products could be CH_2CH + Cl, and that H_2 is a secondary fragment produced when the vinyl radical absorbs a *probe* photon. This hypothesis was tested by varying the delay between the two laser pulses. If Cl were a primary product, the heavy fragments would fly out of the field of view much more slowly than if H_2 were produced directly by the photolysis laser. As shown in Fig. 7, $D(0, t)$ corresponds closely to the decay curve calculated for H_2 as the primary product.

VI. MULTIPHOTON DOPPLER PROFILES

The results of the previous sections may be generalized to include the case where more than one photon is absorbed. In discussing multiphoton effects, it is important to distinguish between resonant and nonresonant excitation of intermediate virtual states. We treat first the case of direct (prompt) non-

[3]Since H_2 is a molecular fragment, a rigorous treatment requires correcting $D(0, t)$ for vector correlation effects. In the present case, where Q-branch lines of a two-photon Σ–Σ transition were used to detect the fragment, this correction would have only a small effect.

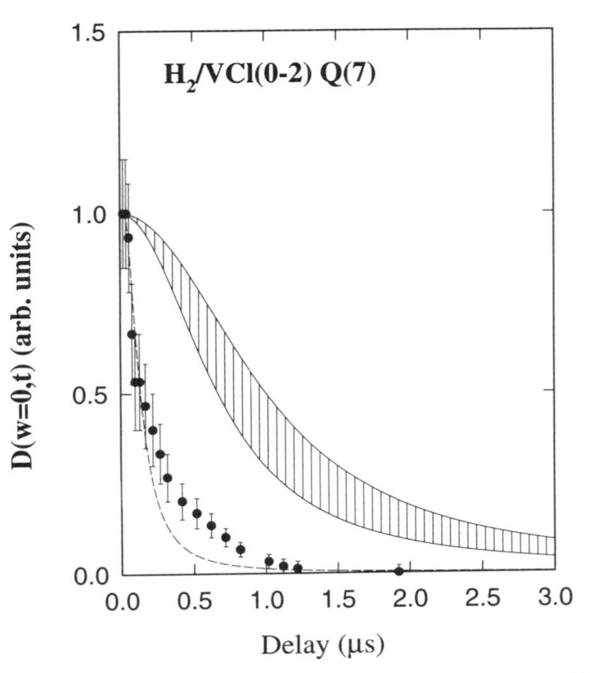

Figure 7. Delay-time scan of H_2 ($v'' = 2$, $J'' = 7$) produced in the 193-nm photodissociation of vinyl chloride. The abscissa is the delay between the pump and probe lasers, while the ordinate is the H_2 signal at the center of the Doppler profile. The filled circles are experimental data with 1σ error bars. The shaded area was calculated assuming that H_2 is produced by secondary photodissociation of the vinyl radical. The dashed line was calculated assuming that H_2 flies out of the field of view immediately after absorbing one photon from the pump laser, with a kinetic energy equal to the mean value calculated from the magic angle Doppler profile.

resonant photodissociation by two photons. The intermediate state vector is denoted by $|J_i, K_i, M_i\rangle$, and the rate of absorption of both photons is proportional to the product of matrix elements [75],

$$f_{E,\mu}(\theta) \propto \langle J'', K'', M''|\boldsymbol{\mu} \cdot \mathbf{E}|J_i, K_i, M_i\rangle^2 \cdot \langle J_i, K_i, M_i|\boldsymbol{\mu} \cdot \mathbf{E}|J', K', M'\rangle^2 \tag{6.1}$$

If both transitions are either parallel or perpendicular, the right side of Eq. (6.1) is proportional to $\cos^4 \theta$, whereas if one transition is parallel and the other is perpendicular it is proportional to $\sin^2 \theta \cos^2 \theta$. (Recall that θ is defined as the angle between **E** and **μ**.) It follows that for the $\|$, $\|$ and $\perp$,

$\perp$ cases

$$f_{E,\mu}(\theta) = \frac{1}{80\pi^2}\left[1 + \frac{20}{7} P_2(\cos\theta) + \frac{8}{7} P_4(\cos\theta)\right] \tag{6.2}$$

while for the $\parallel$, $\perp$ and $\perp$, $\parallel$ cases

$$f_{E,\mu}(\theta) = \frac{2}{15\pi^2}\left[1 + \frac{5}{7} P_2(\cos\theta) - \frac{12}{7} P_4(\cos\theta)\right] \tag{6.3}$$

Transforming to space-fixed coordinates and projecting out the velocity along the probe direction, we obtain

$$D(w) = \frac{1}{2v}[1 + \beta_{2,\mathrm{eff}} P_2(w/v) + \beta_{4,\mathrm{eff}} P_4(w/v)] \tag{6.4}$$

where

$$\beta_{s,\mathrm{eff}} = \beta_s P_s(\cos\theta_p) \tag{6.5}$$

for $s = 2$ or 4. The values of β_s for the various types of transitions are listed in Table I.

These results are readily generalized to the absorption of n photons, with Eq. (6.4) containing terms up to $\beta_{2n,\mathrm{eff}}P_{2n}(w/v)$. The anisotropy parameters for three identical photons are listed in Table II.

We consider next the effect of resonant absorption of one of the intermediate photons. In this case the molecule becomes oriented and/or aligned (orientation being induced by circularly polarized light and alignment by both circularly and linearly polarized light) in the intermediate discrete state. Memory of the orientation and alignment is retained by the final state and may be observed in the Doppler profile. The general form of the Doppler

TABLE I
Anisotropy Parameters for Two-photon Doppler Profiles in the Short Lifetime Limit

Transition	β_2	β_4
$\parallel$, $\parallel$	20/7	8/7
$\parallel$, $\perp$ or $\perp$, $\parallel$	5/7	−12/7
$\perp$, $\perp$	−10/7	3/7

TABLE II
Anisotropy Parameters for Three-photon Doppler Profiles in the Short Lifetime Limit

Transition	β_2	β_4	β_6
$\parallel, \parallel, \parallel$	10/3	24/11	16/33
$\parallel, \parallel, \perp$; $\parallel, \perp, \parallel$; $\perp, \parallel, \parallel$	5/3	−16/11	−40/33
$\perp, \perp, \parallel$; $\perp, \parallel, \perp$; $\parallel, \perp, \perp$	0	−21/11	10/11
$\perp, \perp, \perp$	−5/3	9/11	−5/33

profile produced by the absorption of n photons is given by

$$D(w) = \frac{1}{2v} \sum_{i=0}^{2n} \beta_i P_i(w/v) \tag{6.6}$$

where even values of i are generated by linearly polarized light, and both even and odd values are produced by circularly polarized light. The case of resonant absorption of the first photon followed by direct dissociation by a second photon was treated classically by Sander and Wilson [76], semiclassically by Chen and Yeung [77], and quantum mechanically by Singer, Freed, and Band [78].

We consider next the case of multiphoton excitation of a discrete (predissociated) upper level with nonresonant absorption of the intermediate photons. This problem is a generalization of single-photon predissociation considered in Section IV, and it may be treated semiclassically in the limit of axial recoil, using Eq. (4.12) as a starting point. For two-photon absorption we follow closely Chen and Yeung's (CY) treatment of the rotational line strength [79]. The molecule is initially in a state $|J''K''M''\rangle|0n''\rangle$, passes through virtual states $|J_iK_iM_i\rangle|in_i\rangle$, and ends up in state $|J'K'M'\rangle|fn'\rangle$, where n'', n_i, and n' designate vibrational and electronic quantum numbers, and the energies of the intermediate states are E_{i0}. The matrix elements of the components of the transition operator $\mathbf{r}$ in the molecular frame are given by

$$\mu_{fi}^{q'} = \langle fn''|r_{q'}|in_i\rangle \tag{6.7}$$

and

$$\mu_{i0}^{q''} = \langle in_i|r_{q''}|0n''\rangle \tag{6.8}$$

Assuming the closure relation for the intermediate states (Eq. (7) in CY) we obtain

$$f_{E,a}(\theta_s) = \sum_{M',M''} |\langle J''K''M''| \sum_{q',q''} \sum_{i,n_i} D^{1*}_{0q'} D^{1*}_{0q''} |J'K'M'\rangle|^2 \cdot \frac{\mu^{q'}_{fi}\, \mu^{q''}_{i0}}{E_{i0} - \hbar\omega} |D^{J'}_{M'K'}|^2 \tag{6.9}$$

Using the properties of the rotational matrices to evaluate the integrals $\langle J''K''M''|D^{1*}_{0q'}D^{1*}_{0q''}|J'K'M'\rangle$ and an inverse Clebsch–Gordan series to evaluate the sums over M' and M'', we obtain for the angular distribution of the fragments,

$$f_{E,a}(\theta_s) = \frac{1}{2}\, \delta_{J'J''}\delta_{K'K''} g_0 - 2\sqrt{\frac{10}{3}}\, \delta_{J'J''}\, \delta_{K'K''}\, g_1 S_1(\theta_s) + \frac{10}{3}\, g_2 S_2(\theta_s) \tag{6.10}$$

The quantities g_i,

$$g_0 = \left[\sum_{q'q''} \sum_{i,n_i} \begin{pmatrix} 1 & 1 & 0 \\ -q' & -q'' & 0 \end{pmatrix} \frac{\mu^{q'}_{fi}\, \mu^{q''}_{i0}}{E_{i0} - \hbar\omega} \right]^2 \tag{6.11}$$

$$g_1 = \sum_{q'q''} \sum_{i,n_i} \begin{pmatrix} 1 & 1 & 0 \\ -q' & -q'' & 0 \end{pmatrix} \begin{pmatrix} 1 & 1 & 2 \\ -q' & -q'' & 0 \end{pmatrix} \left(\frac{\mu^{q'}_{fi}\, \mu^{q''}_{i0}}{E_{i0} - \hbar\omega} \right)^2 \tag{6.12}$$

$$g_2 = (2J'' + 1) \left[\sum_{q'q''} \sum_{i,n_i} \begin{pmatrix} 1 & 1 & 2 \\ -q' & -q'' & q' + q'' \end{pmatrix} \cdot \begin{pmatrix} J'' & 2 & J' \\ K'' & -q' - q'' & K' \end{pmatrix} \frac{\mu^{q'}_{fi}\, \mu^{q''}_{i0}}{E_{i0} - \hbar\omega} \right]^2 \tag{6.13}$$

contain all the necessary information describing the virtual intermediate states. The angular information is contained in the quantities

$$S_1(\theta_s) = (-1)^{J'+K'}(2J' + 1)^{1/2} \begin{pmatrix} J' & 2 & J' \\ K' & 0 & K' \end{pmatrix}^2 P_2(\cos\theta_s) \tag{6.14}$$

and

$$S_2(\theta_s) = \sum_{k''} \langle J''k''2\ K - k''|J'K'\rangle^2 |D^2_{K'-k'',0}|^2$$

$$= \tfrac{3}{8} [\langle J''K' - 2, 22|J'K'\rangle^2 + \langle J''K' + 2, 2 - 2|J'K'\rangle^2] \sin^4 \theta_s$$

$$+ \tfrac{3}{2} [\langle J''K' - 1, 21|J'K'\rangle^2 + \langle J''K' + 1, 2 - 1|J'K'\rangle^2]$$

$$\cdot \sin^2 \theta_s \cos^2 \theta_s + \langle J''K', 20|J'K'\rangle^2 [P_2(\cos \theta_s)]^2 \quad (6.15)$$

For all transitions other than the Q-branch with $K' = K''$, only the last term in Eq. (6.10) contributes to the angular distribution. The angular factor for this term may be written as

$$S_2(\theta_s) = \tfrac{1}{5} (A + B + C) + \tfrac{1}{7} (-2A + B + 2C) P_2(\cos \theta_s)$$

$$+ \tfrac{3}{35} (A - 4B + 2C) P_4(\cos \theta_s) \quad (6.16)$$

where

$$A = \langle J''K' - 2, 22|J'K'\rangle^2 + \langle J''K' + 2, 2 - 2|J'K'\rangle^2 \quad (6.17)$$

$$B = \langle J''K' - 1, 21|J'K'\rangle^2 + \langle J''K' + 1, 2 - 1|J'K'\rangle^2 \quad (6.18)$$

and

$$C = \langle J''K', 20|J'K'\rangle^2 \quad (6.19)$$

Equation (6.16) may be recast in the form

$$S_2(\theta_s) = \tfrac{1}{5} (A + B + C) [1 + \beta_2 P_2(\cos \theta_s) + \beta_4 P_4(\cos \theta_s)] \quad (6.20)$$

where the anisotropy parameters have the values

$$\beta_2 = \frac{5}{7} \frac{-2A + B + 2C}{A + B + C} \quad (6.21)$$

and

$$\beta_4 = \frac{3}{7} \frac{A - 4B + 6C}{A + B + C} \quad (6.22)$$

TABLE III
Anisotropy Parameters for Two-photon Doppler Profiles in the Long Lifetime, High-J Limit

Rotational Branch	β_2	β_4
Q	−5/7	27/28
P, R	−5/14	−9/14
O, S	5/7	9/56

The values of β_2 and β_4 in the high-J limit are listed in Table III for all five rotational branches. The case of the Q-branch for $K' = K''$ can be calculated in a straightforward manner, provided that the relative magnitudes of g_0, g_1, and g_2 are known or treated as parameters.

These equations may be combined with the results of the previous sections to calculate Doppler profiles under various conditions. For example, magic angles may be used to eliminate either $\beta_{2,\text{eff}}$ (at $\theta_v = 54.74°$) or $\beta_{4,\text{eff}}$ (at $\theta_v = 30.56°$ or $70.12°$), but not both simultaneously.

It is also possible to use the VADS technique with multiphoton excitation. We illustrate the latter approach for $Cl(^2P_{3/2})$ atoms produced in the predissociation of the $F^1\Delta$ state of HCl [80]. Since the predissociation lifetime of this state is much longer than the rotational period, the Doppler profile is in the long lifetime limit. Figure 8 shows the VADS profile for the $Q(8)$ transition. The minimum in the center of the Doppler profile is expected for a $\perp$, $\perp$ transition.

VII. INFLUENCE OF ROTATIONAL POLARIZATION ON DOPPLER LINE SHAPES

A. Rotational Polarization

In the previous sections we presented a semiclassical derivation of the distribution of *atomic* photofragments (or more generally, ensembles of molecules with an isotropic distribution of angular momentum vectors). The arguments concerning velocity anisotropy of *molecular* photofragments are no different; the number density of particles with a velocity projection w is correctly given by Eq. (3.4). What has been neglected so far is that our ability to detect the fragments by optical means may not be independent of the Doppler shift. Just as the translational anisotropy of fragments stems from the directional interaction of the parent molecule transition dipole moment $\boldsymbol{\mu}$ with the polarized dissociation light, so the detection of fragments depends on the alignment of the *fragment* transition dipole moment $\boldsymbol{\mu}_f$ with the probe beam polarization. Just as the forces that generate fragment velocities display the anisotropy of the photoselection and subsequent disso-

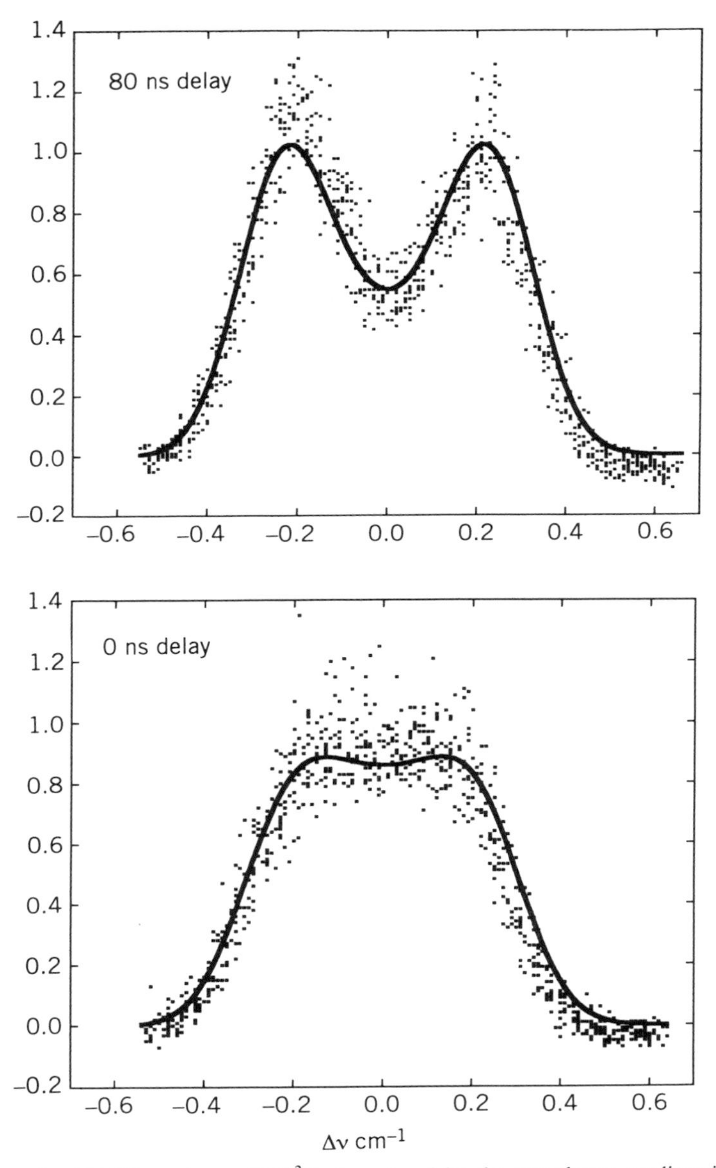

Figure 8. VADS spectrum of $Cl(^2P_{3/2})$ produced by the two-photon predissociation of the $F\,^1\Delta$ state of HCl.

ciation, so do the torques that generate fragment rotation reflect this anisotropy. The anisotropy of the fragment rotation is referred to as rotational polarization.

The sensitivity of the detection probability to the direction of rotation axis can be readily seen in the semiclassical interpretation of the M-dependence of the matrix elements of the dipole operator: $|\langle J'M'\Lambda'|\boldsymbol{\mu} \cdot \mathbf{E}|J''M''\lambda''\rangle|^2$. For light linearly polarized along the laboratory axis $\hat{\mathbf{Z}}$, which is also used to define the space-fixed projection quantum number M, the transition probability is proportional to $\langle J'M', 10|J''M''\rangle^2$. For a Q-branch transition ($J' = J''$, $M' = M''$), this squared Clebsch–Gordon coefficient is $M^2/J(J + 1) = \cos^2 \theta$, with θ being the semiclassical angle between $\mathbf{J}''$ and $\hat{\mathbf{Z}}$. For R- or P-branch transitions ($J' = J'' \pm 1$, $M' = M''$), the high-J limits of the squared Clebsch–Gordan coefficients are both $\frac{1}{2}(J^2 - M^2)/J(J + 1) = \frac{1}{2} \sin^2 \theta$. We see, then, that Q-branch transitions selectively excite large M states ($\mathbf{J}$ parallel to the electric field), while P- and R-branch transitions selectively excite small M values ($\mathbf{J}$ perpendicular to the electric field) [18]. The degree and sign of rotational alignment can thus be measured by comparing either the relative intensities of Q- and R-branch transitions from the same initial state, or by measuring the relative intensity of a single transition as a function of the probe laser polarization. One can also appreciate that in scanning across a Doppler-broadened absorption line, the distribution of $\mathbf{J}$ directions may change with the selected velocity ensemble, imparting an additional variation in the Doppler profile beyond that derived solely from the velocity distribution. These effects were first seen and explained nearly 10 years ago [14–17]. We note that a related effect in the Doppler spectroscopy of atomic fragments from electron impact dissociation of simple molecules had been observed several years earlier by Hatano and coworkers [81]. Unequal populations of magnetic sublevels of atomic fragments were shown to affect the Doppler-broadened atomic emission lines in a way similar to that observed for molecular photofragmentation.

The anisotropic distributions of molecular rotation can be described in several different ways. Axially symmetric distributions can be described by the distribution of a classical angle θ between $\mathbf{J}$ and the symmetry axis $\hat{\mathbf{Z}}$, or by an expansion of this distribution in Legendre polynomials, $P_n(\cos \theta)$. Quantum mechanically, the relative populations of the M_J states of a given J contain this same information. Alternatively, the M_J distribution can also be described by its Legendre moments. For non–axially symmetric distributions, a classical distribution function of an azimuthal angle ϕ as well as θ is required to describe the anisotropy of $\mathbf{J}$. An expansion of this angular distribution would require the inclusion of spherical harmonics, Y_{JM_J}, $\neq 0$, unlike the axially symmetric case, where spherical harmonics with $M_J = 0$

(Legendre polynomials) suffice. Quantum mechanically, a non–axially symmetric ensemble of **J** is characterized by definite phase differences, or coherences, between the M_J states. The M_J populations alone do not fully describe this anisotropy, which can, however, be reprsented by the density matrix $\rho_{MM'}(J) = \langle JM|\rho|JM'\rangle$, or its state multipole expansion. Recent reviews by Orr-Ewing and Zare [2, 3] include a clear and elementary discussion of rotational polarization in its many guises. The point of the present discussion is simply to alert the reader to the connections between the different languages used to describe rotational polarization.

Houston and coworkers developed density matrix [14, 82] and semiclassical [82] formulations to describe the effects of **v**–**J** correlations on laser-induced fluorescence Doppler line shapes. Simultaneously, Dixon [15] extended the more modern and elegant spherical tensor formulation of angular momentum polarization of Fano and Macek [83] and Greene and Zare [19, 20] to Doppler-resolved intensity measurements. The formulation of Dixon, while forbidding to the novice, provides systematic and powerful tools to generalize the analysis to arbitrary detection schemes, using multiple photons of arbitrary polarization and arbitrary experimental geometries. The Doppler analysis of more complex sample ensembles, such as photofragments from chiral parent molecules [84] and bimolecular reaction products [36, 41], has been successfully treated, requiring only minor extensions of Dixon's original work. While it is not our intent in this chapter to provide a full derivation of the rotational alignment corrections to the Doppler profiles, it may be helpful to include here a short discussion of the bipolar harmonics and bipolar moments, which appear as the dynamic observables in the Doppler spectroscopy of photofragments and bimolecular reaction products. We will feature a graphical presentation of the bipolar harmonics, and show a simple example of the analysis for one-photon absorption line shapes.

We have seen that the Doppler profile of a photofragment depends not only on the velocity distribution, as described in the earlier sections of this chapter, but also its correlated angular momentum distribution. The final state of a photofragment with a particular recoil speed has a translational anisotropy described by a distribution function of $\hat{\mathbf{v}}$ and a rotational anisotropy for each J state described by a density matrix $\rho_{MM'}(J)$, which may be an explicit function of the direction of fragment translation. The formulation simplifies considerably if the rotational anisotropy is described instead by a semiclassical probability distribution function of the direction $\hat{\mathbf{J}}$. Extension to a fully quantum theory of photofragment vector correlations has recently been presented [85]; here, the differences are small when fragment **J** greatly exceeds the total angular momentum of the initial parent molecule. The semiclassical correlated distributions of directions $\hat{\mathbf{v}}$ and $\hat{\mathbf{J}}$ can be expanded in bipolar harmonics, using the body-fixed direction of the parent transition moment as the reference axis. The bipolar harmonics form a natural basis for problems concerning the correlations of two vectors, and consist of appropriately weighted products of spherical harmonics of two directions given in spherical coordinates by angles $\theta_a\phi_a$ and $\theta_b\phi_b$. They are defined [86] as

$$B_{KQ}(k_1k_2;\ \theta_a\phi_a\theta_b\phi_b) = \sum_{q_1}\sum_{q_2}(-1)^{K-Q}[K]^{1/2}\begin{pmatrix}k_1 & K & k_2\\ q_1 & -Q & q_2\end{pmatrix} \times C_{K_1q_1}(\theta_a\phi_a)\ C_{k_2q_2}(\theta_b\phi_b) \tag{7.1}$$

where $[X] \equiv (2X + 1)$, and $C_{kq}(\theta\phi)$ is a modified spherical harmonic [86].

We adopt here the notation of Dixon [6], using unprimed variables $\theta_t\phi_t$ and $\theta_r\phi_r$ for the direction of translation and rotation in the space fixed frame (cf. $\theta_s\phi_s$) and primed variables $\theta_t'\phi_t'$ and $\theta_r'\phi_r'$ in the body-fixed frame (cf. $\theta_M\phi_M$). The desired probability distribution function can be expanded in the body-fixed basis,

$$P(\theta_t'\phi_t'\theta_r'\phi_r') = \sum_K\sum_Q\sum_{k_1}\sum_{k_2}\frac{[k_1]\ [k_2]}{16\ \pi^2}\ b_Q^K(k_1k_2)\ B_{KQ}(k_1k_2;\ \theta_t'\phi_t'\theta_t'\theta_r'\phi_r') \tag{7.2}$$

and then transformed to give the distribution of space-fixed translational and rotational directions,

$$P(\theta_t\phi_t\theta_r\phi_r) = \frac{1}{16\pi^2}\left[\sum_k [k]^2 b_0^0(kk)\ B_{00}(kk;\ \theta_t\phi_t\theta_r\phi_r) + \frac{2}{5}\sum_{k_1}\sum_{k_2}[k_1]\ [k_2]b_0^2(k_1k_2)B_{20}(k_1k_2;\ \theta_t\phi_t\theta_r\phi_r)\right] \tag{7.3}$$

where $b_0^k(k_1k_2)$ are the bipolar moments. The restriction to $K = 0$, 2 and $Q = 0$ in Eq. 7.3 arises from the symmetry properties of one-photon excitation, and the factor of 2/5 arises from the transformation to space-fixed coordinates. Dixon has derived [15] the variation of the rotational alignment parameters as functions of the relative Doppler shift, w/v (the projection of $\hat{\mathbf{v}}$ on the direction of Doppler resolution), in terms of renormalized bipolar moments, denoted $\beta_0^K(k_1k_2)$. From this result, one can write the Doppler profile as a function of the probing geometry, spectroscopic transition, and a small number of $\beta_0^K(k_1k_2)$ parameters, which are the dynamically relevant quantities. The details can be seen in the original work [15] and its initial application to 1 + 1 (one probe photon, one detected photon) laser-induced fluorescence detection of photofragments [16] induced by single-photon dissociation with linearly polarized light.

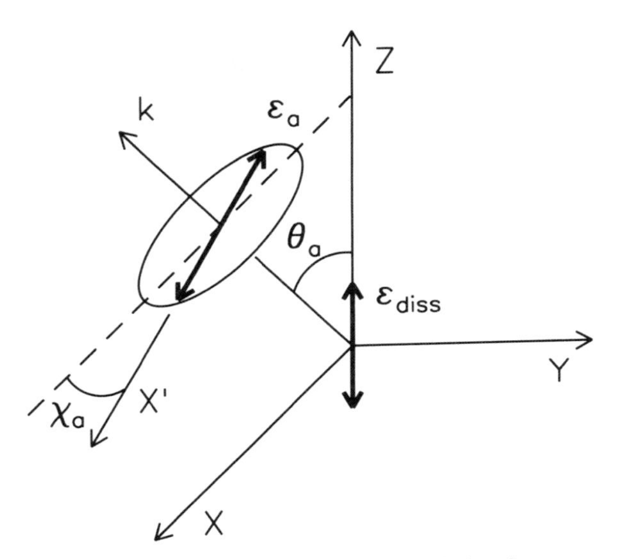

Figure 9. Coordinate system for one-photon probing of photofragments created by linearly polarized light. The laboratory $\hat{\mathbf{Z}}$-axis is defined by the photolysis polarization. The analysis beam makes an angle θ_a with respect to $\hat{\mathbf{Z}}$ and has a plane of polarization rotated by an angle χ_a from the plane of $\hat{\mathbf{Z}}$ and the propagation direction.

The analysis is particularly simple for the case of one-photon detection—either Doppler-resolved chemiluminescence, as originally contemplated by Dixon [15], or direct transient absorption, as first demonstrated by Hall and Wu [87]. In this case, only two angles are required to characterize the analysis process, a polar angle θ_a and an azimuthal angle χ_a, as shown in Fig. 9. The Doppler profile for an ensemble with a single speed, v, can be written as

$$I(w) = \frac{1}{2v}\left[g_0 + g_2P_2\left(\frac{w}{v}\right) + g_4P_4\left(\frac{w}{v}\right)\right] \tag{7.4}$$

with constants g_i [unrelated to those of Eqs. 6.9–6.12] given by

$$\begin{aligned} g_0 &= b_0\beta_0^0(00) + b_1\beta_0^2(02) \\ g_2 &= b_2\beta_0^2(20) + b_3\beta_0^0(22) + b_4\beta_0^2(22) \\ g_4 &= b_5\beta_0^2(42) \end{aligned} \tag{7.5}$$

The b_i multipliers are constants that consist of a geometrical factor and an angular momentum factor, both of which depend on the probe transition and the number of photons involved in the detection scheme. For single-photon absorption, they take on the following simple forms:

$$
\begin{aligned}
b_0 &= 1 \\
b_1 &= h^{(2)} \left[\tfrac{3}{5} \sin^2 \theta_a \cos 2\chi_a - \tfrac{2}{5} P_2(\cos \theta_a)\right] \\
b_2 &= 2 P_2(\cos \theta_a) \\
b_3 &= -h^{(2)} \\
b_4 &= h^{(2)} \left[\tfrac{6}{7} \sin^2 \theta_a \cos 2\chi_a + \tfrac{4}{7} P_2(\cos \theta_a)\right] \\
b_5 &= \frac{h^{(2)}}{35} \left[9 \sin^2 \theta_a \cos 2\chi_a - 36 P_2(\cos \theta_a)\right]
\end{aligned} \tag{7.6}
$$

The geometrical factors are simple functions of the experimental angles θ_a and χ_a; and the angular momentum factors, $\pm h^{(2)}$ in this case, have the simple form below for P-, Q-, and R-branch absorption transitions,

$$
\begin{aligned}
\mathrm{Q}\uparrow\ h^{(2)} &= +1 \\
\mathrm{R}\uparrow\ h^{(2)} &= -J/(2J+3) \\
\mathrm{P}\uparrow\ h^{(2)} &= -(J+1)/(2J-3)
\end{aligned} \tag{7.7}
$$

The generalization to a distribution of velocities analogous to Eq. (3.5) is then

$$
D(w) = \tfrac{1}{2} \int_{|w|}^{\infty} [g_0 + g_2 P_2(w/v) + g_4 P_4(w/v)] v f(v)\, dv \tag{7.8}
$$

where it must be considered that the bipolar moments may themselves be functions of velocity.

We will shortly apply these equations to an example of photodissociation in order to get past the somewhat frightening notation and see the simple geometrical nature of the vector correlations. Once this has been recognized, the more complex cases involving two or more probe photons are seen to differ only in the number of nonvanishing, higher order contributions to Eqs. (7.4) and (7.5), and a somewhat more complex calculation required to obtain the b_i constants [15, 88].

B. The Bipolar Harmonics Visualized

The renormalized bipolar moments themselves are a measure of the overlap between the joint probability distribution functions, $P(\theta_t'\phi_t'\theta_r'\phi_r')$, and the orthogonal generating functions, $B_0^K(k_1k_2;\ \theta_t'\phi_t'\theta_r'\phi_r')$. These functions of four angles can be graphed as polar plots of $B_0^K(k_1k_2\theta_r'\phi_r')$ on the surface of a sphere corresponding to the range of values for $\theta_t'\phi_t'$ or $\hat{\mathbf{v}}$. This representation breaks the symmetry between the directions **v** and **J** in the bipolar harmonics, in exchange for being able to visualize a function of four coordinates. Figures 10–14 illustrate the five lowest-order, even renormalized bipolar harmonics. In each case the large sphere drawn in outline depicts the set of recoil directions with respect to the body-fixed direction of the parent molecule's transition moment, $\boldsymbol{\mu}$, pointing toward the north pole of the sphere. The polar plots at each point on the velocity sphere can be considered the contribution to the corresponding bipolar moment due to a photofragment

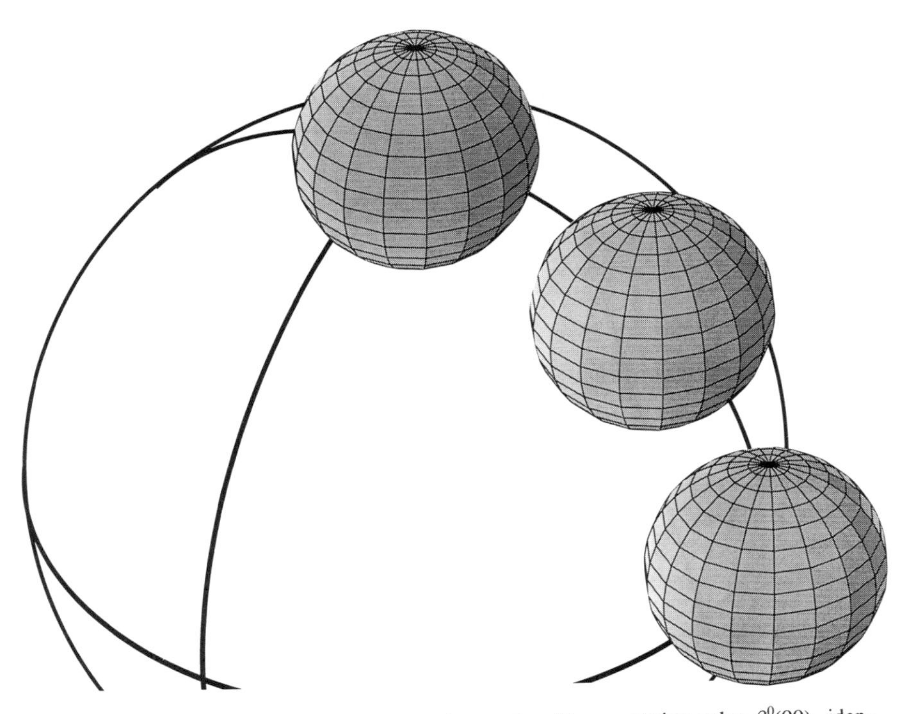

Figure 10. Representation of the bipolar harmonic with expectation value $\beta_0^0(00)$, identically 1.0 for all values of $\hat{\mathbf{v}}$ and $\hat{\mathbf{J}}$, The angular distribution of $\hat{\mathbf{J}}$ for each **v** is depicted by a polar plot of the angular arguments of $\hat{\mathbf{J}}$ positioned on the velocity sphere at a position **v**.

moving along $\mathbf{v}$ with its angular momentum pointed along $\mathbf{J}$. The symmetry of one-photon dipole excitation makes all of these functions symmetric in the north and south hemispheres, and azimuthally symmetric in ϕ_t. The functions are thus plotted along a single meridian from $\theta_t = 0$ to $\theta_t = \pi/2$. The first bipolar moment, $\beta_0^0(00)$, is identically one, corresponding to a normalization of the joint probability distribution in $\hat{\mathbf{v}}$ and $\hat{\mathbf{J}}$. The $\beta_0^0(00)$ weighting function can be viewed as positive unit spheres in $\theta_r'\phi_r'$ uniformly covering the surface of the $\theta_t'\phi_t'$ sphere, as shown in Fig. 10.

Figure 11 shows the generating function of $\beta_0^2(20)$, which is one half of the familiar velocity anisotropy parameter, β. The spheres in $\theta_r'\phi_r'$ space reflect the uniform integration over the distribution of rotational directions, $\hat{\mathbf{J}}$. The grey shaded polar plots depict positive weights, and the white plots depict negative weights, with a magnitude proportional to the radius. Velocities parallel to the transition moment, aligned with the poles of the figure, contribute $+1$ weight; velocities perpendicular to the transition moment,

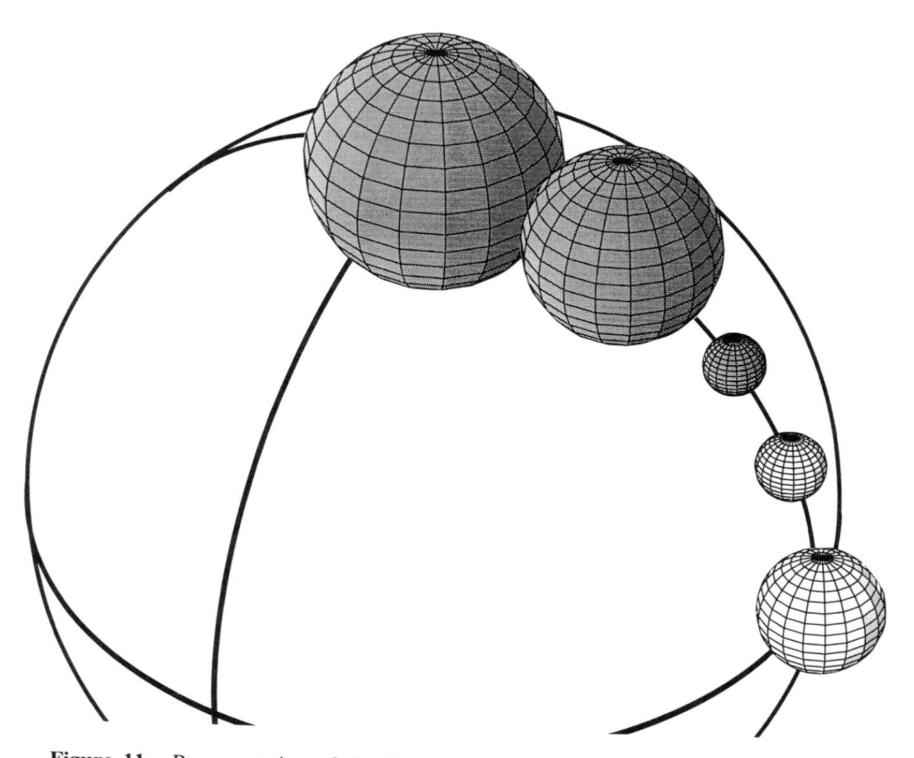

Figure 11. Representation of the bipolar harmonic with expectation value $\beta_0^2(20)$, the $\boldsymbol{\mu}$–$\mathbf{v}$ correlation.

along the equator of the figure, contribute $-\frac{1}{2}$ weight. The zero contribution of magic angle recoil to the $\beta_0^0(22)$ moment is represented by the vanishing diameter of the spheres on a ring at north and south lattitude 35.26°, which separates regions of positive and negative weight.

Figure 12 shows the generating function of $\beta_0^0(02)$, which is 5/4 times the rotational alignment parameter, $A_0^{(2)}$. This bipolar moment depends only on the distribution of the rotational directions, $\theta_r'\phi_r'$, relative to the transition moment of parent molecule, and is independent of the recoil direction, as indicated by a uniform covering of the $\theta_t'\phi_t'$ sphere with identical functions $P_2(\cos\theta_r)$. This function provides $+1$ weight for $\mathbf{J} \parallel \boldsymbol{\mu}$ and $-\frac{1}{2}$ weight for $\mathbf{J} \perp \boldsymbol{\mu}$, independent of $\mathbf{v}$.

Figure 13 depicts the generating function for the lowest-order $\mathbf{v}$–$\mathbf{J}$ correlation moment, $\beta_0^0(22)$. Positive weights for this moment are contributed when $\mathbf{v} \parallel \mathbf{J}$, and negative weights are contributed for $\mathbf{v} \perp \mathbf{J}$, independent of the direction of $\boldsymbol{\mu}$.

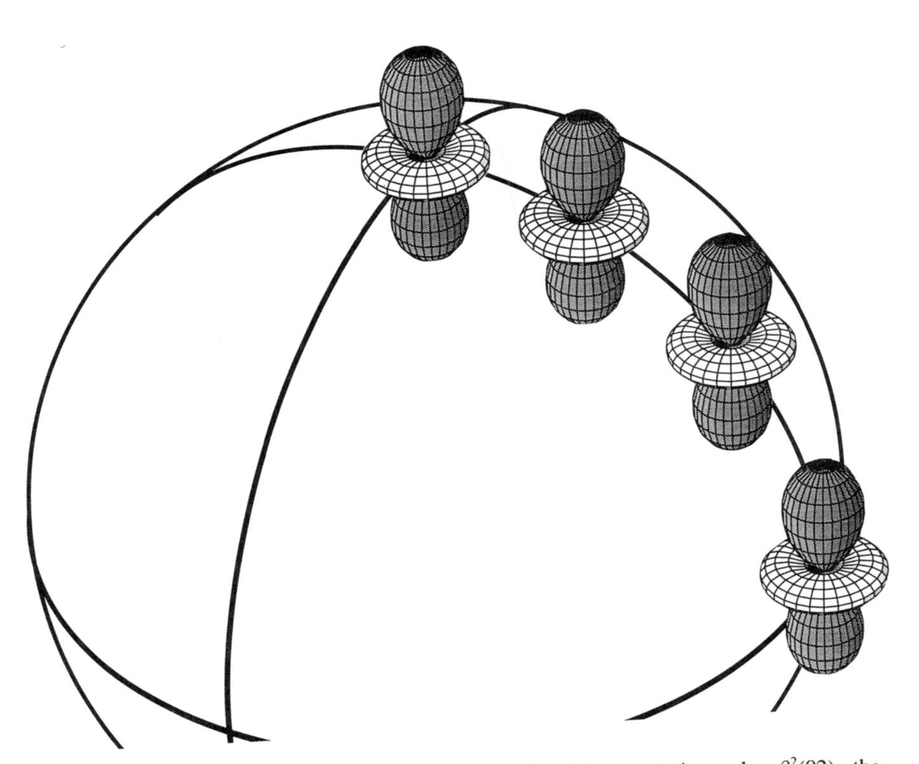

Figure 12. Representation of the bipolar harmonic with expectation value $\beta_0^2(02)$, the $\boldsymbol{\mu}$–$\mathbf{J}$ correlation.

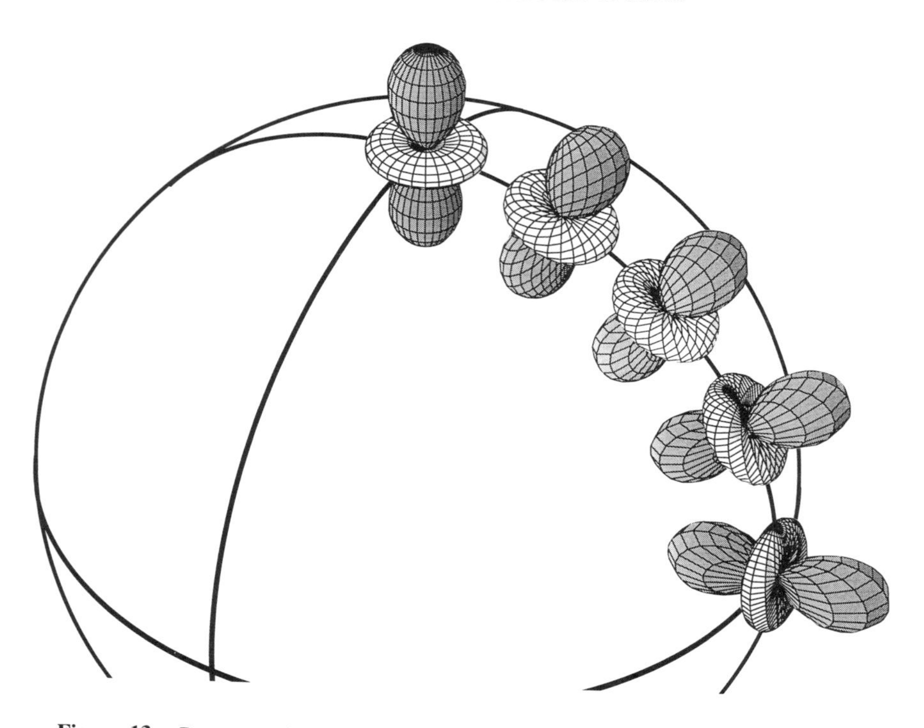

Figure 13. Representation of the bipolar harmonic with expectation value $\beta_0^0(22)$, the **v–J** correlation.

Figure 14 shows the generating function for the lowest-order triple vector correlation moment, $\beta_0^2(22)$. This moment depends on the overlap of the actual joint distribution function with a weighting function of $\theta_r'\phi_r'$ that changes shape with θ_t'. For fragments recoiling parallel to $\boldsymbol{\mu}$, the weighting function is $-P_2(\hat{\mathbf{J}} \cdot \hat{\boldsymbol{\mu}})$, while for perpendicular recoil, $\beta_0^2(22)$ is sensitive to the azimuthal asymmetry of **J** about **v**, as shown. The smooth evolution from the polar to the equatorial shapes with θ_t' displays an amusing pattern of transitions from rings to lobes and back. By now, we have seen enough to imagine the higher-order functions. For $k_2 = 4$, the figures in $\theta_r'\phi_r'$ space will have five lobes instead of 1 or 3, while for $k_1 = 4$ the shapes will change more rapidly with θ_t'. Because of the more highly oscillatory nature of the higher-order bipolar harmonics, the overlap with photofragment velocity and angular momentum distributions typically falls off quickly, and the essential shape of the correlated distribution of $\boldsymbol{\mu}$, **v**, and **J** is well characterized by these first five bipolar moments. The neglect of the $P_4(w/v)$ term in Eq. (7.4) is a common procedure [16], often necessitated by

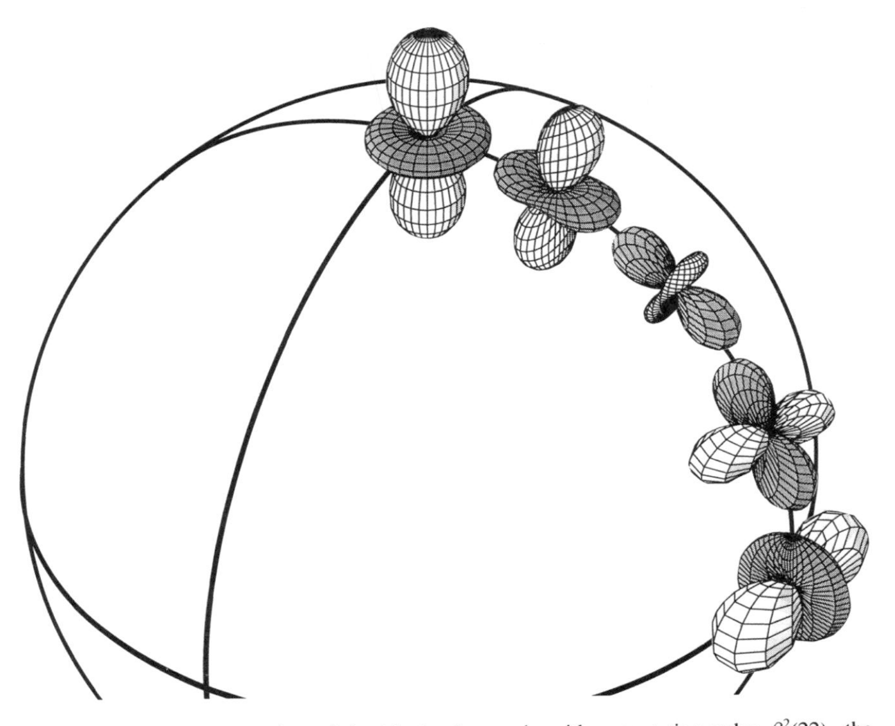

Figure 14. Representation of the bipolar harmonic with expectation value $\beta_0^2(22)$, the **μ–v–J** correlation.

limited experimental precision, but generally justified by its small contribution. When they can be determined, the corresponding higher-order bipolar moments frequently are found to confirm conclusions independently drawn from the lower order moments [87].

Following this introduction, let us make a few qualitative remarks about the influence of these bipolar moments on the one-photon Doppler profiles, recalling Eqs. (7.4–7.6). Neglecting rotational polarization is the same as keeping only terms containing bipolar moments with index $k_2 = 0$. These remaining terms correspond to the previously derived Eqs. (2.10), (3.3), and (3.4), with $g_0 = 1$ and $g_2 = b_2\beta_0^2(20) = 2\ P_2\ (\cos\theta_a)\beta_0^2(20) = \beta\ P_2(\cos\theta_a)$. The effect of rotational alignment, $b_1\beta_0^2(02)$, is to change the integrated intensity but not the shape of the Doppler profile. The effects of the **v–J** correlation, $b_3\beta_0^0(22)$, and the **μ–v–J** correlation, $b_4\beta_0^2(22)$, are to change the shape, but not the intensity, of the Doppler profile, since the $P_2(w/v)$ term integrates to zero. Each of these vector correlations make contributions of opposite sign for Q-branches and P- or R-branches, making them easiest to observe by comparing Doppler profiles for different rotational branches probing the same photofragment states. The effects of the **v–J** correlation, $b_3\beta_0^0(22)$, are independent of the experimental geometry, in keeping with the K = Q = 0 tensor indices of the corresponding bipolar harmonic. If the velocity distribution is isotropic, the $b_3\beta_0^0(22)$ term can still give rise to a nonvanishing $P_2(w/v)$ contribution to the Doppler profile. Similarly, probing at $\theta_a = 54.7°$ $\chi_a = 45°$ cancels the experimental sensitivity to all bipolar moments but $\beta_0^0(22)$.

C. Magic Angle Doppler Spectroscopy for Molecular Photofragments

To illustrate the use of the bipolar moment analysis on photofragment Doppler spectra, we show Doppler spectra of CN radical fragments from the 193-nm photodissociation of ethyl thiocyanate, C_2H_5SCN. The example is taken from the work of Wu and Hall [87] and shows both strong velocity anisotropy and rotational polarization. The speed distribution for each detected CN state, indicative of the correlated C_2H_5S internal energy distribution, is extracted from a composite magic angle Doppler profile.

In single-photon absorption spectroscopy, access to the only two relevant analysis angles, θ_a and χ_a shown in Fig. 9, can be attained in a single, orthogonal, pump-probe geometry, with the aid of polarization optics in each beam. The photolysis polarization selects θ_a, and the probe polarization selects χ_a. The importance of the probe laser polarization and the rotational branch used for detecting Doppler spectra can be seen in Fig. 15, which shows magic angle Doppler spectra of the same ($v = 0$, $N = 69$, $J = 69.5$) state of CN produced by the 193-nm photodissociation of C_2H_5SCN. According to Eqs. (3.4) and (3.5), derived in the absence of rotational polarization of the photofragments, all these Doppler profiles should be identical, since $\beta_{\text{eff}} = 0$ when θ_a is a magic angle for any value of χ_a. We see, instead, that the Doppler profiles depend upon the probe laser polarization, which indicates a nonzero value of $\beta_0^2(22)$, and that the Doppler profiles depend upon the rotational branch selected to probe the photofragment, which indicates a nonzero value of $\beta_0^0(22)$.

From an inspection of the b_i constants in Eq. (7.6), we see that the ideal "magic angle" Doppler spectrum, that is, the one whose derivative is related to the speed distribution through Eq. (5.1), is not observed in any single line for any geometry. There does exist, however, an experimental geometry, $\theta_a = 54.7°$ and $\chi_a = 45°$, at which sensitivity to all rotational polarization except for the lowest order **v–J** correlation vanishes exactly. The effects of this **v–J** correlation can in turn be eliminated by constructing a weighted average of Q- and P- or R-branch lines, measured in this geometry, with weights proportional to $|1/h^{(2)}|$. Since b_1 vanishes in this geometry also, any variation in the observed integrated intensities of the Q- and R-branch

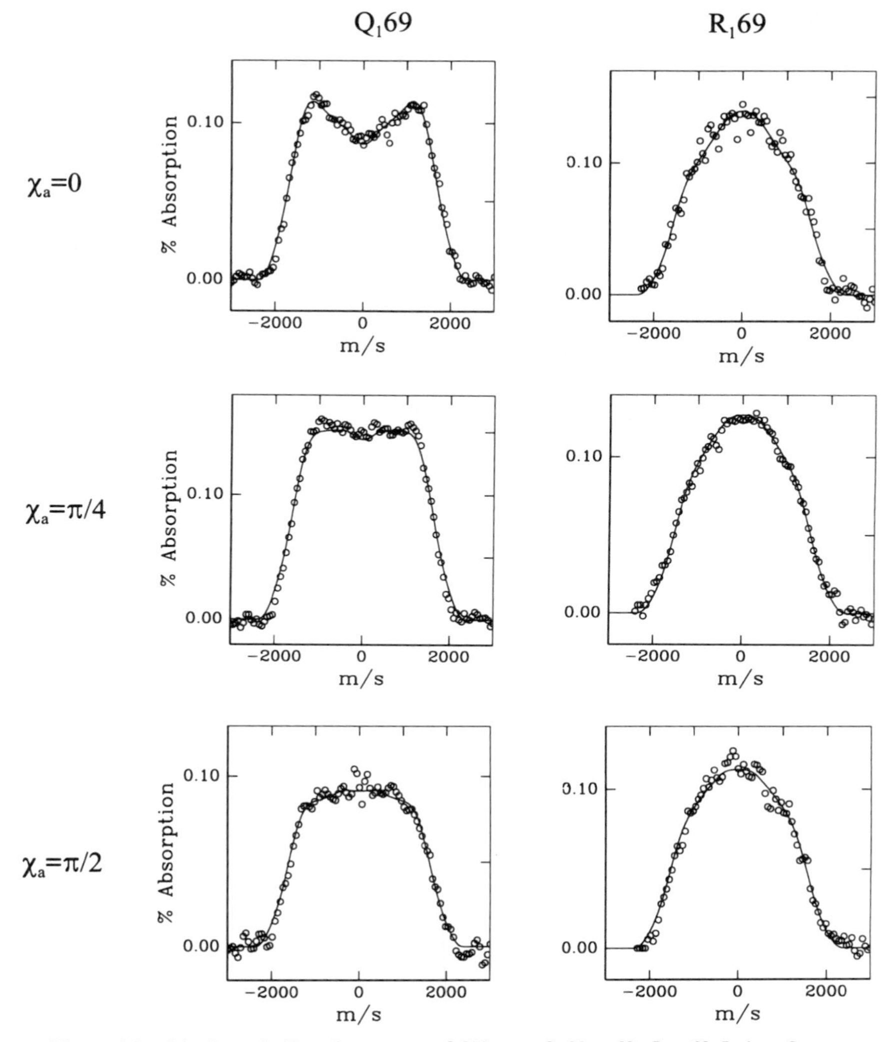

Figure 15. Magic angle Doppler spectra of CN, $v = 0$, $N = 69$, $J = 69.5$ photofragments from the 193-nm photolysis of C_2H_5SCN, obtained by transient absorption spectroscopy on the A–X (2–0) band [89]. The left column of spectra were measured on the Q_1-branch line and the right column was measured on the R_1-branch line. All lines were measured with the probe beam propagating at magic angle $\theta_a = 54.7°$, but with probe laser polarization angle χ_a varied in 45° steps between the rows.

lines can be attributed to artifactual variations in absolute signal. The observed Doppler line shapes can therefore be normalized to constant area, and added with Q:R weighting of 1:(2 + 3/J) (approaching 1:2 at high J) to give the true $\beta_{\text{eff}} = 0$ Doppler profile.

A similar composite of multiple geometries and probe transitions has been described by Huber [13] and Docker [89] for generating Doppler profiles with $\beta_{\text{eff}} = 0$, using laser-induced fluorescence. In this case, three different geometries are required for each of two transitions, and the geometries require realignment of beams, and not just polarization optics. The analogous REMPI-TOF method for measuring projection of fragment velocity along a flight direction wins the contest for simplicity, however, since the probe laser polarization axis need not be perpendicular to the axis of velocity resolution. This allows a measurement with $\beta_{\text{eff}} = 0$ for a single probe transition, by choosing both the probe polarization and the photolysis polarization axis to be at an angle of 54.7° to the velocity resolution direction [47, 88]. A final trick, frequently useful to measure speed distributions without the complications of rotational polarization, is to use Q-branch lines of two-photon $\Delta\Lambda = 0$ transitions. The two-photon line strength factors are typically dominated by their isotropic term [90]. In this case, the angular momentum factors in the b_i multipliers are small, although not precisely zero, and the weak sensitivity to rotational alignment justifies their neglect [72, 91].

From the 2:1 weighted R- and Q-branch lines measured at $\theta_a = 54.7°$ and $\chi_a = 45°$, as shown in Fig. 15, the isotropic Doppler profile and the speed distribution, $v^2 f(v)$ can be derived using Eq. (5.1). This fragment state of CN has a speed distribution that peaks at 1800 m/s, with a full width at half maximum of about 900 m/s. Energy balance arguments show that the coincident C_2H_5S fragment possesses only about 25% of the total excess energy of fragmentation, despite its possessing the majority of the degrees of freedom in the system.

D. Measurement of Bipolar Moments

To continue with the example in which CN is generated from C_2H_5SCN by photodissociation, we follow a simple generalization of the technique first described by Comes et al. [16], but take advantage of the experimentally determined speed distribution. The analysis proceeds by assuming that each of the five bipolar moments illustrated in Figs. 10–14 is independent of velocity, which is reasonable in this case because of the fast and sharply peaked distribution of velocities. A more general analysis to allow for velocity-dependent bipolar moments has been described by Aoiz et al. [41, 42] building upon elements of analysis from Kinsey [11], Ticktin and Huber [92], and Taajes et al. [10]. If the bipolar moments can be considered velocity independent, their extraction is simple. Once the isotropic speed distribution has been determined, the Doppler profiles can all be fit with a linear combination of two (or three) basis functions, $D_0(w)$, $D_2(w)$, and (optionally) $D_4(w)$. The first is simply the measured isotropic composite Doppler line shape,

$$D_0(w) = \tfrac{1}{2} \int_{|w|}^{\infty} v f(v) \, dv \tag{7.9}$$

The second and third can be constructed with P_2 and P_4 functions, weighted by $f(v)$ according to

$$D_2(w) = \tfrac{1}{2} \int_{|w|}^{\infty} P_2(w/v) \, v f(v) \, dv \tag{7.10}$$

and

$$D_4(w) = \tfrac{1}{2} \int_{|w|}^{\infty} P_4(w/v) \, v f(v) \, dv \tag{7.11}$$

A simple least-squares calculation then fits each Doppler profile shown in Fig. 15, and others measured at $\theta_a = 0$ and $\pi/2$, with parameters g_0, g_2, and (optimally) g_4 in the equation

$$D_{\text{obs}}(w) = g_0 D_0(w) + g_2 D_2(w) + g_4 D_4(w) \tag{7.12}$$

which can be seen to be the same as Eq. (7.8). If the laser resolution or the thermal velocity distribution of the parent molecule cannot be neglected, these equations must be slightly modified to include a convolution, which will tend to wash out the g_4 term and selectively decrease the anisotropy of the slower fragments [87]. The integrated intensities are proportional to $g_0 = 1 + b_1 \beta_0^2(02)$, allowing $\beta_0^2(02)$ to be determined most easily by comparing intensities at different values of χ_a. Each measured Doppler profile contributes a g_2/g_0 ratio from the fit, which, along with $\beta_0^2(02)$ and the associated constant $b_0, \ldots, b_4$ from Eq. (7.6), gives one equation in three unknowns: the bipolar moments $\beta_0^2(20)$, $\beta_0^0(22)$, and $\beta_0^2(22)$. If g_4 is included in the linear fit (its contribution will generally be small), the value of g_4/g_0 gives an independent estimate of $\beta_0^2(42)$ for each measured line. Solving a set of overdetermined linear equations then leads to experimental values for a set of velocity-averaged bipolar moments needed to characterize the shape of the vector correlations between and among $\boldsymbol{\mu}$, $\mathbf{v}$, and $\mathbf{J}$.

Qualitative interpretation of such a set of bipolar moments is generally accomplished by comparison with the set of bipolar moments derived from limiting Cartesian arrangements of $\boldsymbol{\mu}$, $\mathbf{v}$ and $\mathbf{J}$ [16, 89, 93]. Setting $\boldsymbol{\mu}$ along the body-fixed axis $\hat{\mathbf{z}}$, $\mathbf{v}$, and $\mathbf{J}$ can be each aligned along $\hat{\mathbf{x}}$-, $\hat{\mathbf{y}}$-, or $\hat{\mathbf{z}}$-directions, giving nine extreme arrangements. One arrangement, $\boldsymbol{\mu}\|\mathbf{v}\|\mathbf{J}$ (A), is unique, the other eight occur in pairs: $\boldsymbol{\mu}\|\mathbf{v}\perp\mathbf{J}$ (B), $\boldsymbol{\mu}\|\mathbf{J}\perp\mathbf{v}$ (C), $\boldsymbol{\mu}\perp\mathbf{v}\|\mathbf{J}$ (D) and $\boldsymbol{\mu}\perp\mathbf{v}\perp\mathbf{J}$ (E), differing only in the labeling of $\hat{\mathbf{x}}$ and $\hat{\mathbf{y}}$. These five different limiting arrangements have been labeled as cases A–E by Dixon, [15] who compiled for each of these five cases a large set of bipolar moments, later extended to higher orders by Docker [88]. Dixon and others [16, 89, 94] have invoked linear combinations of bipolar moments to evaluate body-fixed $\hat{\mathbf{x}}$, $\hat{\mathbf{y}}$, and $\hat{\mathbf{z}}$ components of $\langle \mathbf{J}^2 \rangle$ for $\mathbf{v}\perp\boldsymbol{\mu}$, in order to provide an interpretation of measured bipolar moments in the photodissociation of H_2O_2 and CH_3ONO. Black [93] has considered linear combinations of cases B, C, and E to fit CN Doppler profiles from the dissociation of ICN. Only those three cases for which $\mathbf{v}\perp\mathbf{J}$ were included in this analysis, since angular momentum conservation in the triatomic dissociation precludes significant components of $\mathbf{J}$ along the relative velocity.

Hall and Wu [87] have extended and formalized this interpretation, based on "effective bipolar moments" that are written as linear combinations of all five limiting cases. The five-column table of bipolar moments for these five limiting cases can be considered as a rectangular transformation matrix of dimensions $5 \times \infty$, which gives an unbounded number of nonvanishing bipolar moments required to describe the sharply peaked correlations of the five limiting orthogonal cases, or any linear combination of them. The linear combinations of these five cases can be considered as a Cartesian representation of the same information carried by the five lowest-order bipolar moments—the ones depicted in Figs. 10–14. Indeed, the 5×5 transformation submatrix obtained by ignoring bipolar moments with k_1 or $k_2 > 2$,

$$\begin{bmatrix} \beta_0^0(00) \\ \beta_0^2(02) \\ \beta_0^2(20) \\ \beta_0^0(22) \\ \beta_0^2(22) \end{bmatrix} = \begin{bmatrix} 1 & 1 & 1 & 1 & 1 \\ 1 & -\frac{1}{2} & 1 & -\frac{1}{2} & -\frac{1}{2} \\ 1 & 1 & -\frac{1}{2} & -\frac{1}{2} & -\frac{1}{2} \\ 1 & -\frac{1}{2} & -\frac{1}{2} & 1 & -\frac{1}{2} \\ -1 & \frac{1}{2} & \frac{1}{2} & \frac{1}{2} & -1 \end{bmatrix} \times \begin{bmatrix} a \\ b \\ c \\ d \\ e \end{bmatrix} \tag{7.13}$$

turns out to have a simple inverse, relating the truncated bipolar and the Cartesian representations,

$$\begin{bmatrix} a \\ b \\ c \\ d \\ e \end{bmatrix} = \frac{1}{9} \times \begin{bmatrix} 1 & 2 & 2 & 2 & 2 \\ 2 & -2 & 4 & -2 & -2 \\ 2 & 4 & -2 & -2 & 2 \\ 2 & -2 & -2 & 4 & 2 \\ 2 & -2 & -2 & -2 & -4 \end{bmatrix} \times \begin{bmatrix} \beta_0^0(00) \\ \beta_0^2(02) \\ \beta_0^2(20) \\ \beta_0^0(22) \\ \beta_0^2(22) \end{bmatrix} \tag{7.14}$$

The vector [a b c d e] can be interpreted as the partial contributions to the full joint probability distribution of the five respective limiting cases, denoted A–E. The incomplete subspace of **μ**, **v**, and **J** correlations spanned by these five orthogonal limits can be visualized in the same way we have graphed the bipolar harmonics in Fig. 16. In this figure double-ended arrows are labelled A–E to indicate the body-fixed direction of **J**, plotted on the velocity sphere for **v** either parallel to **μ** (A and B) or perpendicular to **μ** (C, D, and E). As before, the vertical pole of the figure is the body-fixed parent transition dipole. The dual degeneracy of cases B–E can be seen in the symmetry with respect to exchange of $\hat{\mathbf{x}}$ and $\hat{\mathbf{y}}$. The lengths of the double-pointed arrows in Fig. 16 depict the [a b c d e] distribution obtained with Eq.

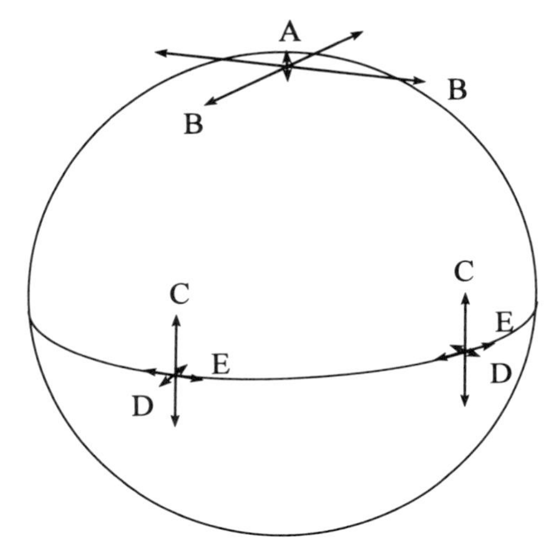

Figure 16. Cartesian representation of the correlated distribution of vectors **μ v J** for a high-**J** CN photofragment state with **v** = 0 in the 193-nm photodissociation of C_2H_5SCN. The visualization scheme is the same as for the bipolar harmonics in Figs. 10–14. The arrows labeled A–E indicate relative probabilities of finding **μ**, **v**, and **J** in the relative body-fixed orientations depicted by the position and direction of the arrows.

(7.14) from the set of bipolar moments observed for the high-J CN states of C_2H_5SCN photodissociation. In this case, we see a dominance of case B, with **v** parallel to **μ** and **J** perpendicular to **v**. There is virtually no amplitude with **J** parallel to **v** (neither A nor D), and the minor perpendicular velocity component is characterized by a moderately strong preference for **J** to be parallel to **μ** (C > E). This pattern of vector correlations can be related to the excited states and exit channel dynamics of the photodissociation [87]. The variety of correlated **μ, v, J** distributions that can be so represented in this four-dimensional space of normalized A–E fractions is quite large, and can represent many physically sensible distributions. While lacking the formal elegance of the spherical tensors, the geometrical connection with excited state symmetries, transition state geometries, and exit channel torques may be somewhat easier to grasp intuitively.

In conclusion, Doppler spectroscopy has developed into one of the most versatile tools available for studying the dynamics of elementary chemical reactions. One of its chief advantages is its experimental simplicity. In many cases it provides information from a bulb experiment that would otherwise require a molecular beam apparatus. It is possible to use Doppler spectroscopy to measure (1) the differential cross sections of photofragments and bimolecular reaction products, (2) the three-dimensional velocity distributions of atoms and molecules, and (3) the correlation between velocity and rotational angular momentum of molecular products. Despite the power of this technique, Doppler spectroscopy maintains the simplicity and elegance envisioned by Zare and Herschbach over 30 years ago [5].

ACKNOWLEDGMENTS

Work by GEH at Brookhaven National Laboratory was performed under contract with the U.S. Department of Energy and supported by its Division of Chemical Sciences DE-AC02-76CH00016). Support by the National Science Foundation (CHE-9408801) is gratefully acknowledged for work done by R.J.G. at the University of Illinois at Chicago. RJG wishes to thank Mr. Rohana Liyanage, Dr. Guoxin He, Dr. Yibo Huang, Professor Neil Shafer-Ray, and Dr. Hongkun Park for many helpful discussions.

REFERENCES

1. J. P. Simons, *J. Phys. Chem.* **91,** 5378 (1987); P. L. Houston, *J. Phys. Chem.* **91,** 5388 (1987); M. P. Docker, A. Hodgson, and J. P. Simons, in M. N. R. Ashfold and J. E. Baggott, Eds., *Molecular Photodissociation Dynamics*, Royal Society of Chemistry, London 1987; P. L. Houston, *Acct. Chem. Res.* **22,** 309 (1989); G. E. Hall and P. L. Houston, *Ann. Rev. Phys. Chem.* **40,** 375 (1989).
2. A. J. Orr-Ewing and R. N. Zare, *Ann. Rev. Phys. Chem.* **45,** 315 (1994).
3. A. J. Orr-Ewing and R. N. Zare, in A. Wagner and K. Liu, Eds., *Chemical Dynamics and Kinetics of Small Radicals*, World Scientific Publications, Singapore, 1995, p. 936.

4. M. Brouard and J. P. Simons, in A. Wagner and K. Liu, Eds., *Chemical Dynamics and Kinetics of Small Radicals*, World Scientific Publications, Singapore, 1995, p. 795.
5. R. N. Zare and D. R. Herschbach, *Proc. IEEE* **51,** 173 (1963).
6. J. Solomon, *J. Chem. Phys.* **47,** 889 (1967).
7. G. E. Busch and K. E. Wilson, *J. Chem. Phys.* **56,** 3638 (1972).
8. R. W. Diesen, J. C. Wahr, and S. E. Adler, *J. Chem. Phys.* **50,** 3635 (1969).
9. R. Schmiedl, H. Dugan, W. Meier, and K. H. Welge, *Z. Phys. A.* **304,** 137 (1982).
10. J. I. Cline, C. A. Taatjes, and S. R. Leone, *J. Chem. Phys.* **93,** 6543 (1990); C. A. Taatjes, J. I. Cline, and S. R. Leone, *J. Chem. Phys.* **93,** 6554 (1990).
11. J. L. Kinsey, *J. Chem. Phys.* **66,** 2560 (1977).
12. Z. Xu, B. Koplitz, and C. Wittig, *J. Chem. Phys.* **87** 1062, (1987); Z. Xu, B. Koplitz, and C. Wittig, *J. Chem. Phys.* **90,** 2692 (1989).
13. M. Dubs, U. Brühlmann, and J. R. Huber, *J. Chem. Phys.* **84,** 3106 (1986).
14. G. E. Hall, N. Sivakumar, P. L. Houston, and I. Burak, *Phys. Rev. Lett.* **56,** 1671 (1986).
15. R. N. Dixon, *J. Chem. Phys.* **85,** 1866 (1986).
16. K.-H. Gericke, S. Klee, F. J. Comes, and R. N. Dixon, *J. Chem. Phys.* **85,** 4463 (1986).
17. M. P. Docker, A. Hodgson, and J. P. Simons, *Chem. Phys. Lett.* **128,** 264 (1986).
18. D. A. Case, G. M. McClelland, and D. R. Herschbach, *Mol. Phys.* **35,** 541 (1978).
19. C. H. Greene and R. N. Zare, *Ann. Rev. Phys. Chem.* **33,** 119 (1982).
20. C. H. Greene and R. N. Zare, *J. Chem. Phys.* **78,** 6741 (1983).
21. W. D. Phillips, J. A. Serri, D. J. Ely, D. E. Pritchard, K. R. Way, and J. L. Kinsey, *Phys. Rev. Lett.* **41,** 937 (1978); E. J. Murphy, J. H. Brophy, G. S. Arnold, W. L. Dimpfl, and J. L. Kinsey, *J. Chem. Phys.* **70,** 5910 (1979); J. A. Serri, C. H. Becker, M. B. Elbel, J. L. Kinsey, W. P. Moskowitz, and D. E. Pritchard, *J. Chem. Phys.* **74,** 5116 (1981); J. A. Serri, J. L. Kinsey, and D. E. Pritchard, *J. Chem. Phys.* **75,** 663 (1981).
22. P. L. Jones, U. Hefter, A. Mattheus, J. Witt, K. Bergmann, W. Müller, W. Meyer, and R. Schinke, *Phys. Rev. A* **26,** 1283 (1982).
23. J.-M. L'Hermite, G. Rahmat, and R. Vetter, *J. Chem. Phys.* **93,** 434 (1990).
24. B. Girard, N. Billy, G. Gouédard, and J. Vigué, *J. Chem. Phys.* **95,** 4056 (1991).
25. A. G. Suits, P. de Pujo, O. Sublemontier, J.-P. Visticot, J. Berlande, J. Cuvellier, T. Gustavsson, J.-M. Mestdagh, P. Meynadier, and Y. T. Lee, *Phys. Rev. Lett.* **67,** 3070 (1991).
26. R. A. Gottscho, R. W. Field, R. Bacis and S. J. Silvers, *J. Chem. Phys.* **73,** 599 (1980).
27. N. Smith, T. P. Scott and D. E. Pritchard, *J. Chem. Phys.* **81,** 1229 (1984).
28. A. J. McCaffery, K. L. Reid, and B. J. Whitaker, *Phys. Rev. Lett.* **61,** 2085 (1988); C. P. Fell, A. J. McCaffery, K. L. Reid, and A. Ticktin, *J. Chem. Phys.* **95,** 4948 (1991); K. L. Reid and A. J. McCaffery, *J. Chem. Phys.* **95,** 4958 (1991).
29. J. Park, N. Shafer, and R. Bersohn, *J. Chem. Phys.* **91,** 7861 (1989).
30. J. F. Hershberger, J. Z. Chou, G. W. Flynn, and R. E. Weston, Jr., *Chem. Phys. Lett* **149,** 51 (1988); J. F. Hershberger, S. A. Hewitt, S. K. Sarkar, G. W. Flynn, and R. E. Weston, Jr., *J. Chem. Phys.* **91,** 4636 (1989); F. A. Kahn, T. G. Kruetz, G. W. Flynn, and R. E. Weston, Jr., *J. Chem. Phys.* **92,** 4876 (1990).

31. G. K. Chawla, G. C. McBane, P. L. Houston, and G. C. Schatz, *J. Chem. Phys.* **88,** 5481 (1988); G. C. McBane, Ph.D. Thesis, Cornell University, Ithaca, NY 1990.
32. N. E. Shafer, A. J. Orr-Ewing, W. R. Simpson, H. Xu, and R. N. Zare, *Chem. Phys. Lett.* **212,** 155 (1993).
33. G. W. Johnson, S. Satyapal, R. Bersohn, and B. Katz, *J. Chem. Phys.* **92,** 206 (1990); A. Chattopadhyay, S. Tasaki, R. Bersohn, and M. Kawasaki, *J. Chem. Phys.* **95,** 1033 (1991); B. Katz, J. Park, S. Satyapal, S. Tasake, A. Chattopadhyay, W. Yi, and R. Bersohn, *Faraday Discuss. Chem. Soc.* **91,** 73 (1991).
34. G. E. Hall, *Proc. of the Twelfth Combustion Res. Conf.*, Tahoe City, CA, 1990, 122.
35. N. E. Shafer, Ph.D. Thesis, Columbia University, New York, 1990.
36. F. Green, G. Hancock, A. J. Orr-Ewing, M. Brouard, S. P. Duxon, P. A. Enriquez, R. Sayos, and J. P. Simons, *Chem. Phys. Lett.* **182,** 268 (1991).
37. F. Green, G. Hancock, A. J. Orr-Ewing, *Faraday Discuss. Chem. Soc.* **91,** 79 (1991).
38. M. Brouard, S. P. Duxon, P. A. Enriquez, and J. P. Simons, *J. Phys. Chem.* **95,** 8169 (1991).
39. D. S. King, D. G. Sauder, and M. P. Casassa, *J. Chem. Phys.* **97,** 5919 (1992).
40. M. Brouard, S. P. Duxon, P. A. Enriquez, and J. P. Simons, *J. Chem. Phys.* **97,** 7414 (1992).
41. F. J. Aoiz, M. Brouard, P. A. Enriquez, and R. Sayos, *J. Chem. Soc. Faraday Trans.* **89,** 1427 (1993).
42. M. Brouard, S. Duxon, P. A. Enriquez, and J. P. Simons, *J. Chem. Soc. Faraday Trans.* **89,** 1435 (1993).
43. W. R. Simpson, A. J. Orr-Ewing, and R. N. Zare, *Chem. Phys. Lett.* **212,** 163 (1993).
44. M. L. Costen, G. Hancock, A. J. Orr-Ewing, and D. Summerfield, *J. Chem. Phys.* **100,** 2754 (1994).
45. N. E. Shafer, H. Xu, R. P. Tuckett, M. Springer, and R. N. Zare, *J. Phys. Chem.* **98,** 3369 (1994).
46. H. L. Kim, M. A. Wickramaaratchi, X. Zheng, and G. E. Hall, *J. Chem. Phys.* **101,** 2033 (1994).
47. M. Mons and I. Dimicoli, *Chem. Phys. Lett.* **131,** 298 (1986); M. Mons and I. Dimicoli, *Chem. Phys.* **130,** 307 (1988); M. Mons and I. Dimicoli, *J. Chem. Phys.* **90,** 4037 (1989).
48. G. E. Hall, N. Sivakumar, R. Ogorzalek, G. Chawla, H.-P. Haerri, and P. L. Houston, *Faraday Discuss. Chem. Soc.* **82,** 13 (1986).
49. R. Ogorzalek-Loo, G. E. Hall, H. P. Haerri, and P. L. Houston, *J. Phys. Chem.* **92,** 5 (1988).
50. J. F. Black and I. Powis, *Chem. Phys.* **125,** 375 (1988).
51. H. J. Hwang and M. A. El-Sayed, *Chem. Phys. Lett.* **170,** 161 (1990).
52. D. W. Chandler and P. L. Houston, *J. Chem. Phys.* **87,** 1445 (1987).
53. T. Kinugawa and T. Arikawa, *J. Phys. Chem.* **96,** 4801 (1992).
54. H. Ni, J. M. Serafin, and J. J. Valentini, *Chem. Phys. Lett.*, **244,** 207 (1995).
55. A. Sanov, C. R. Bieler, and H. Reisler, *J. Phys. Chem.*, **99,** 13637 (1995).
56. R. N. Zare, *Angular Momentum*, Wiley, New York, 1988, p. 95.
57. C. Jonah, *J. Chem. Phys.* **55,** 1915 (1971).

58. S. Mukamel and J. Jortner, *J. Chem. Phys.* **61,** 5348 (1974).
59. R. Schmiedl, H. Dugan, W. Meier, and K. H. Welge, *Z. Phys. A* **304,** 137 (1982).
60. S. Yang and R. Bersohn, *J. Chem. Phys.* **61,** 4400 (1974).
61. G. Gerber and R. Möller, *Phys. Rev. Lett.* **55,** 814 (1985).
62. R. N. Zare, *Berichte der Bunsen-Geselschaft* **86,** 422 (1982).
63. T. J. Butenhoff, K. L. Carleton, R. D. van Zee, and C. B. Moore, *J. Chem. Phys.* **94,** 1947 (1991).
64. S. J. Singer, K. F. Freed, and Y. B. Band, *J. Chem. Phys.* **79,** 6060 (1983).
65. M. Glass-Maujean and J. A. Beswick, *Phys. Rev. A* **36,** 1170 (1987).
66. M. Glass-Maujean and J. A. Beswick, *J. Chem. Soc., Faraday Trans. 2* **85,** 983 (1989).
67. L. D. A. Siebbeles, J. M. Schins, W. J. van der Zande, and J. A. Beswick, *Chem. Phys. Lett.* **187,** 633 (1991).
68. W. J. van der Zande, F. Vitalis, and L. D. A. Siebbeles, *Chem. Phys.* **186,** 205 (1994).
69. E. Flemming, G. Reichardt, H. Schmoranzer, and O. Wilhelmi, *Phys. Lett. A* **192,** 52 (1994).
70. T. Sato, T. Kinugawa, T. Arikawa, and M. Kawasaki, *Chem. Phys.* **165,** 173 (1992).
71. Y. Huang, G. He, Y. Yang, S. Hashimoto, and R. J. Gordon, *Chem. Phys. Lett.* **229,** 621 (1994).
72. G. He, Y. Yang, Y. Huang, S. Hashimoto, and R. J. Gordon, *J. Chem. Phys.* **103,** 5484 (1995).
73. R. N. Dixon, J. Nightingale, C. M. Western, and X. Yang, *Chem. Phys. Lett.* **151,** 328 (1988).
74. Y. Huang, Y. Yang, G. He, S. Hashimoto, and R. J. Gordon, *J. Chem. Phys.*, **103,** 5476 (1995).
75. R. N. Zare, *Mol. Photochem.* **4,** 1 (1972).
76. R. K. Sander and K. R. Wilson, *J. Chem. Phys.* **63,** 4242 (1975).
77. K. Chen and E. S. Yeung, *J. Chem. Phys.* **72,** 4723 (1980).
78. S. J. Singer, K. F. Freed, and Y. B. Band, *J. Chem. Phys.* **81,** 3064 (1984).
79. K. Chen and E. S. Yeung, *J. Chem. Phys.* **69,** 43 (1978).
80. R. Liyanage, Y. Yang, R. J. Gordon, and R. W. Field, *J. Chem. Phys.* **103,** 6811 (1995).
81. N. Kouchi, K. Ito, Y. Hatano, and N. Oda, *Chem. Phys.* **70,** 105 (1982); Y. Hatano, *Comments At. Mol. Phys.* **13,** 259 (1983).
82. G. E. Hall, N. Sivakumar, D. Chawla, P. L. Houston, and I. Burak, *J. Chem. Phys.* **88,** 3682 (1988).
83. U. Fano and J. H. Macek, *Rev. Mod. Phys.* **45,** 553 (1973).
84. R. Uberna, R. D. Hinchliffe, and J. I. Cline, *J. Chem. Phys.* **103,** 7934 (1995).
85. L. D. A. Siebbeles, M. Glass-Maujean, O. S. Vasyutinskii, J. A. Beswick, and O. Roncero, *J. Chem. Phys.* **100,** 3610 (1994).
86. D. M. Brink and G. R. Satchler, *Angular Momentum*, 3rd ed., Clarendon Press, Oxford, 1993, p. 55.
87. G. E. Hall and M. Wu, *J. Phys. Chem.* **97,** 10911 (1993).
88. M. P. Docker, *Chem. Phys.* **135,** 405 (1989).

89. M. P. Docker, A. Ticktin, U. Brühlmann, and J. R. Huber, *J. Chem. Soc., Faraday Trans. 2*, **85,** 1169 (1989).

90. R. G. Bray and R. M. Hochstrasser, *Mol. Physics* **31,** 1199 (1976).

91. Y. Sato, Y. Matsumi, M. Kawsaki, and R. Bersohn, *J. Phys. Chem.*, **99,** 16307 (1995).

92. A. Ticktin and J. R. Huber, *Chem. Phys. Lett.* **156,** 372 (1989).

93. J. F. Black, *J. Chem. Phys.* **98,** 6853 (1993); **100,** 5392 (1994).

94. M. Brouard, M. T. Martinez, J. O'Mahony, and J. P. Simons, *J. Chem. Soc. Faraday Trans. 2* **85,** 1207 (1989).

VIBRATIONAL PREDISSOCIATION DYNAMICS OF VAN DER WAALS COMPLEXES: PRODUCT ROTATIONAL STATE DISTRIBUTIONS

MARSHA I. LESTER

Department of Chemistry, University of Pennsylvania, Philadelphia, PA 19104-6323

CONTENTS

Advances in Chemical Physics, Volume XCVI, Edited by I. Prigogine and Stuart A. Rice.
ISBN 0-471-15652-3

I. INTRODUCTION

The photodissociation dynamics of van der Waals (vdW) complexes have been extensively investigated for more than 20 years [1, 2]. Studies of these model systems, however, represent only a small subset of the enormous number of photochemical investigations carried out in recent times [3–5]. The great interest in the photochemistry of vdW complexes stems, in part, from the fact that their dissociation dynamics have been shown to be highly nonstatistical due to the small number of states involved. In addition, the weak binding between the constituent molecules (monomers) in these complexes, typically due to dispersion, induction, and electrostatic interactions, means that only a small amount of energy is required to induce dissociation. The absorption of an infrared photon corresponding to one vibrational quantum of the monomer usually provides sufficient energy to break the weak intermolecular bond. Alternatively, the complex can be prepared in an excited electronic state of the monomer with one or more quanta of vibrational excitation. Again, the monomer vibrational energy is used to break the intermolecular bond and dissociation takes place on the excited electronic state surface. In both cases, the dissociation dynamics takes place on a single electronic potential energy surface.

The principle mechanisms for dissociation of a weakly bound complex are vibrational predissociation and internal rotational predissociation. In vibrational predissociation, a photon is used to excite a vibrational mode of a monomer constituent within the complex. Energy redistribution within the complex slowly transforms the monomer vibrational excitation into translational motion along the vdW bond, resulting in bond breakage. The vibrational predissociation process is a half-collision analogue of vibrational to translational energy transfer ($V \rightarrow T$) in molecular collisions. In some cases, excess energy (beyond that required to break the intermolecular bond) can flow into product rotation and/or vibration, which can be envisioned as analogues of vibration to rotation ($V \rightarrow R$) and vibration to vibration ($V \rightarrow V$) energy transfer processes. Van der Waals complexes can also break apart through transfer of rotational excitation from the monomer (prepared by infrared or optical excitation) to translational motion along the vdW bond. This process is similar to rotational to translational ($R \rightarrow T$) inelastic energy transfer in collisions. In order for either of these predissociation processes to occur, the internal energy released from the monomer must exceed the intermolecular binding energy. This review will focus exclusively on the vibrational predissociation dynamics of weakly bound complexes in their ground and excited electronic states.

The main experimental observables of the dissociation event are the rotationally resolved absorption spectrum of the complex, the lifetime of the vibrationally excited complex, and the final state distribution of the product molecules. These observations can provide information on the structure of the complex, the characteristics of the intermolecular potential energy surface, and the fragmentation mechanism [6]. To date, the primary source of information on the vibrational predissociation dynamics of weakly bound complexes has come from measurements of the homogeneous line widths associated with excitation of various vibrational modes (intramolecular and/or intermolecular vibrations) [7, 8]. These line widths provide a measure of the vibrational predissociation lifetimes. For a few systems, direct time-domain measurements of the lifetimes have also been achieved [9–11]. The lifetime data gives information on the rate of the dissociation process, specifically, the rate of energy transfer from the initially prepared state to the dissociation coordinate.

More detailed information on the dynamical process can be obtained from the final state distribution of the products. The product state distribution reveals the final outcome of the dissociation event and, in some cases, the dynamical pathway(s) leading to dissociation. Only recently, however, have experimental methods been developed that can provide information on the final state distribution of the fragments for these predissociation events. These methods include various laser pump-probe schemes and measurements of photofragment angular distributions. Application of these methods has been quite limited to date. Only a few systems have been studied at the state-to-state level of detail that requires measurement of both the initially prepared state of the complex and the final state distribution of the photofragments. The final state distributions determined for these systems will be the primary subject of this review.

The final state distribution of the fragments reveals the partitioning of excess energy (after bond breakage) over product rotational, vibrational, and translational degrees of freedom. Experimental measurements of final state distributions permit detailed questions about the dynamics of the vibrational predissociation process to be addressed: Is the excess energy statistically distributed over product channels, as might be expected if the initial excitation is fully randomized in the complex prior to vibrational predissociation? Does the final state distribution contain some dynamical information that reflects the dissociation event and the underlying potential energy surface? Does the final state distribution of the fragments depend on the initially prepared state of the complex?

The large body of data compiled on the lifetimes of weakly bound complexes has shown that vibrational predissociation is a highly nonstatistical process [1, 7, 8]. The vibrational predissociation lifetimes are strongly mode-specific, as illustrated by the HF dimer, where the decay rate has been shown to be 20 times faster following excitation of the "hydrogen-bonded" HF stretch as compared to that occurring upon excitation of the "free" HF vibration [12, 13]. As a result, the final distributions following vibrational predissociation of these complexes are also expected to be highly nonstatistical.

Through the examples discussed in this review, we will explore how the initially prepared state, the coupling between the initial and final states, and the available product channels influence the vibrational predissociation dynamics. We will examine product rotational distributions and discover what these distributions tell us about dissociation pathways and the anisotropy of the underlying potential energy surface. We will revisit the propensity rules proposed to explain the rates of vibrational predissociation in systems where final state distributions are now also available. Finally, to the extent possible, comparisons will be made between the experimental measurements and theoretical calculations of these observables. This will yield additional insight into the vibrational predissociation dynamics as well as provide a stringent test of *ab initio* and semiempirical intermolecular potential energy surfaces (PES).

II. THEORETICAL BACKGROUND

In vibrational predissociation, the coupling between the motion along the intramolecular vibrational coordinate r and the intermolecular coordinate R (due to the interaction potential) determines the rate of energy exchange and therefore the lifetime of the complex. This coupling is usually very weak for vdW complexes and the lifetimes of the resonance states are very long. As a result, theoretical treatments of vibrational predissociation typically decouple the intermolecular motion along R from the intramolecular motion r of the monomer. This permits zero-order states to be defined for the initially prepared quasibound state and the final state. The vibrationally excited complex is represented as a bound state with a wavefunction $\chi_1(R)$, while the final state is represented by a one-dimensional continuum wavefunction $\chi_0(R; E)$ evaluated at the same total energy E. The vibrational to translational coupling term responsible for vibrational predissociation can also be expressed as $V_I(R)$. Using the Golden Rule [14], the rate of decay of the initially prepared state, Γ, can then be evaluated as follows:

$$\Gamma \sim \frac{4\pi^2}{h^2} |\langle \chi_1 | V_I(R) | \chi_0 \rangle|^2 \tag{2.1}$$

The dissociation rate depends on the overlap of the bound state wavefunction, the oscillatory continuum wavefunction, and the coupling potential $V_I(R)$.

The rate of dissociation depends crucially on the translational energy that is available to the fragments after dissociation (E_{trans}). The larger E_{trans}, the more rapidly the continuum wavefunction oscillates, and the poorer the overlap with the bound state wavefunction. Thus, the rate of dissociation will slow as E_{trans} increases. The well-known *energy gap* and *momentum gap* laws, advanced by Beswick and Jortner [14] and Ewing [15], are derived from the Golden Rule expression by modeling the intermolecular stretching motion as a Morse oscillator in R to obtain approximate expressions for the wavefunctions. These laws clearly show the dependence of the decay rate Γ on the translational energy release E_{trans}:

$$\Gamma \sim \exp\left[-\frac{2\pi^2(2mE_{trans})^{1/2}}{\alpha h}\right] \tag{2.2}$$

where m is the reduced mass of the complex and α is the range parameter of the Morse oscillator. The amount of energy flowing into translational degrees of freedom will be decreased if a product molecule is formed with rotational (or vibrational) excitation. Internal excitation of the molecular fragments will *decrease* the energy available to translation and thereby *increase* the rate of predissociation.

The *energy gap* and *momentum gap* models predict that the highest accessible rotational states will have the largest probability, since by energy conservation the corresponding translational energy release will be minimized. This condition yields the best overlap between the bound and oscillatory continuum wavefunctions. These models were devised to explain the wide range of vibrational predissociation lifetimes observed [14, 15]. The prediction that the highest accessible product rotational states will be populated can now be tested by direct examination of final state distributions. The population of highly excited product rotational state (j) will require significant anisotropy of the intermolecular potential to enable the large Δj changing transitions [5]. This requirement is contrary to the usual expectation that the vdW interaction is weak and that the intermolecular potential varies only slowly with angle.

A high degree of rotational excitation has been observed following vibrational predissociation in some systems, e.g., $(HF)_2$ and NeICl. As will be illustrated through this review, the final product rotational distributions do not simply follow the *energy gap* law but rather show interesting dynamical features that reveal detailed information about the vibrational predissociation dynamics.

The product rotational distributions observed following vibrational predissociation of vdW complexes exhibit some of the same trends seen upon photodissociation of covalently bound molecules. In direct dissociation of stable molecules, the product rotational distributions are controlled by two factors: (i) the initial bound state wavefunction of the parent molecule and (ii) final state interactions as the fragments separate [5]. In the Franck–Condon limit, the final rotational state distribution simply reflects the bound state wavefunction of the parent molecule. In this case, the square of the bending wavefunction of the parent molecule is mapped directly onto the rotational state distribution of the fragments, with the final state distribution determined by the free rotor expansion coefficients of the bound state wavefunction. At the other extreme, the final rotational state distribution will be completely determined by interactions between the fragments in the exit channel. Intermediate situations can be described using the *Rotational Reflection Principle* [5]. The product rotational state distribution reflects the square of the bending wavefunction of the parent molecule, with the mapping mediated by the anisotropy of the dissociative potential energy surface along the route sampled by the fragments as they separate. The torque on the fragments along the dissociation path is expressed as the rotational excitation function, the amount of rotational angular momentum generated in the trajectory as a function of initial bending angle.

Although this approach was developed for direct dissociation, the same two factors are important in determining final state distributions in the fragmentation of van der Waals complexes. In vibrational predissociation, however, the initial bound state wavefunction is that of the vibrationally excited complex prior to fragmentation. In this review, the product rotational distributions will be classified according to the criterion described above for direct dissociation. Note, however, that in vibrational predissociation the coupling potential may depend strongly on angle with the coupling strongest for collinear geometries and smallest for perpendicular configurations [16]. Therefore, the angular dependence of the coupling term may also influence the product rotational distribution in vdW complexes.

III. PHOTOFRAGMENT ANGULAR DISTRIBUTIONS

The earliest experiments that provided information on the partitioning of energy in the photofragments from vibrational predissociation relied on measurements of fragment angular distributions. In these studies on H_2O [17, 18] and HF [18] clusters, ethylene dimers [19], and SF_6 dimers [20], the angular distributions were peaked near $0°$, indicating that only a small fraction of the available energy was released as relative translational energy. This result is consistent with the *energy gap* and *momentum gap* laws [14, 15] which have been proposed for these systems. Unfortunately, the angular resolution in these experiments was insufficient to resolve the individual internal states of the fragments.

More recently, R. E. Miller and coworkers have measured photofragment angular distributions using a rotatable molecular beam apparatus with angular resolution adequate to determine the rotational state distribution of the photofragments [21–30]. Their method involves the use of an F-center laser to photodissociate the complex of interest in a molecular beam, while the angular distribution of the fragments is determined by means of a bolometer detector cooled with liquid helium. The photofragment angular distribution is measured by rotating the molecular beam source about the photolysis point. Although the laser provides extremely high resolution for the excitation step, the angular resolution of the photofragments is quite limited. Nevertheless, the angular resolution of this apparatus was found to be sufficient to resolve the individual rotational states of the two HF photofragments (j_1, j_2) from the HF dimer. The experimental method will be illustrated using the HF dimer; however, the photofragment angular distribution method is applicable to any system in which the density of final states is very low. In the limit of high density of final states, however, the angular distribution becomes structureless and final state distributions cannot be determined uniquely. To date, the systems investigated at the state-to-state level of detail with this method include $(HF)_2$ [21–25], H_2–HF [26], N_2–HF [25, 27, 28], Ar–CO_2 [29, 30], and CO_2–HF [28]. The final state distributions for each of these systems in turn will be discussed below.

A. Vibrational Predissociation of HF Complexes

1. Experimental Method

Upon excitation of one of the HF stretching vibrations in $(HF)_2$, the dimer predissociates on the ground vibrational state potential energy surface, and excess energy flows into product translation and rotation. Each combination of rotational quantum numbers for the two HF fragments corresponds to a unique rotational energy (E_{rot}), giving a specific (quantized) value for the translational energy of the fragments (E_{trans}) in the center-of-mass frame of reference [21–25],

$$E_{trans} = h\nu + E_{init} - E_{rot} - D_0 \tag{3.1}$$

where E_{init} is the initial internal energy of the dimer in its ground state and D_0 is the dissociation energy of the dimer. (This energy conservation relationship can be generalized to accommodate rotational and/or vibrational excitation of the fragments by replacing E_{rot} with the total internal energy of the two fragments, E_{frag}.) The result is that each (j_1, j_2) rotational channel gives rise to a Newton sphere with a well defined radius, v_{HF}. The Newton spheres corresponding to the major product channels following ν_1 excitation (“free” HF stretch) of the HF dimer, namely (1, 11), (4, 10), and (3, 10), are shown in Fig. 1. The origin of the Newton spheres lies at the head of the HF dimer velocity vector. In the laboratory frame (origin at the tail of the HF dimer velocity vector), the fragments will be scattered over a range of angles from 0° (along the HF dimer velocity vector) to the laboratory angle corresponding to the tangent to the Newton sphere for that (j_1, j_2) product channel. The larger the kinetic energy release, the larger the Newton sphere and therefore the larger the range of angles accessed.

Since the lifetime of the HF dimer is on the order of nanoseconds [12, 13], the dimer undergoes many rotational periods prior to dissociation. As a result, the angular distributions will be spherically symmetric (isotropic) in the center-of-mass frame. Therefore, the maximum fragment flux for each (j_1, j_2) product channel will be seen by the detector at the laboratory angle corresponding to the tangent to the Newton sphere for that channel [21, 22]. At this laboratory angle, there will be a large range of center-of-mass angles contributing to the signal. The amplitude of each peak in the angular distribution is proportional to the probability of dissociation into a specific product rotational channel.

The (j_1, j_2) correlated product rotational distribution of the HF products were extracted from the photofragment angular distribution using the following procedure: First, the angular distributions were converted to functions of the translational energy release in the dissociation process, after

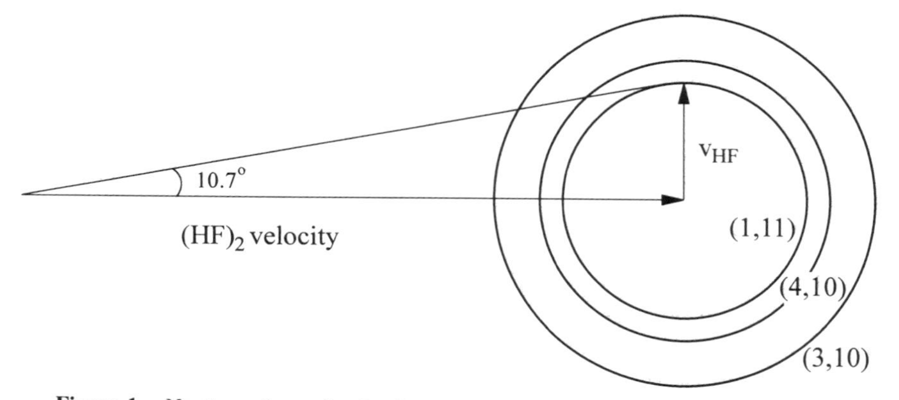

Figure 1. Newton spheres for the three major product channels following ν_1 excitation (“free” HF stretch) of the HF dimer. The radius of each sphere is defined by the HF fragment velocity (v_{HF}) in the center-of-mass frame of reference. The transformation from the center-of-mass to laboratory frame yields peaks in the angular distribution at laboratory scattering angles that are tangent to one of the Newton spheres. This is illustrated for the $(j_1, j_2) = (1, 11)$ channel which appears in the photofragment angular distribution at 10.7°.

taking into account the experimental geometry, initial velocity distribution of the dimer, and laser polarization in a Monte Carlo averaging procedure. Next, the HF dimer dissociation energy (D_0) was adjusted until the peaks in the angular distributions were consistent with the possible (j_1, j_2) exit channels, resulting in a value of D_0 = 1062 ± 1 cm^{-1} [21, 22]. Finally, the translational energy release was converted into E_{rot} via conservation of energy and the (j_1, j_2) rotational distribution for the two HF fragments was obtained from E_{rot}. The deconvolution procedure was found to be most effective at low kinetic energies where the product rotational channels are well separated in energy.

2. *HF Dimer*

The HF dimer, a prototype hydrogen-bonded complex, has been extensively investigated from both experimental and theoretical perspectives. The ground state structure of the dimer and its tunneling dynamics were first determined by Klemperer and coworkers using a molecular beam electric resonance method [31] and subsequently via microwave studies [32–34]. The infrared spectrum of the dimer recorded by Pine and coworkers showed large differences in the line widths (and corresponding vibrational predissociation lifetimes) for the "free" HF vibration (ν_1) as compared to the "hydrogen-bonded" HF stretch (ν_2) [12]. Excitation of the "hydrogen-bonded" HF stretch results in a decay rate 20 times faster than that observed upon excitation of the "free" HF vibration. The strong intramolecular mode dependence of the vibrational predissociation lifetime was attributed to the stronger coupling of the "hydrogen-bonded" HF vibration to the intermolecular hydrogen-bonding coordinate than for the "free" HF stretch. R. E. Miller and coworkers have since investigated the vibrational predissociation dynamics of the HF dimer at the state-to-state level of detail by measuring the correlated rotational state distributions for the two HF fragments for various initially prepared states of the dimer [21–25]. These final state distributions and a comparison of the experimental results with theoretical calculations of the product rotational distributions are detailed in this section.

The energetically available HF product rotational channels (j_1, j_2) following excitation of the "free" HF stretch (ν_1) are depicted in Fig. 2. The experimentally obtained product rotational distribution following excitation of the HF dimer on the ν_1 $^RR_0$ $(0)^+$ transition is given in Fig. 3 [22]. More than 50% of the HF fragments appear in three product channels: (j_1, j_2) = (1, 11), (4, 10), and (3, 10), with translational recoil energies of 183.5, 255.5, and 419.4 cm^{-1}, respectively. These most probable product channels have relatively low translational energy release, consistent with a simple *energy gap* model. *Energy gap*, however, is not the only criterion controlling the product rotational distribution. For example, the (5, 10) channel, with

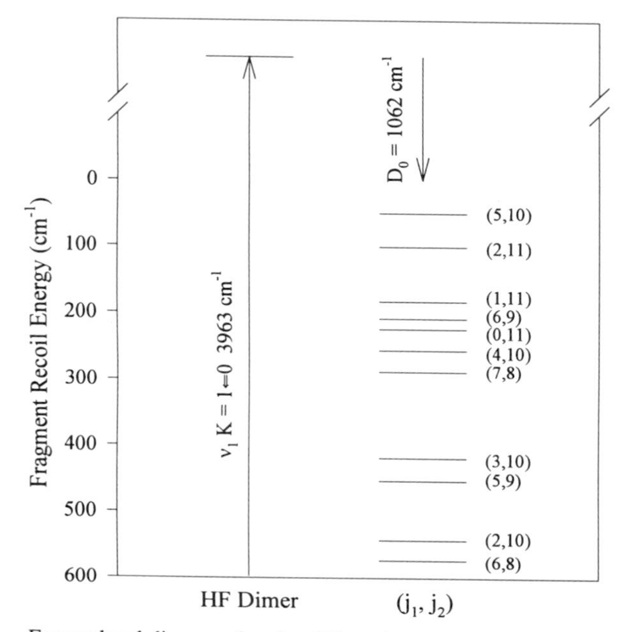

Figure 2. Energy level diagram showing HF product rotational channels (j_1, j_2) that may be populated following ν_1 excitation of the HF dimer ("free" HF stretch) on the $K = 1 \leftarrow 0$ sub-band at 3963 cm^{-1}. The energy scale gives the translational recoil of the HF fragments associated with each (j_1, j_2) channel.

only 50.9 cm^{-1} of translational energy release, is significantly less populated than the (1, 11) channel. Furthermore, comparison of the HF product distribution with a statistical distribution (phase-space theory) shows that the final state distribution resulting from vibrational predissociation of the HF dimer is highly nonstatistical. The most strongly preferred channels are those that have one fragment that is highly rotationally excited (high j) and the other with low rotational excitation (low j), while the channels with $j_1 \approx j_2$ are suppressed. The correlation between high j and low j can be rationalized by geometric considerations if the dissociation occurs via a sudden repulsive force along the intermolecular bond. In this circumstance, the torque on the proton acceptor will be small, since the repulsive force is mainly directed towards its center of mass (near the F atom). By contrast, a large torque will be exerted on the H atom of the proton donor due to the fact that this H atom lies far from the corresponding center of mass. This impulsive picture of the dissociation process suggests that the product rotational state distributions are principally determined by final state interactions as the fragments separate.

If dissociation were to occur via a statistical process, the initial excitation

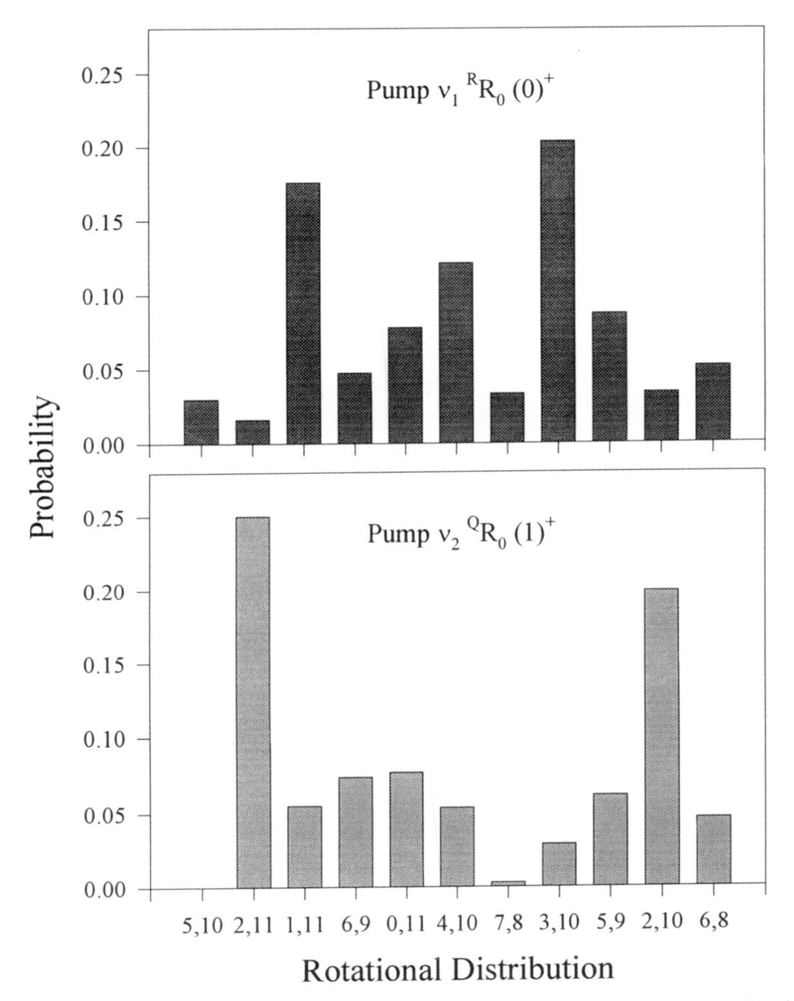

Figure 3. Rotational distributions for the HF fragments (j_1, j_2) following vibrational predissociation of the HF dimer. *Top panel:* excitation of the "free" HF stretch (ν_1). *Lower panel:* excitation of the "hydrogen-bonded" HF stretch (ν_2). The final state distributions were extracted from photofragment angular distributions as described by Bohac, Marshall, and Miller [22].

would be fully randomized over the intermolecular modes before dissociation. This would likely favor product channels where $j_1 \approx j_2$, contrary to the experimental observations.

The product rotational distribution following excitation of the "hydrogen-bonded" HF vibration (ν_2) is also shown in Fig. 3 [22]. Since the ν_2 level of the dimer lies nearly 100 cm^{-1} lower in energy than ν_1, one product channel $(j_1, j_2) = (5, 10)$ is closed. In addition, the (2, 11) channel becomes very nearly resonant with the total energy available to the photofragments. This near resonant channel as well as the (2, 10) channel ($E_{trans} = 447.6$ cm^{-1}) dominate the final state distribution. Again, the final state distribution is highly nonstatistical, and the propensity to favor high j − low j product channels is observed. Thus, even though vibrational predissociation proceeds approximately 20 times faster for ν_2 as compared to ν_1 excitation, both product rotational distributions appear to be governed by the forces acting on the fragments as they separate.

The photofragment angular distributions were also found to exhibit a strong laser polarization dependence, with a high correlation between the direction of the transition moment in the HF dimer ($\boldsymbol{\mu}$) and the recoil velocity vector of the fragment ($\mathbf{v}_{HF}$) [23]. The laser polarization effect could be observed, despite the long lifetime of the HF dimer, because single rotational levels of the dimer prepared by the pump laser do not evolve on the time scale of molecular rotation. The initial alignment of the dimer is therefore retained until fragmentation occurs. The best fit for the data was obtained with an anisotropy parameter of $\beta = 1.8$, analogous to the limiting situation ($\beta = 2$) in which fragments are preferentially ejected along the inertial a-axis of the dimer (along the F–F axis). The highly peaked fragment distribution along the a-axis is consistent with the impulsive mechanism proposed for photodissociation of the HF dimer (see Fig. 11 of [22]).

The final state distributions [22] as well as the predissociation lifetimes [12, 13], are found to be essentially independent of the initial rotational state J of the dimer. Some memory of the initially prepared wave function, however, is carried over to the product rotational distribution [22]. The two tunneling states associated with the ν_1 transition, separated in energy by only 0.3 cm^{-1}, exhibit noticable differences in their product rotational distributions. Small differences between the two tunneling states were also observed in the total rate of dissociation. These differences between the symmetric and antisymmetric tunneling states are not a result of energy considerations, but rather are probably due to a symmetry effect.

A more striking influence of the initially prepared level in the HF dimer on the fragmentation dynamics is seen upon excitation of combination bands composed of an intermolecular vibration and the "free" HF stretch (ν_1) [24]. The product rotational distributions have been measured following excitation of combination bands with F–F stretching (ν_4) and *trans*-bending (ν_5) vibrations. The additional energy associated with the intermolecular vibrations (127.5 and 166.5 cm^{-1} for $\nu_1 + \nu_4$ and $\nu_1 + \nu_5$, respectively) enables three additional final rotational states to be energetically accessible, namely $(j_1, j_2) = (7, 9)$, (8, 8), and (3, 11). These three additional open channels are seen in the angular distributions from both combination bands. They do not dominate over higher kinetic energy channels, indicating again that the *energy gap* in not the most important factor influencing the dissociation dynamics. The additional energy derived from the ν_4 F–F intermolecular stretch flows directly into kinetic energy of the fragments, increasing the average kinetic energy release by 150 cm^{-1} as compared to excitation of the ν_1 fundamental. The intermolecular stretch seems to be effectively decoupled from the bending motion of the complex and therefore from the fragment rotational degrees of freedom. In contrast, excitation of the ν_5 intermolecular bend does not influence the kinetic energy release. Rather, the additional energy appears as rotational excitation of the fragments, particularly in the (7, 9) and (3, 11) channels. The correlation between high j and low j is somewhat weaker for the $\nu_1 + \nu_5$ combination band than that found for the "free" HF fundamental, presumably because the intermolecular bending excitation increases the probability of finding the dimer in the centrosymmetric C_{2h} geometry, which favors fragments with similar HF rotational excitation.

From a theoretical perspective, much of the effort has focused on deriving potential energy surfaces for the HF dimer. The PES is highly anisotropic and has relatively strong intermolecular interactions. Several *ab initio* calculations that give quantitative agreement with spectroscopic observables have been carried out [35]. In addition, a number of semiempirical surfaces have been developed that give excellent agreement with spectral data [36]. Since the spectroscopic measurements probe the PES primarily in the vicinity of the potential minimum, these comparisons show the validity of these potentials in the region of the minimum. Dynamical experiments sample a broader region of the PES and therefore can provide a stringent test of the PES in regions beyond the potential minimum.

The first attempt at calculating the state-to-state vibrational predissociation dynamics of the HF dimer was made by Halberstadt et al. [37], using close coupling methods and freezing the rotational degree of freedom associated with the proton acceptor. Using this atom-diatom approximation, the HF fragment was found to be preferentially produced in the highest energetically accessible rotational state. More recently, Zhang and Zhang [38] have performed four-dimensional quantum dynamics calculations of the vibrational predissociation of the HF dimer, employing the Quack and Suhm surface [36] as well as the PES of Bunker et al. [35]. The calculated lifetimes and product state distributions were found to be in qualitative agreement with experiment. Both PESs indicate that the low–kinetic energy, high–rotational energy channels are favored in the fragmentation process, with about 90% of the available energy going into HF rotation, in accord with experimental and the previous theoretical findings. On closer examination, the lifetimes and product rotational distributions computed for the two PESs are found to differ substantially from each other. This shows the high sensitivity of the vibrational predissociation dynamics of the HF dimer to the details on the underlying PES. Further refinements of these PESs will be required as the calculated product distributions agree better with one another than with the experimental results. The implication is that both potentials underestimate the anistropy of the potential in the regions sampled during the fragmentation process.

The influence of initial intermolecular excitation on the fragmentation dynamics has also been examined theoretically [39]. The quantum dynamics calculations of Bacic, Zhang, and coworkers have shown that intermolecular bending excitation (ν_5) in conjunction with the "free" HF (ν_1) fundamental vibration increases the product rotational energy by an amount that is similar to the (calculated) bending frequency (161 cm^{-1}) [39]. The theoretical calculations predict that the intermolecular bending excitation emerges almost completely as rotational excitation of the HF products, in accord with experimental findings [24]. The HF dimer combination band involving intermolecular stretching excitation ($\nu_1 + \nu_4$) results in an increase in the average translational energy release by about 150 cm^{-1}. The calculations indicate a rather direct flow of the intermolecular stretching excitation (ν_4) into translational energy of the HF fragments, again as observed experimentally [24]. The agreement between theory [39] and experiment [24] is quite good in terms of these average quantities, i.e., product rotation and translation. There is also qualitative agreement regarding the dominant open product rotational channels; however, significant differences are found between the computed and observed product rotational distributions.

Final state distributions have also been computed for the bend (ν_5) and stretch (ν_4) combination bands of the "hydrogen-bonded" HF stretch (ν_2) [39], which have not been experimentally determined. For $\nu_2 + \nu_4$ excitation, the increase in the translational energy of the products is predicted to be comparable to the calculated ν_4 intermolecular stretching frequency of 130 cm^{-1}. This implies nearly complete energy transfer of the intermolecular stretch to product translation as was also observed following $\nu_1 + \nu_4$ excitation. By contrast, a significant difference is predicted between the product distribution for the combination band involving intermolecular bend (ν_5) with the "hydrogen-bonded" HF stretch (ν_2) and the band seen with ν_5 excitation in combination with the "free" HF stretch (ν_1). With $\nu_2 + \nu_5$ excitation, the theoretical calculations indicate that a large fraction of the energy contained in the ν_5 bending vibration is converted to *translational* energy of the products, while ν_5 excitation in combination with ν_1 resulted in product *rotational* excitation. This suggests that the intermolecular coupling between the intermolecular bend and HF strech is stronger for the "hydrogen-bonded" HF (ν_2) than for "free" HF (ν_1).

3. H_2/D_2–HF

High-quantity *ab initio* calculations have been performed on H_2/D_2–HF to derive a PES for this system [40]. The resultant H_2/D_2–HF PES is highly anisotropic and is dominated by quadrupole–dipole and quadrupole–quadrupole interactions. The minimum energy configuration is T-shaped, with the H side of the HF molecule pointing towards the midpoint of the H_2 molecule. The very encouraging agreement between the bound-state spectra predicted from this PES and available spectroscopic data [41], demonstrate the accuracy of the *ab initio* potential within the region of the PES that is accessed spectroscopically. Vibrational predissociation of H_2/D_2–HF provides a way to examine the dynamics taking place on this PES more fully [41, 42].

Photofragment angular distributions have been measured for the HF fragments following excitation of ortho and para D_2–HF in the HF stretching region [26]. The complexes derived from ortho D_2 ($j_{D_2} = 0$) and para D_2 ($j_{D_2} = 1$) could be selectively prepared by exciting a $\Sigma \leftarrow \Sigma$ band at 3950 cm^{-1} for ortho D_2-HF and a $\Pi \leftarrow \Pi$ band at 3948 cm^{-1} for para D_2-HF. Both experiment and theory indicate that the D_2 fragments are produced in with one quantum of vibrational excitation, following vibration to vibration energy transfer ($V \rightarrow V$) across the weak intermolecular bond from the initially excited HF stretch, as illustrated in Fig. 4. After accounting for the energy required to break the D_2–HF bond (D_0 is assumed to be 68 cm^{-1} for both ortho and para D_2–HF), the remaining $\sim$900 cm^{-1} of energy can be partitioned between HF rotation, D_2 rotation, and translational recoil of the fragments. For para D_2, the most probable product channels are found to be $(j_{HF}, j_{D_2}) = (4, 3)$ and (3, 3). For the ortho D_2 complex, the nearly resonant (3, 4) channel is by far the most probable channel, with nearly 50% of the products being produced in this channel. Note that vibrational predissociation of the ortho and para complexes can give rise to only even or only odd j D_2 products, respectively. In both cases, the most probable channels are those with $j_{HF} \approx j_{D_2}$. This propensity is quite different than that found for the HF dimer where the preferred channels had a correlation between high j and low j for the HF fragments.

For ortho D_2–HF, product rotational distributions based on the *ab initio* potential of Clary [40], which were predicted in advance of experimental measurements, are found to be in almost quantitative agreement with experimental results. The agreement between experiment and theory is also quite good for para D_2–HF, with both experiment and theory showing the preference for populating states with $j_{HF} \approx j_{D_2}$ in low rotational levels. Similar results were obtained by Zhang, Zhang, and Bacic using a time-dependent wavepacket approach [43] as compared to Clary's time-dependent close-coupling method [40]. The dynamics are dominated by the V–V coupling, which is expected to be electrostatic in nature.

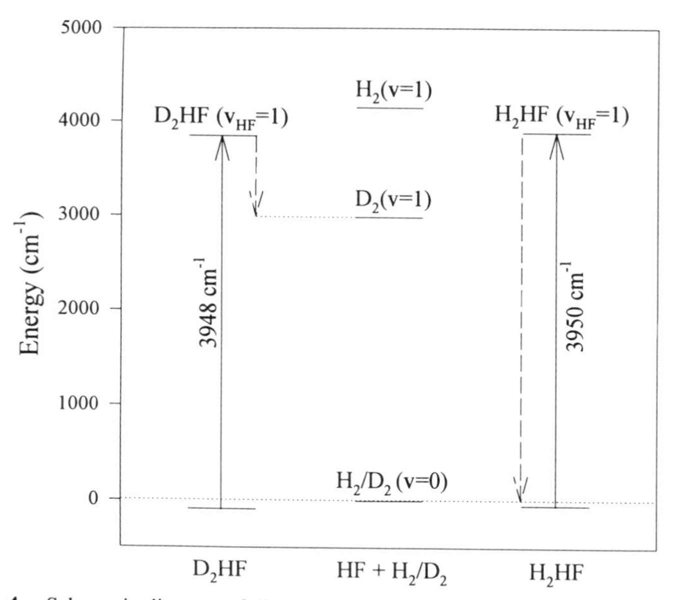

Figure 4. Schematic diagram of dissociation pathways for H_2/D_2–HF complexes following HF vibrational excitation. For simplicity, only the Π states of the complex are shown. The D_2–HF complexes may dissociate via the near resonant $V \rightarrow V$ pathway, yielding D_2 ($v = 1$) fragments with approximately 900 cm^{-1} of excess energy to be disposed in rotation and translation. The H_2–HF complexes can only dissociate via the inefficient $V \rightarrow R/T$ pathway with approximately 3900 cm^{-1} of excess energy. Figure adapted from Lovejoy, Nelson, and Nesbitt [41].

For H_2–HF, only the infrared transition associated with the ortho complex ($j_{H_2} = 1$) have been observed to date (Π ← Π transition at 3950 cm^{-1}), and, as a result, photofragment angular distributions have been obtained only for the ortho complex [26]. The vibrational predissociation dynamics for H_2–HF are quite different from those for D_2–HF, since the V–V channel is energetically closed (see Fig. 4). For H_2–HF, the ~3900 cm^{-1} of excess energy released from the HF vibration (beyond that required to break the H_2–HF bond, namely 48 cm^{-1}) flows into translational recoil and/or rotational excitation of the fragments. Since both the H_2 and HF fragments are formed without vibrational excitation, many rotational channels are open. In fact, so many channels are open that the relatively unstructured photofragment angular distribution could not be assigned to specific (j_{HF}, j_{H_2}) channels. The highest probabilities were observed for states with high rotational energy, consistent with the minimization of energy in translational degrees of freedom [14, 15]. An average kinetic energy release of 985 cm^{-1} was deduced from the observed angular distribution.

Theoretical calculations of Zhang, Zhang, and Bacic [43] for the vibrational predissociation of H_2–HF yield quite different results. They predict a significantly longer predissociation lifetime than observed experimentally [41, 42] (1600 ns vs. 27 ns) and a much larger average kinetic energy release of 2210 cm^{-1}. Furthermore, the calculations indicate that product channels with large kinetic energy release should have the highest probability, as compared to the product channels observed experimentally, which have significantly smaller translational energies. This suggests that the anisotropy of the PES is too weak, most likely at short range in the region of the repulsive wall. The deficiency may be due to high-order terms in the anisotropy of the potential that are needed to accommodate large rotational excitation of the fragments [26], since this same potential works quite well for D_2–HF where rotational excitation of the photofragments is small. As a result, the H_2–HF complex is forced to dissociate via states of lower rotational excitation and corresponding higher translational energies. This could also lead to slower predissociation based on *energy gap* arguments.

4. N_2–HF

The photofragment angular distribution method has also been applied to the N_2–HF system following HF vibrational excitation [27]. The dissociation pathways are illustrated in Fig. 5. In this case, however, the structure observed in the angular distribution was insufficient to assign a unique set of final states. This is primarily a result of the large number of open channels available, corresponding to rotational states associated with the N_2 fragment. In addition, each product channel gives rise to two peaks in the angular distribution, since the two fragments with different masses scatter with different angles in the laboratory frame. To overcome these problems, a pump-probe method was introduced in which a second F-center laser was used to probe the final states of the HF fragments. These pump-probe experiments are carried out with angular resolution of the photofragments, simultaneously giving information on the rotational and translational distributions of the HF fragments. Furthermore, a dc electric field was used to orient the linear N≡N–HF complex in the laboratory frame [25, 28]. Since the complex dissociates preferentially along its axis (determined from laser polarization studies), the two fragments recoil to different sides of the apparatus. This approach enables the angular distributions of the N_2 and HF fragments to be detected independently. With this additional information, the final state distribution could be determined uniquely.

Surprisingly, the probe laser detected only two rotational states of the HF product with appreciable population: only $j_{HF} = 7$ (at small scattering angles) and $j_{HF} = 12$ (at relatively large scattering angles) were observed with essentially equal probability [27]. Since the change in kinetic energy release

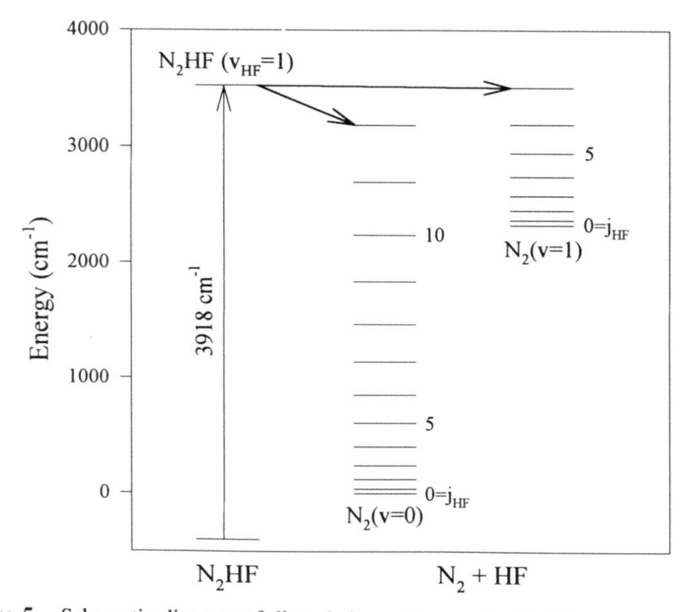

Figure 5. Schematic diagram of dissociation pathways for N_2–HF complexes following HF vibrational excitation. HF fragments are produced exclusively in $j_{HF} = 12$ and $j_{HF} = 7$, the highest energetically available HF rotor levels associated with N_2 the $v = 0$ and $v = 1$ channels [27]. The rotational states associated with the N_2 fragment are not shown.

is relatively small compared to the large energy difference between these two HF rotor levels, the balance of energy must be carried away by the N_2 fragment as vibrational excitation. Thus, HF ($j_{HF} = 7$) products are produced with N_2 ($v = 1, j$), while HF ($j_{HF} = 12$) fragments are generated with N_2 ($v = 0, j$) products (see Fig. 5). Finally, the N_2 rotational distributions were extracted by fitting the N_2 and HF photofragment angular distributions, after taking into account laser polarization dependence ($\beta = 2$), dc electric field strength, and the N_2–HF dissociation energy (398 cm^{-1}). The resultant distributions are displayed in Fig. 6 [27].

Energy gap appears to be an important factor in determining which final states are important. For the N_2 $v = 0$ and $v = 1$ channels, only 1% and 5%, respectively, of the total available energy appears in translation. Most of the excess energy appears as HF rotation and/or N_2 vibration, with 90% in HF rotation for the N_2 ($v = 0$) channel and 32.7% in HF rotation when N_2 is vibrationally excited (accommodating 66% of the available energy). N_2 rotation accounts for only 5% and 0.3% of the total available energy for the N_2 $v = 0$ and $v = 1$ channels, respectively.

The HF rotational distribution is highly nonstatistical as only two rotor states, $j_{HF} = 7$ and 12, are populated. These HF rotor states are the highest

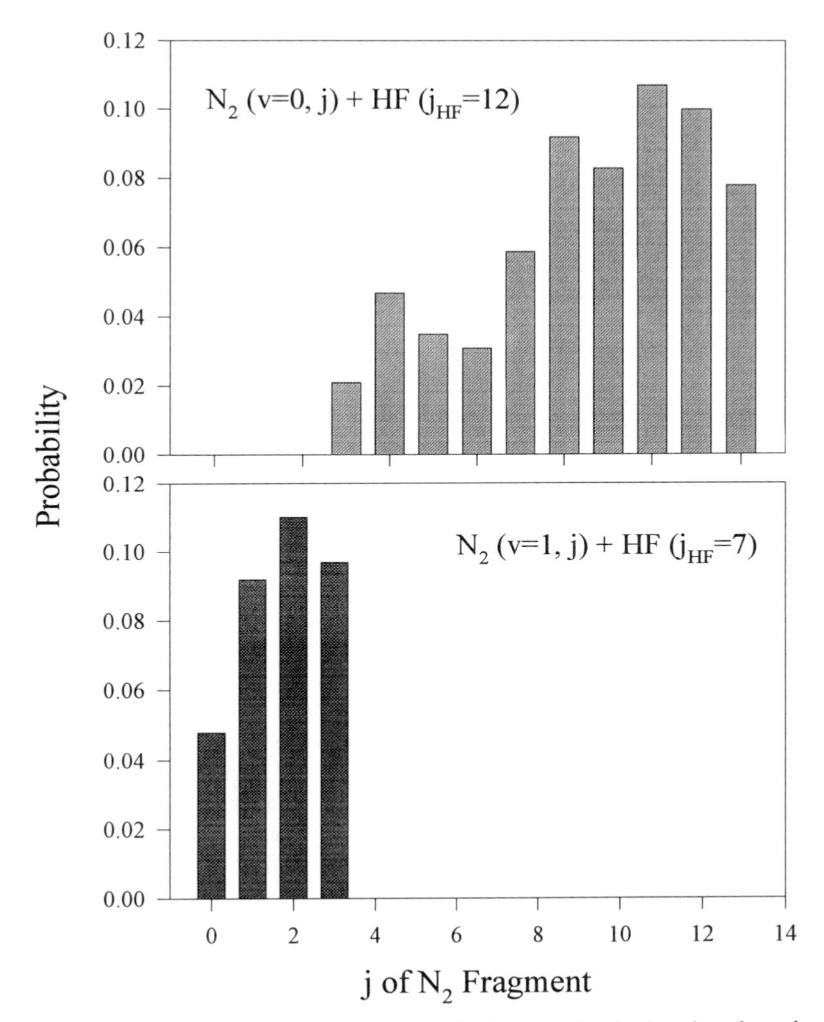

Figure 6. N_2 product rotational distributions for the two vibration/rotation channels observed following vibrational predissociation of N_2–HF. *Upper panel:* N_2 rotational level probabilities associated with the N_2 ($v = 0, j$) + HF ($j_{HF} = 12$) channel. *Lower panel:* N_2 product states occurring with the N_2 ($v = 0, j$) + HF ($j_{HF} = 7$) channel. In each case, only a single HF rotor state is populated. Figure adapted from data of Bemish et al. [27].

energetically available levels associated with the N_2 $v = 1$ and $v = 0$ channels, respectively. This same result emerges for NO–HF complexes following HF vibrational excitation (Section IV.A.4), where again only two HF rotor states are populated (corresponding to the NO $v = 0$ and $v = 1$ channels) [44]. The inverted rotational distributions for N_2 with $v = 0$ and

$v = 1$ (Fig. 6) are again consistent with minimization of energy in translational recoil. The rotational excitation of the N_2 fragment is distributed nearly statistically. Both the HF and N_2 rotational distributions are significantly "hotter" than those derived from the initial bending motions in the complex (Franck–Condon limit), suggesting that both fragments experience a significant torque in the dissociation process.

B. Vibrational Predissociation of CO_2 Complexes

The vibrational predissociation dynamics of Ar–CO_2 has also been measured at the state-to-state level of detail. This system, like D_2–HF, examines the role of product vibrational degrees of freedom in the dissociation dynamics. In this case, energy is transferred from a high-frequency CO_2 vibration to lower-energy CO_2 modes ($V \rightarrow V$). The T-shaped Ar–CO_2 complex [45] is prepared in states of different energies within the $(10^01)/(02^01)$ Fermi diad, using an F-center laser operating at 2.7 μm; line width measurements indicate that vibrational predissociation lifetimes are in excess of 50 ps [46, 47]. The final state distributions are determined by measuring the angular distributions of the photofragments [29, 30]. The dominant dissociation channels correspond to near-resonant intramolecular V–V energy transfer processes, and the $(21^10)/(13^10)/(05^10)$ CO_2 triad was found to constitute the most important product vibrational channels. Excitation of the upper member of the $(10^01)/(02^01)$ diad leads to production of a CO_2 fragment in the upper member of the product triad, the highest energetically available vibrational level. For excitation of the lower member of the diad, the angular distribution suggests that the middle member of the product triad is the one principally populated. These assignments of product vibrational channels also enabled the dissociation energy for Ar–CO_2 to be refined at 166 cm^{-1} [29]. In Ar–CO_2, the vast majority of the excess energy flows into fragment vibration; the balance of the energy goes into rotational excitation of CO_2 and translational recoil.

The product rotational distributions exhibit a strong propensity for formation of the CO_2 fragments in odd j states [30]. Dissociation from the upper member of the diad yields CO_2 fragments preferentially in final rotational states with $j = 5$, 7, and 9 (in the upper member of the product vibrational triad); while dissociation from the lower member of the diad produces ~80% of the fragments in the $j = 13$ and 15 rotor states (of the middle member of the product triad). These CO_2 rotational distributions are shown in Fig. 7 [30]. Product rotational states up to $j = 10$ and $j = 16$ are energetically available for the upper and lower members of the initially prepared diad, respectively. Thus, product channels that minimize the energy release to translationals are most populated, subject to the additional constraint that odd j's are strongly favored over even j's. This propensity for odd j's begins to break down with increasing rotational excitation of the

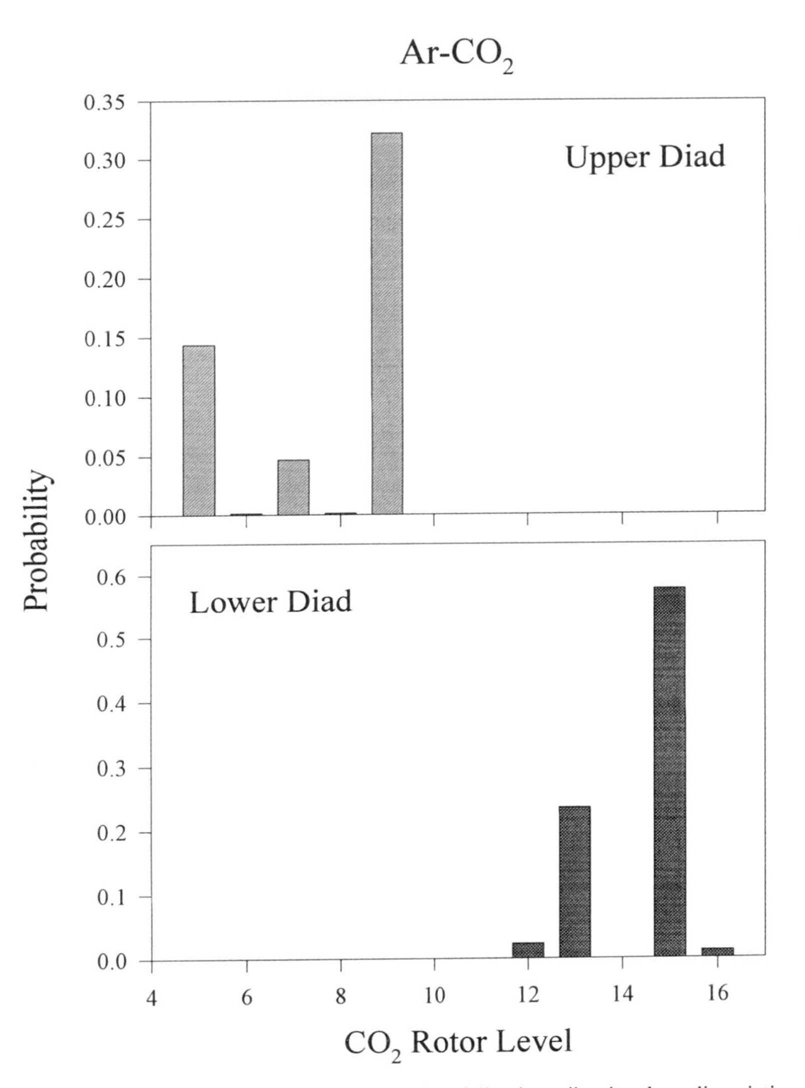

Figure 7. CO_2 fragment rotational distribution following vibrational predissociation of Ar–CO_2 from the upper member of the $(10^01)/(02^01)$ Fermi diad (*upper panel*) and lower member of the diad (*lower panel*). The CO_2 fragments are produced in near resonant vibrational levels with the upper and middle member of the $(21^10)/(13^10)/(05^10)$ CO_2 triad principally populated in the upper and lower panels, respectively. A strong propensity is observed for formation of the CO_2 fragments in *odd* rotor states. Figure adapted from Bohac et al. [30].

Ar–CO_2 complex. The propensity can be understood if one considers the symmetry properties of the CO_2 molecule [30]. The initially prepared state of the Ar–CO_2 complex in the $(10^01)/(02^01)$ Fermi diad can be described by a superposition of states with only odd values of j; as a result, the propensity to produce final rotational states with odd j's can be seen as a preference to retain the character of the initially prepared state. A similar propensity has been predicted theoretically [48] and observed experimentally [49] for inelastic collisions of CO_2 in which the vibrational angular momentum of the CO_2 molecule changes from zero to a nonzero value, or vice versa.

The dependence of the angular distribution on laser polarization shows a strong vector correlation between the transition moment of the Ar–CO_2 complex ($\boldsymbol{\mu}$) and the recoil velocity vector of the fragments ($\mathbf{v}$), indicating that the center-of-mass angular distributions are highly peaked along the inertial a-axis of the parent complex [29, 30]. Again, this effect relaxes towards an isotropic distribution with increasing rotational excitation of the parent complex.

Predissociation of CO_2–HF has also been examined upon infrared excitation of the HF stretch at 3910 cm^{-1}, using a probe laser to detect HF fragments in concert with bolometric detection of the photofragment angular distribution [28]. Only a single HF rotational level is populated, namely $j = 6$. If the CO_2 (00^01) vibration is excited in parallel with this HF rotor state, then nearly all of the available energy would be accommodated (the CO_2–HF dissociation energy is ~ 700 cm^{-1}). As in N_2–HF, this suggests that vibrational predissociation proceeds via a V–V energy transfer process across the weak intermolecular bond,

$$CO_2\ (00^00)\text{—HF}(v = 1) \longrightarrow CO_2\ (00^01) + \text{HF}(v = 0, j = 6) \quad (3.2)$$

with the minimal balance of energy going into CO_2 rotation and relative translational motion.

IV. PUMP-PROBE METHODS IN THE GROUND ELECTRONIC STATE

An alternative approach for obtaining final-state distributions following vibrational predissociation on the ground electronic state surface involves the use of a pump-probe laser scheme. King and coworkers have utilized an extremely powerful infrared pump–UV probe technique to investigate NO-containing complexes, specifically, NO dimer [9, 50, 51], NO–C_2H_4 [52], and NO–HF [44]. In these studies, infrared excitation is used to initiate the vibrational predissociation process, while the NO fragment is probed by laser-induced fluorescence (LIF). The internal energy, kinetic energy, and rate of appearance of the NO fragments are measured. Similar pump-probe schemes have been devised to examine the HCl fragments produced upon vibrational predissociation of acetylene–HCl and HCl dimers [53, 54]. In this case, the internal state distribution of the HCl products is probed using resonance-enhanced multiphoton ionization.

A. Vibrational Predissociation of NO Complexes

1. Experimental Method

The infrared pump–UV probe scheme used to investigate the vibrational predissociation dynamics of NO complexes is shown in Fig. 8 [9, 50, 51]. A frequency-doubled dye laser operating at ~ 226 nm on the NO $A\ ^2\Sigma^+ - X\ ^2\Pi_{1/2,3/2}$ 0–0 transition is utilized to probe the NO fragments by laser-induced fluorescence (LIF). Two different kinds of experiments were performed: frequency- and time-resolved measurements.

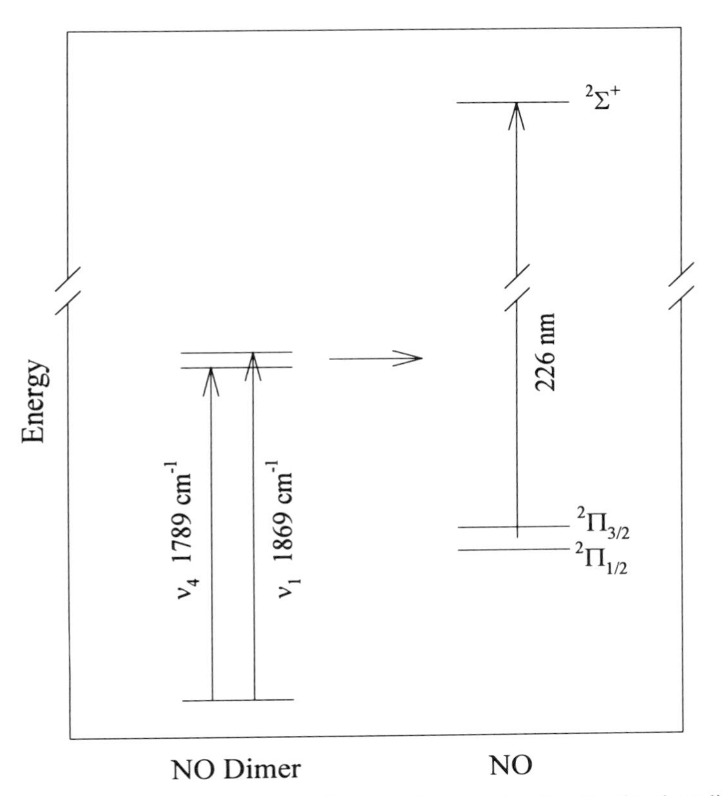

Figure 8. Infrared pump–UV probe scheme used to examine the vibrational predissociation dynamics of NO dimers. The internal energy, kinetic energy, and rate of appearance of the NO fragments are probed in frequency- and time-domain experiments by laser-induced fluorescence on NO $A\ ^2\Sigma^+ - X\ ^2\Pi$ transitions.

In frequency domain experiments [50, 51], LIF measurements at a resolution of 0.3 cm^{-1} enabled the relative population of NO fragments formed in rotational (j), spin-orbit (Ω), and lambda-doublet (Λ) states to be ascertained. The kinetic energy of the NO fragments was also determined in the frequency domain by Doppler spectroscopy. For these experiments, the UV probe frequencies were obtained from a pulse-amplified single-mode continuous-wave ring laser which was frequency doubled and then sum frequency mixed with 1064 nm pulses. The laser bandwidth from this laser system was 0.017 cm^{-1}. Internal energy distributions and Doppler profiles of fragments were obtained by fixing the pump laser on a specific rovibrational transition of the complex and then scanning the probe laser through NO fragment transitions. Suitable collection optics, spectral filters, and a spatial mask allowed detection of fluorescence from a region only 1.5 mm in length, where the pump laser, probe laser, and molecular beam overlap. The fluorescence signal was then detected by a photomultiplier tube.

High-resolution scans revealed the Doppler profiles for each internal state of the NO fragment, which in turn gives the translational energy distribution of the NO fragment. A single recoil speed results in a maximum Doppler shift of $\pm\Delta\nu_D$ about the transition rest frequency of ν_0. Molecules with such a δ-function speed distribution will have a Doppler profile which can be expressed as [55]

$$g(\nu) = (2\Delta\nu_D)^{-1}\,[1 + \beta_{\text{eff}}\,P_2((\nu - \nu_0)/\Delta\nu_D)] \tag{4.1}$$

where β_{eff} is an effective anisotropy that accounts for flux anisotropy and other vector correlations. Here, $P_2(x) = (3x^2 - 1)/2$. The experimentally observed Doppler profiles must also be convoluted with a Gaussian experimental resolution function, $\Delta\nu_G$. By fitting the observed Doppler profiles to the $g(\nu)$ expression, the recoil speed and spatial anistropy can be determined. Thus, the Doppler profiles contain exactly the same information as the photofragment angular distributions discussed in the previous section.

Finally, the time-domain experiments enabled the vibrational predissociation lifetimes of the complexes to be measured in real time [9, 51]. These experiments used a picosecond laser system consisting of a mode-locked cw YAG laser, which synchronously pumped two dye lasers (ω_1, ω_2). The dye laser pulses were subsequently amplified and mixed using nonlinear optical crystals to generate the necessary infrared pump ($\omega_1 - \omega_2$) and UV probe ($2\omega_1$ + 1064 nm) pulses. The UV probe pulse was variably delayed in time with respect to the infrared pump pulse.

2. *NO Dimer*

Spectroscopic studies of the NO dimer have shown that the equilibrium geometry in the ground electronic state is a trapezoidal structure with C_{2v}

symmetry [56]. Since the dissociation energy of the dimer is $\sim 800\ cm^{-1}$, infrared excitation of either of the two NO stretching fundamentals in the dimer (ν_1 at 1869 cm^{-1} or ν_4 at 1789 cm^{-1}) will lead to predissociation of $(NO)_2$, as shown in Fig. 8. The ν_1 and ν_4 vibrations of the dimer are shifted 7 cm^{-1} and 87 cm^{-1}, respectively, towards lower energy of the monomer fundamental (1876 cm^{-1}). Both state- and time-resolved measurements of the NO products have been carried out in order to elucidate the final state distribution of the products and the rate of dissociation [9, 50, 51].

In the frequency-domain experiments, Casassa et al. pumped the ν_1 vibration of the NO dimer using the frequency-doubled output of a pulsed CO_2 laser operating on the 10.6 μm $P(30)$ transition [50, 51]; for ν_4 excitation, infrared pulses were generated by difference frequency mixing of 532 nm light with a dye laser [50, 51]. A CO laser has also been used for ν_4 excitation by Brechignac and coworkers [57]. In addition, high-resolution infrared spectra of the NO dimer have been obtained by Matsumoto et al. [58] by direct absorption using a diode laser.

Following infrared excitation of the NO dimer, the internal state distributions of the NO photofragments have been determined from laser-induced fluorescence spectra [50, 51]. For ν_1 excitation, the rotational population of the NO fragments followed a Boltzmann distribution with rotational temperatures of 101 K for the lower $^2\Pi_{1/2}$ spin–orbit state and 112 K for the upper $^2\Pi_{3/2}$ spin–orbit state. The NO rotational distribution from ν_4 excitation is qualitatively the same with rotational temperatures of 90 K and 71 K for the lower and upper spin–orbit states, respectively. In both cases, the relative population in the spin–orbit states is substantially different than that expected if the spin–orbit and rotational degrees of freedom were equilibrated. The $^2\Pi_{3/2}$ state receives significantly more population than expected, implying a strong spin correlation in the predissociation process.

Doppler profile measurements of the NO products were obtained following ν_1 excitation, but were not possible with ν_4 excitation, due to inadequate infrared pump laser intensity. The Doppler profiles were "top hat" in shape, indicative of a narrow kinetic energy distribution. The NO Doppler profiles were analyzed to obtain the average kinetic energy and effective anisotropy parameters. No vector correlations were detected ($\beta_{eff} = 0$), indicating that the angular flux of NO fragments was isotropic. This analysis yielded an average kinetic energy of 400 cm^{-1}, which by conservation of energy also gives an NO dimer dissociation energy of 800 cm^{-1}.

The average product internal energies were surprisingly small, with only 135 cm^{-1} and 102 cm^{-1} (per NO fragment) for ν_1 and ν_4 excitation, respectively. This accounts for less than 25% of the available energy. The overwhelming majority of the excess energy appeared as relative translational energy of the NO fragments (75%), in stark contrast to theoretical predictions based on *energy gap* or *momentum gap* arguments.

Time-resolved measurements show a dramatic change in the appearance of the NO fragments following excitation of the ν_1 and ν_4 fundamentals [9, 51]. Predissociation from the antisymmetric stretching mode (ν_4) occurs in 40 ps, while predissociation from the symmetric stretch (ν_1) is more than 20 times slower with a lifetime of 880 ps! The mode specificity of the predissociation lifetime is quite surprising, given that the initially prepared ν_1 and ν_4 levels have similar product state distributions. Line broadening observed in high-resolution infrared spectra of $(^{14}NO)_2$ and $(^{15}NO)_2$ confirm the predissociation rates obtained in the picosecond pump-probe experiments [58].

Two mechanisms have been proposed to explain the predissociation dynamics observed for the NO dimer [51]. One involves a nonadiabatic transition from the initially excited vibrational state to a repulsive electronic surface. This is plausible since the $^2\Pi$ configurations of the NO monomers combine to form several low-lying electronic states in the dimer. A crossing to a repulsive surface could increase the rate of predissociation with minimal rotational excitation of the fragments. It is unlikely to explain, however, the mode-specific lifetimes. The other proposed mechanism is based on a mode-dependent coupling between the NO stretching vibrations and the dissociation coordinate. In comparing ν_4 and ν_1, the significantly larger spectral shift of the ν_4 vibration from the NO fundamental is indicative of a stronger coupling with the intermolecular modes for ν_4 than ν_1 [56]. The significantly faster predissociation of the NO dimer upon ν_4 excitation as compared to ν_1 excitation is also consistent with the mode-specific coupling model [9, 51]. Matsumoto et al. [58] have suggested that a combination of these two mechanisms is operative in predissociation of the NO dimer. They propose that the rate of predissociation is enhanced by an electronically nonadiabatic transition to a dissociative surface, while the vibrational potential coupling is responsible for the mode specificity.

The NO dimer predissociation dynamics has also been investigated following infrared overtone excitation of the $\nu_1 + \nu_5$ band at 3626 cm^{-1} [59]. Excitation spectra of the overtone transition, recorded by fixing the probe laser on a NO $A \leftarrow X$ transition and scanning the pump laser, revealed spectral line broadening that corresponds to a 34 ps predissociation lifetime. LIF spectra of the NO products showed that nearly 97% of the dissociation events yield one NO fragment in the $v = 0$ state and the other in the $v = 1$ state. Like the ν_1 and ν_5 fundamentals, overtone excitation results in minimal rotational energy release, with each NO fragment receiving less than 3% of the energy available after vibrational energy redistribution. The most plausible origin of the NO rotational state distribution is transformation of the initial bending (and/or torsional) wave function of the NO dimer to product rotation (Franck–Condon limit). Of the remaining energy (approximately 1000 cm^{-1}), nearly all goes into translational recoil of the fragments

(85%), as was found upon fundamental excitation of the dimer. Almost 75% of the dissociations yield one NO fragment in the lower $^2\Pi_{1/2}$ and the other in the upper $^2\Pi_{3/2}$ spin-orbit state. This provides further evidence that low-lying electronic states of the dimer may play a role in the predissociation process. A nonadiabatic process likely occurs at some point along the reaction coordinate, affecting the spin–orbit distribution of the fragments.

3. *NO–C$_2$H$_4$*

Final state distributions have also been determined for the NO fragments following infrared excitation at 953 cm^{-1} (using a CO_2 laser) of the ν_7 mode of ethylene in NO–C_2H_4 complexes [52]. LIF spectra of the NO fragments revealed a Boltzmann-like rotational distribution with a temperature of 75 K, corresponding to an average NO fragment rotational energy of 52 cm^{-1}. Doppler profiles of the NO fragments showed a kinetic energy release of approximately 105 cm^{-1} in each of the two fragments. The NO velocity vectors were isotropically distributed in space about the photolysis point. Assuming an NO–C_2H_4 intermolecular bond dissociation energy of ~400 cm^{-1}, energy balance requires that more than 50% of the available energy appear in ethylene rotation (300 cm^{-1}). Only 10% of the excess energy is observed in NO rotation, while nearly 40% of the excess energy is deposited in translational recoil.

4. *NO–HF*

The final state distribution following vibrational predissociation of NO–HF (v = 1) was determined by King and coworkers using Doppler spectroscopy of the NO fragments [44]. Prior to these photodissociation dynamics measurements, NO–HF had been characterized by infrared spectroscopy using the molecular beam optothermal technique [60]. The HF $v = 0 \rightarrow 1$ transition in NO–HF was identified at 3877 cm^{-1}, which is 84 cm^{-1} to the low energy side of the fundamental HF transition. Spectroscopic analysis yielded an equilibrium geometry in which the hydrogen atom in HF points towards the center of mass of NO, with an NO–HF angle of $30 \pm 15°$.

King and coworkers [44] generated infrared pump pulses at 2.3 μm by difference frequency mixing of a dye laser with 532 nm light followed by optical parametric amplification. The high-resolution (0.017 cm^{-1} bandwidth) probe laser at 226 nm was used to detect the internal state and kinetic energy distributions of the NO fragments by LIF.

The Doppler profile for each NO (v, j) fragment state probed was consistent with a *single* recoil speed for the NO fragment, and by conservation of linear momentum, for the HF cofragment as well. No vector correlations were observed between the transition dipole of the dimer ($\boldsymbol{\mu}$) and the NO fragment velocity vector ($\mathbf{v}$), nor between the fragment velocity and angular momentum (**j**) vectors. Furthermore, the recoil energy for the NO fragment was inversely related to the NO product internal energy, such that the sum of the NO ($v = 0, j$) fragment internal energy and its corresponding recoil energy was a constant (280 cm^{-1}). This indicates that all NO ($v = 0, j$) fragments have HF (j_0) cofragments in a single rotational level. Similar results were found for the NO ($v = 1, j$) fragments that were formed (constant energy of 1988 cm^{-1} with a NO vibrational energy component of 1876 cm^{-1}); that is, a single HF (j_1) cofragment rotor state is populated for all NO ($v = 1, j$) fragments.

The NO ($v = 0, j$) and NO ($v = 1, j$) fragments were formed with similar probability, as were the two spin–orbit states. The probability of populating a given NO ($v = 0, j$) rotor level was found to increase with internal energy up to a cutoff around 275 cm^{-1}. The NO ($v = 1, j$) rotational distribution was relatively flat out to an energy of about 100 cm^{-1}. Assuming an NO–HF dimer bond dissociation energy of 448 cm^{-1}, energy conservation requires that the NO ($v = 0, j \leq 11.5$) fragments be matched with an HF product in the $j_0 = 12$ state and NO ($v = 1, j \leq 6.5$) fragments have an HF cofragment in the $j_1 = 8$ rotor state.[1] Thus, the picture that emerges is a final state distribution for the NO–HF dimer [44] which shows a remarkable similarity to the N_2–HF system (see Section III.A.4) [27].

Again, the HF rotational distribution is highly nonstatistical with only two levels populated. These HF rotor levels, $j_1 = 8$ and $j_0 = 12$, are the highest energetically available levels associated with the NO $v = 1$ and $v = 0$ channels, respectively. The inverted rotational distribution for the NO ($v = 0$) fragment is further evidence of minimization of energy in translational recoil. The highest populated NO ($v = 1$) rotor level, $j = 6.5$, has a recoil energy of only 15 cm^{-1}, clearly showing the propensity for near-resonant exit channels that minimize the *energy gap*.

B. Vibrational Predissociation of HCl Complexes

Two groups have recently used the resonance-enhanced multiphoton ionization (REMPI) technique to probe the internal state distribution of the HCl fragments resulting from vibrational predissociation of acetylene–HCl and $(HCl)_2$ [53, 54]. In both cases, the HCl fragment internal energy distribution is probed via 2 + 1 ionization through the $V\,^1\Sigma$–$X\,^1\Sigma$ transition using photons at approximately 238 nm [61–63], and ion signals are collected on the HCl^+ mass channel of a time-of-flight mass spectrometer.

Naaman et al. [53] have examined the vibrational predissociation of acet-

[1]King and coworkers [44] also suggest a second possibility for the bond dissociation energy, namely 1769 cm^{-1}, which would be consistent with HF final states $j_0 = 9$ and $j_1 = 2$; however, this seems to be an unreasonably high value of the NO–HF intermolecular bond strength.

ylene–HCl following vibrational excitation of the symmetric CH stretch in the acetylene moiety at 3270 cm^{-1}. Previous studies of C_2H_2–HCl have shown that the complex is T-shaped, with the HCl hydrogen-bonded to the π cloud of the acetylene [64, 65]. High-resolution infrared spectra revealed that the C_2H_2–HCl absorption lines were homogeneously broadened, corresponding to predissociation lifetimes of 150–750 ps; C_2H_2-DCl lines, on the other hand, were sharp [64]. To explain the anomalously short predissociation lifetime in C_2H_2–HCl it was proposed that there was a near-resonant V–V channel in which the vibrational excitation initially localized in C_2H_2 was transferred across the intermolecular bond to HCl. Deuterium substitution of the HCl subunit increases the energy gap between the C_2H_2 and DCl vibrations by 800 cm^{-1}, thereby slowing the rate of predissociation. Using an infrared pump (difference frequency mixing of 1.06 μm with a dye laser) and REMPI probe scheme, Naaman and coworkers [53] have since shown that HCl is formed in $v = 1$ with little rotational excitation ($j = 0$ to 3). Less than 10% of the available energy was deposited in relative translation of the fragments, consistent with an *energy gap* model. Measurements of the branching ratio between HCl $v = 1$ and $v = 0$ as well as the HCl ($v = 0, j$) rotational distribution would further elucidate the mechanism for vibrational predissociation in this system.

Valentini and coworkers [54] have used stimulated Raman to excite the "hydrogen-bonded" HCl stretch (ν_2) of $(HCl)_2$ at 2853.8 cm^{-1}, and REMPI to examine the rotational distribution of the HCl fragments. This pump-probe method is illustrated in Fig. 9. Spectroscopic studies of the HCl dimer have shown that it has a nearly L-shaped equilibrium structure with a dissociation energy of $\sim$430 cm^{-1} [66]. Serafin et al. [54] have directly observed the HCl photofragments in the $j = 10$ to 14 rotor levels. The probability is largest for $j = 12$ and 14. The highest energetically available rotor level is $j = 14$, and, by energy conservation, this would imply that the highest energetically accessible level for the other HCl fragment would be $j = 4$. HCl photofragments in $j = 0$ to 9 could not be detected because the signals are overwhelmed by REMPI detection of the HCl monomer. The portion of the HCl rotational distribution observed is consistent with a statistical distribution in which every energetically accessible state of the fragments is equally probable and all open product channels are counted. As was found for the HF dimer [21, 22], most of the available energy appears as product rotation and there is minimal translational recoil energy. When further experimental results are available, it will be interesting to compare the correlated j_1, j_2 distribution for the HF dimer with that obtained for $(HCl)_2$. This may reveal the importance of the rapid tunneling and large zero-point motions in $(HCl)_2$ on the vibrational predissociation dynamics [54].

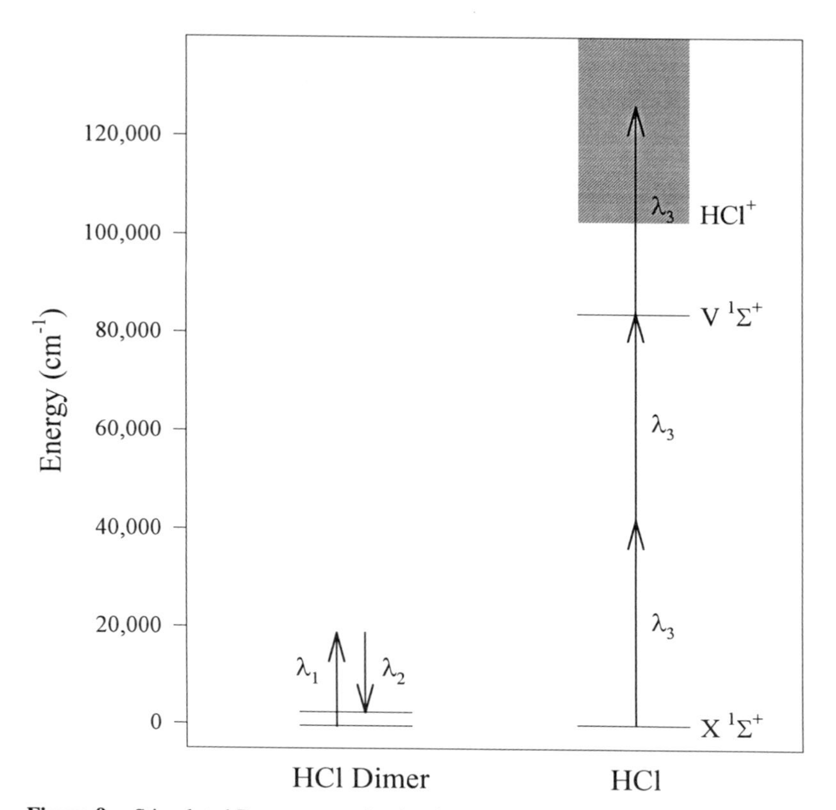

Figure 9. Stimulated Raman scattering has been used to excite the HCl dimer "hydrogen-bonded" HCl stretch (ν_2) at 2853 cm^{-1} (λ_1 = 532 nm and λ_2 = 627 nm). Following vibrational predissociation, the internal state distribution of the HCl fragments is probed by 2 + 1 resonance-enhanced multiphoton ionization with 238 nm photons (λ_3).

V. PREDISSOCIATION DYNAMICS ON EXCITED ELECTRONIC STATE SURFACES

A. Halogen–Rare Gas Systems

The spectroscopy and vibrational predissociation dynamics of halogen–rare gas complexes have been extensively investigated for the past 20 years. A great variety of experimental methods have been applied to these systems. These range from high-resolution laser-induced fluorescence studies [67, 68] to picosecond time-resolved pump-probe measurements [11]. With each new experiment has come a deeper understanding of the underlying intermolecular potential energy surfaces and the dynamics taking place on these surfaces. Great advances have also been made on the theoretical side; *ab initio* potential energy surfaces [69, 70] and three-dimensional quantum dynamics calculations [71, 72] are now available for several of these systems. This extensive body of literature on halogen–rare gas complexes has been reviewed many times [67, 68, 73], most recently in 1990 by Janda and Bieler [74]. As a result, in this review only three systems will be discussed, specifically NeICl, $NeCl_2$, and NeIBr. These systems are highlighted here because they illustrate the wide range of dynamical behavior that has been observed in the halogen–rare gas family. In addition, theoretical calculations performed on each of these complexes will be compared and contrasted with experimental results. The focus of this review will continue to be product rotational distributions and the nature of the predissociation dynamics that can be derived from these results.

1. Experimental Method

The product rotational distributions of the halogen fragments following vibrational predissociation of halogen–rare complexes have been evaluated using a pump-probe technique, which was first demonstrated in 1986 on NeICl [10, 75]. The pump-probe scheme is illustrated in Fig. 10 for excitation of NeICl to the ICl $A\,^3\Pi_1$ electronic state with 23 quanta of ICl stretch:

$$\text{PUMP: NeICl } (X, v'' = 0, J'') + h\nu_1 \longrightarrow \text{NeICl } (A, v' = 23, J') \tag{5.1}$$

$$\text{VP } (\Delta v = -1)\text{: NeICl } (A, v' = 23, J') \longrightarrow \text{Ne} + \text{ICl } (A, v' = 22, j) \tag{5.2}$$

$$\text{PROBE: ICl } (A, v' = 22, j) + h\nu_2 \longrightarrow \text{ICl } (\beta, v, j \pm 1) \tag{5.3}$$

A tunable dye laser prepares NeICl in the $A\,^3\Pi_1$ or $B\,^3\Pi_{0+}$ electronic state in a specific rotational level(s), J', with one or more quanta of ICl vibrational excitation, v'. Energy transfer from the ICl stretch to intermolecular vibrational degrees of freedom results in vibrational predissociation (VP) of the complex. Typically, one or two quanta of ICl vibration contain sufficient energy to break the intermolecular bond without changing the electronic state of the ICl molecule. The excess energy is partitioned between translational recoil of the fragments and rotational excitation (j) of the ICl molecule. The final state distribution of the ICl fragments is then probed with a second tunable dye laser that promotes the ICl products to an ion-pair electronic state, $E(\Omega = 0^+)$ or $\beta(\Omega = 1)$, from which the total fluorescence is collected. Similar pump-probe schemes have since been demonstrated for complexes of Cl_2 [76–79], IBr [80, 81], and I_2 [11] with rare gas atoms.

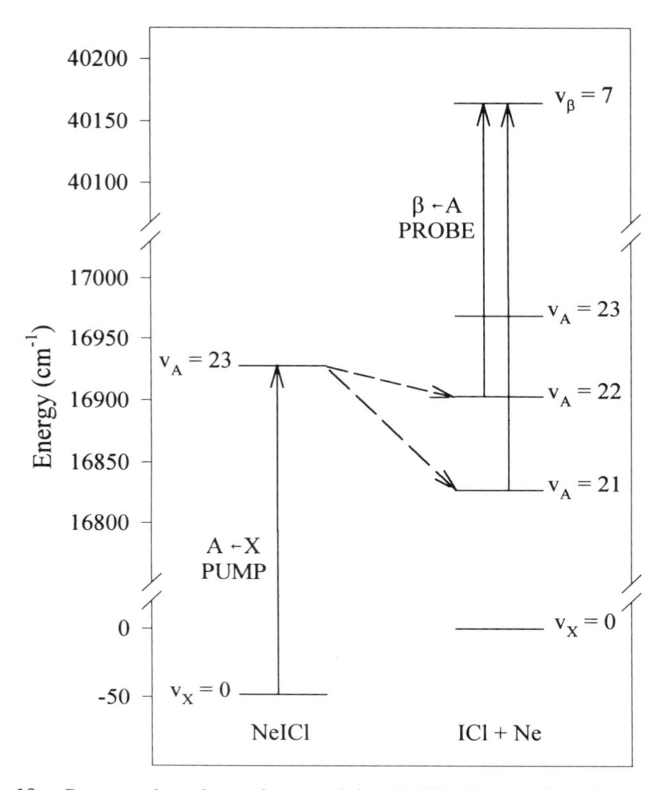

Figure 10. Pump-probe scheme for examining NeICl vibrational predissociation dynamics in the $A\ ^3\Pi_1$ electronic state with 23 quanta of ICl vibrational excitation. The rotational distributions of the ICl fragments produced in the $v = 22$ and 21 vibrational levels of the $A\ ^3\Pi_1$ state ($\Delta v = -1$ and -2 channels) are probed by excitation to an ion-pair electronic state (β, $v_\beta = 7$) and collecting the induced fluorescence.

2. *NeICl*

The vibrational predissociation dynamics of NeICl complexes have been examined following excitation to the lowest intermolecular level correlating with ICl in the $B\ ^3\Pi_{0+}$ ($v' = 2$) state as well as a series of ICl vibrational levels, $v' = 11$ to 23, in the $A\ ^3\Pi_1$ electronic state [10, 75, 82–84]. The NeICl excitation features are positioned a few wavenumbers to the high energy side of the corresponding bare ICl molecule transitions; the features exhibited asymmetric band contours but had insufficient rotational structure at the laser resolution of approximately 0.1 cm^{-1} to permit the structure of the complex to be determined. The excitation laser was fixed at the peak of the NeICl feature and the probe laser was scanned to determine the product rotational distributions by laser-induced fluorescence. Populations were recovered from the unsaturated spectral line intensities by taking into account the appropriate Franck–Condon factors, Hönl–London factors, and degeneracies.

The nascent rotational distribution of the ICl (B, $v' = 1$, j) fragments following vibrational predissociation of NeICl via the $\Delta v = -1$ channel is shown in Fig. 11 [75, 85]. The rotational distribution is highly inverted, peaking at $j = 17$, and narrow in breadth with a full width at half-maximum (FWHM) of at most 12 rotational levels. The lowest 8 rotational levels have negligible populations. ICl rotation accounts for up to 40% of the 140 cm^{-1}

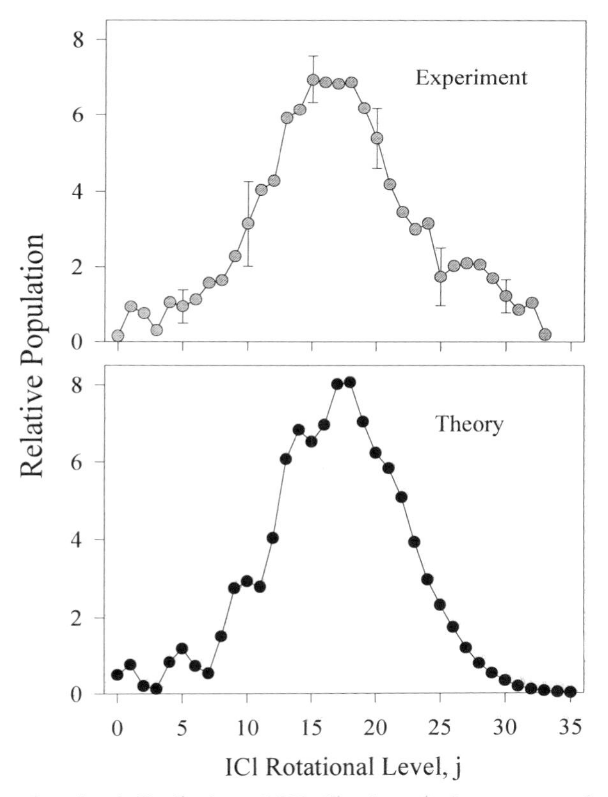

Figure 11. Rotational distribution of ICl (B, $v' = 1$) fragments produced following vibrational predissociation of NeICl via the $\Delta v = -1$ channel. A strongly peaked distribution was observed experimentally (top panel) which has been attributed to a rotational rainbow [75, 85]. Three-dimensional quantum dynamics calculations of Roncero et al. [87] also yield a highly inverted final state distribution (lower panel) arising from final state interactions as the Ne and ICl fragments separate.

of energy available to products, with an average rotational energy of nearly 20% of the energy available to products. A strongly peaked rotational distribution has also been observed for HeICl (B, $v' = 2$), with the primary peak in the distribution occurring at lower j ($j = 8$) [75, 86].

The strongly peaked rotational state distribution at high j has been attributed to a direct scattering event as the fragments separate [75]. Classically, one would expect high angular momentum transfer to the ICl fragments if the Cl side of the ICl molecule experiences a strong torque as the rare gas atom recoils away. More generally, there will be regions of the intermolecular potential energy surface that result in maximum rotational angular momentum transfer (at a maxima of Schinke's classical excitation function [5]). In the vicinity of such a location, many orientations of the Ne atom with respect to ICl will lead to the same degree of rotational excitation of the ICl fragments. This causes a peak in the ICl product rotational distribution at the maximum j, a phenomenon known in scattering as a rotational rainbow. Semiclassical [86] and three-dimensional quantum mechanical [16, 87] calculations have confirmed this interpretation of the ICl product distributions.

The ICl rotational distributoins produced upon vibrational predissociation of NeICl prepared in a series of ICl vibrational levels in the $A\ ^3\Pi_1$ electronic state were also examined in order to determine the role of available energy on the product distribution [83, 84]. The energy available to products (vibrational energy released from the ICl vibration minus the NeICl bond dissociation energy) via the $\Delta v = -1$ channel ranged from 110 to 26 cm^{-1} for the A state levels investigated, primarily due to the anharmonicity in the ICl potential. The vibrational predissociation lifetime also decreases by nearly three orders of magnitude over the range of levels studied, from 3 ± 2 ns at $v' = 14$ to less than 50 ps at $v' = 23$ [10, 83]. This trend is consistent with the *energy gap* model, which predicts a decrease in vibrational predissociation lifetime with decreasing energy available to product translation. As found for excitation of NeICl to the $B\ ^3\Pi_{0+}$ ($v' = 2$) state, the ICl product rotational distributions measured following excitation to $A\ ^3\Pi_1$ state vibrational levels were peaked, with the maximum populated j in the $\Delta v = -1$ channel varying from 13 to 4 as the initial ICl vibrational level was changed from $v' = 11$ to 23 [84]. For NeICl prepared in $v' = 23$, the $\Delta v = -2$ channel was also open and showed substantially greater rotational excitation than the $\Delta v = -1$ channel [83]. Of the 101 cm^{-1} available to fragments in the $\Delta v = -2$ channel, up to 70% of the energy appears in rotation (levels up to $j = 34$ are populated) with a broad distribution over ICl rotor levels. The product rotational distributions exhibit a trend of increasing rotational excitation of the ICl fragments with the amount of energy available. On average, a constant fraction of the available energy ($\sim 17\%$)

has been transferred to rotation. Furthermore, the peaks of the rotational distributions were shown to scale with available energy. These results are again suggestive of an inelastic collision event between the Ne and ICl fragments as they separate.

The vibrational predissociation dynamics of NeICl (B, $v' = 2$) has been investigated theoretically in a full three-dimensional quantum mechanical study [87]. The results from this calculation have become a benchmark to test other theoretical methods and approximations [88–90]. A model intermolecular potential energy surface was derived for NeICl based on atom–atom pairwise interactions with the Ne–I and Ne–Cl potential parameters scaled to fit the experimental data (dissociation energy, lifetime, and final rotational distribution) [87]. The PES predicts a bent equilibrium geometry for NeICl at 140° (angle between intermolecular axis and ICl bond) with the Ne atom closer to the Cl side of ICl. The lowest intermolecular level of NeICl (B, $v' = 2$) is computed to have a predissociation lifetime of 2.5 ns and an ICl (B, $v' = 1$) product rotational distribution that is shown in Fig. 11 [87]. The peak of the distribution occurs at $j = 18$ with a breadth of 12 units in j. The agreement between experiment and theory is remarkably good considering that a model PES was used in the calculations.

The initial quasibound NeICl wavefunction was also decomposed in terms of free rotor states for comparison with the final state distribution [87]. This initial state decomposition would give the final state distribution if no torques act on the fragments as they separate. The decomposition of the initial quasibound wave function exhibits peaks at $j = 1, 5, 9, \ldots$ and minima at $j = 3, 7, 11, \ldots$, arising from the initial geometry of the complex; similar results were obtained when the vibrational coupling that induces predissociation was included. This distribution is strikingly different from that observed experimentally and in the full dyanmics calculation, a finding which demonstrates that the highly inverted rotational distribution is due to final state interactions as the fragments separate. Some residual structure can be seen in the calculated final state distribution at low j (Fig. 11), which can be attributed to the initial quasibound wavefunction (total angular momentum, $J = 0$). Since the equilibrium geometry of NeICl is bent at 140°, the initial wavefunction will have its maximum at angles that induce maximum angular momentum transfer, resulting in a rotational rainbow. Similar results have been obtained by Lipkin et al. [90] using the complex coordinate method, which treats the photodissociation as a half-collision process.

3. $NeCl_2$

The product state distributions observed following vibrational predissociation of $NeCl_2$ are quite different from those seen for NeICl. In part, this can be expected, since $NeCl_2$ has a T-shaped equilibrium geometry in both the

X and *B* electronic states, with a slight shortening of the intermolecular bond length (0.05 Å) upon electronic excitation [91], while NeICl has a bent structure. The predissociation lifetimes of the excited vibrational levels have been shown to range between 300 and 65 ps for $v' = 8$ to 13 [91]. The Cl_2 product state distributions have been measured for $NeCl_2$ prepared in a series of initial vibrational levels in the *B* electronic state, $v' = 6$ to 13, using the pump-probe scheme described above [76, 77]. As shown in Fig. 12, the Cl_2 rotational distributions for the $\Delta v = -1$ channels are peaked at low j ($j =$ 2 to 5) with nearly 80% of the population in $j \leq 12$. The remaining population (~20%) is spread over higher rotational levels, terminating at the highest rotational level permitted by conservation of energy and angular momentum for initial $v' > 8$. For example, $NeCl_2$ (B, $v' = 11$) predissociation results in an abrupt termination of the rotational distribution at $j =$ 23. For $6 < v' < 10$, the rotational distributions show a bimodal structure, which has been attributed to a quantum interference effect in the case of $NeCl_2$ [77].

The rotational product distributions show a remarkable similarity for $v' = 6$ to 13, in spite of the fact that the energy available to products decreases from 142 to 66 cm^{-1} over this vibrational range [77]. On average, 12% to 17% of the available energy is transferred to Cl_2 rotation, the balance is released as translational recoil. Surprisingly, the $\Delta v = -2$ channels (observed for $v' = 11$ and 13) exhibit rotational distributions similar to the $\Delta v = -1$ channel, even though the $\Delta v = -2$ channel releases three times the amount of energy to products as the $\Delta v = -1$ channel. The average rotational energy of the products from the $\Delta v = -2$ channel is comparable to that of the $\Delta v = -1$ channel, but the percentage of the available energy is much smaller (~7%–8%). This shows that the Cl_2 rotational distributions are essentially independent of the amount of energy released to product degrees of freedom, a situation that is quite different from what would be expected from an impulsive half-collision model [77].

Halberstadt et al. [72] have performed a three-dimensional quantum calculation of the $NeCl_2$ vibrational predissociation dynamics. These calculations use a model PES determined from atom–atom interactions, with potential parameters chosen to reproduce the experimentally observed bond dissociation energy, bond length, and predissociation lifetimes. The calculated product rotational distribution successfully reproduces the general features of the dynamics experiment, as illustrated in Fig. 12 for excitation of $NeCl_2$ to the $v' = 11$ level of the *B* state with no excitation in the intermolecular bending or stretching modes [72]. Since the $NeCl_2$ potential has C_{2v} symmetry, only even j's can result from this initial state; thus, only even j experimental points are shown for comparison in Fig. 12. The theoretical calculation underestimates the population of Cl_2 products at high j,

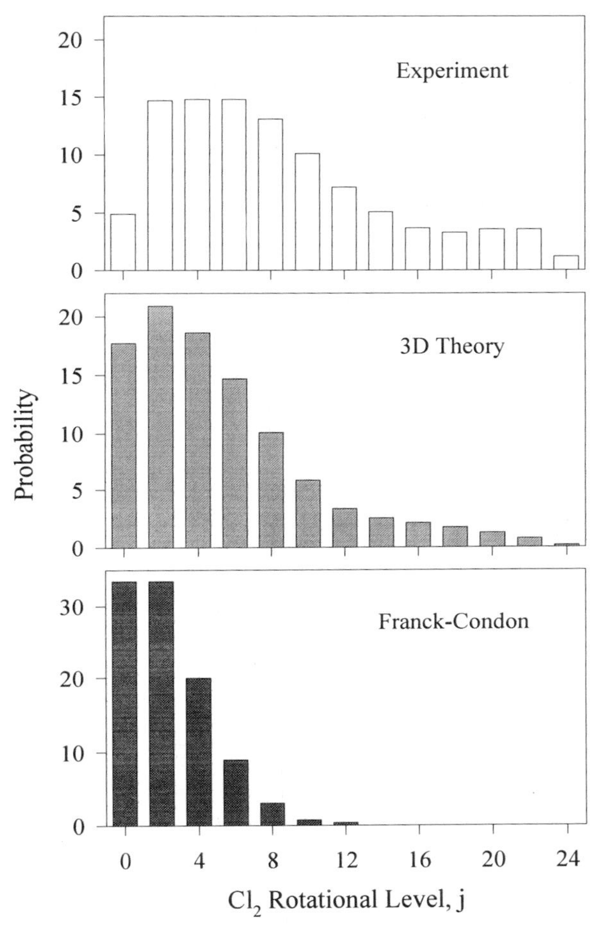

Figure 12. Comparison of experimentally observed and theoretically calculated Cl_2 product rotational distributions [72] following vibrational predissociation of $NeCl_2$ (B, $v' = 11$). The full three-dimensional quantum dynamics calculation of Halberstadt et al. (middle panel) is in good accord with the experimental results of Janda and coworkers (top panel). The Franck–Condon limit (bottom panel) reproduces the low-j peak, but does not account for the high-j tail. Only even j's are shown.

suggesting that the anisotropy of the model potential should be increased. In addition, the $\Delta v = -2$ channel, which accounts for only 1% of the total from the $v' = 11$ level, yields the same product rotational distribution as the $\Delta v = -1$ channel in the close coupling calculation.

Decomposition of the bending wavefunction of the initial state [92] over free rotor basis functions gives the product distribution in the Franck–Condon limit [72]. Rotor levels up to $j = 10$ would contribute to the product distribution in this limit (see Fig. 12). This model predicts the correct position of the low-j peak observed experimentally but does not account for the high-j tail of the distribution. It also predicts that the rotational distribution is independent of energy available for the products. Furthermore, the fact that the $\Delta v = -1$ and -2 channels exhibit similar product distributions can be explained by this model, since the same initial bending wavefunction is projected on the product rotor basis for the two channels. This indicates that the two product channels are directly coupled to the same initial state. Thus, in the case of $NeCl_2$, much of the dynamics can be understood in terms of the Franck–Condon limit, with the high-j tail of the final state distribution arising from the vibrational coupling term and/or final state interactions as the Ne and Cl_2 fragments separate [77].

4. *NeIBr*

Stephenson has examined the fragment rotational distributions following vibrational predissociation of NeIBr prepared in various vibrational levels of the IBr A ($\Omega = 1$) electronic state [81]. A pump-probe method has been used with the probe laser operating on a valence-to-ion pair electronic transition in IBr (such as $G \leftarrow A$) [80, 81]. All of the rotational distributions resulting from the loss of one vibrational quantum peak at relatively low j ($j \leq 6$) with gradually decreasing population towards higher j. The rotational distributions from the $\Delta v = -1$ dissociation pathways extend to their energetic limits, $j = 22$ at $v' = 12$ to $j = 10$ at $v' = 16$. On average, 20% to 30% of the available energy is deposited into rotational degrees of freedom. For the $\Delta v = -2$ channel ($v' = 15$ to 17), the fragments are produced with a significantly higher rotational excitation than that found for the $\Delta v = -1$ process. Again, the distributions extend nearly to the energetic limit. The fraction of total energy available appearing in rotations, however, is on the order of half that observed for the $\Delta v = -1$ pathway.

The product rotational distribution has been evaluated in the Franck–Condon limit based on a model potential for NeIBr. The pairwise additive model potential for NeIBr, constructed in a similar manner as those for NeICl and $NeCl_2$, has a bent equilibrium geometry with a 73° angle between the intermolecular axis and the IBr axis (skewed towards the Br atom). The wavefunction for the lowest bound state of the complex is expanded over IBr free rotor functions to determine the product rotational distributions arising from the initial zero-point bending motion of the complex. The final state distribution predicted for this limiting case severely underestimates the degree of rotational excitation of the products observed experimentally.

The experimental product rotational distributions have been accurately reproduced using a semiclassical model [5, 86] in which classical trajectories were propagated along the model potential, starting from the repulsive wall. The trajectories were then weighted by the initial angular configurations of the complex. These calculations demonstrate that final state interactions (torques) between the recoiling Ne and IBr fragments must be included to account for the rotational excitation of the IBr fragments [81].

B. OH–Ar ($A\ ^2\Sigma^+$) Vibrational Predissociation Dynamics

The vibrational predissociation dynamics of OH–Ar in the excited $A\ ^2\Sigma^+$ electronic state have been investigated at a state-to-state level of detail both experimentally [93–95] and theoretically [96]. The predissociation dynamics have been investigated following preparation of OH–Ar in various intermolecular levels (K, s, b) supported by the OH $A\ ^2\Sigma^+$ $(v = 1)$ + Ar potential: pure intermolecular stretching levels $(K, s, b) = (0, s, 0)$ and angular states $(1, s, 0)$, where K is the projection of the OH–Ar (or, alternatively, the OH) total angular momentum on to the OH–Ar intermolecular axis, s indicates the quanta of intermolecular stretching excitation, and b denotes the amount of bending excitation. Spectroscopic [93, 94, 97–101] and theoretical [102, 103] studies have shown that OH–Ar ($A\ ^2\Sigma^+$) has a linear equilibrium structure in the O–H---Ar configuration with a bond length of 2.9 Å and a bond dissociation energy of 742.9 cm^{-1}. *Ab initio* [102] and semiempirical [97, 104, 105] PESs for OH $A\ ^2\Sigma^+$ + Ar have revealed a highly anisotropic potential with deep minima in the linear O–H---Ar and Ar---O–H configurations and a high barrier at intermediate angles.

Early experiments [93, 94] yielded predissociation lifetimes that were derived from the homogeneous contributions to the line widths of OH–Ar features in fluorescence excitation spectra and *qualitative* information on the rotational distributions of the OH $A\ ^2\Sigma^+$ $(v = 0)$ fragments from dispersed fluorescence spectra. The vibrational predissociation lifetimes, τ_{vp}, were found to be strongly dependent on the initially prepared intermolecular level of the complex [94]. For example, the lifetimes of $(K, s, b) = (0, s, 0)$ levels were shown to decrease rapidly with decreasing intermolecular stretching excitation from 150 ps to about 20 ps for $s = 6$ to 3, whereafter little change in τ_{vp} was discerned. The $K = 1$ levels did not exhibit line broadening at 0.1 cm^{-1} laser resolution, corresponding to lifetimes in excess of 50 ps. Dispersed fluorescence spectra have revealed that OH A–X 0–0 emission is seen exclusively. The resolution of the monochromator ($\sim$60 cm^{-1}) used for the dispersed fluorescence measurements did not enable any *quantiative* information regarding the population of photofragments of OH $A\ ^2\Sigma^+$ $(v = 0)$ rotational levels (n) to be obtained from the spectra; however, some *qualitative* information could be extracted [93, 94]. The OH $A\ ^2\Sigma^+$ $(v = 0)$ product rotational distribution appears to end abruptly at $n = 7$ for complexes prepared in $(0, s, 0)$ intermolecular stretching levels with $s = 3$ to 5. When

OH–Ar is prepared in the (1, 3, 0) and (1, 4, 0) levels, the rotational distribution of OH photofragments extends to higher rotational levels with significant population in $n = 8$ and 9.

In more recent studies of this system [95], fully state-resolved OH $A\ ^2\Sigma^+$ product distributions have been obtained following vibrational predissociation of OH–Ar $A\ ^2\Sigma^+$ ($v = 1$) using a variation of the stimulated emission pumping (SEP) technique [95, 106]. The experimental method is illustrated schematically in Fig. 13 on a potential energy diagram for Ar + OH in its ground $X\ ^2\Pi$ and excited $A\ ^2\Sigma^+$ electronic states. This same approach has been used to examine the OH $A\ ^2\Sigma^+$ product rotational distribution following internal rotational predissociation of OH–Ar [95, 106]. Briefly, the experiment has been carried out as follows: An excimer laser, operating at 193 nm, produces OH $X\ ^2\Pi$ radicals by photolyzing nitric acid entrained in Ar carrier gas either within or just below the exit of a quartz capillary affixed to the end of a pulsed valve. These radicals undergo supersonic expansion with Ar to form OH–Ar complexes; the complexes are excited to various intermolecular vibrational levels supported by the OH $A\ ^2\Sigma^+$ ($v = 1$) + Ar potential using the frequency-doubled output of a Nd:YAG-pumped dye laser operating around 282 nm (arrow 1 of Fig. 13). Spontaneous fluorescence from the OH fragments produced via vibrational predissociation from these levels is collected using a photomultiplier tube. After a 200 ns delay, an excimer-pumped dye laser, operating at 354 nm, is then introduced to probe the OH $A\ ^2\Sigma^+$ fragments by stimulating downward OH $A \rightarrow X$ $0 \rightarrow 1$ rovibrational transitions (arrow 2 of Fig. 13). Each time this laser frequency is resonant with a downward transition originating from an OH rotor level that has been populated as a result of vibrational predissociation, some of the fragments will be driven to the ground state, thereby *decreasing* the overall spontaneous fluorescence and creating a dip in the spectrum.

The amplitude of the fluorescence dip can be directly related to the relative number of OH fragments produced in a specific OH $A\ ^2\Sigma^+$ v, n, j level, N, even though the data is recorded under partially saturated conditions. As derived by Giancarlo et al. [106], the measured intensities of the fluorescence dips, I_{dip}, can be expressed as

$$I_{\text{dip}} \sim -\frac{N}{1 + g_{j'j''}} \left\{ \exp\left[-\frac{\alpha\, S_{j'j''}(1 + g_{j'j''})}{2j' + 1} \right] - 1.02 \right\} \tag{5.4}$$

Here, α is a saturation parameter, dependent on the intensity and duration of the probe laser pulse as well as the Einstein coefficient for stimulated emission, and $g_{j'j''}$ is the ratio of the degeneracy factors for the OH excited and ground state levels. The OH $A\ ^2\Sigma^+$ rotational levels were generally

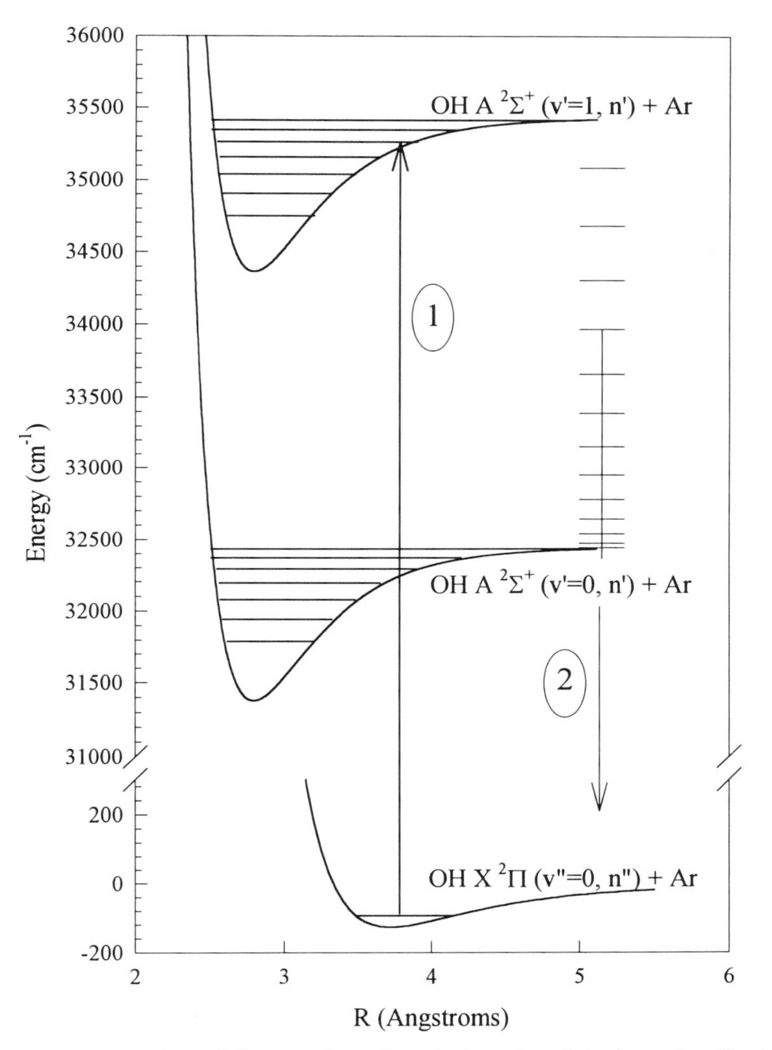

Figure 13. Overview of the experimental method used to investigate the vibrational predissociation dynamics of OH–Ar ($A\ ^2\Sigma^+$). Shown are radial cuts of the OH–Ar potentials in the ground $X\ ^2\Pi$ and excited $A\ ^2\Sigma^+$ electronic states. OH–Ar is excited from the lowest level in the ground electronic state to metastable intermolecular levels correlating with OH $A\ ^2\Sigma^+$ ($v = 1$) + Ar, arrow 1. The rotational distribution of the OH $A\ ^2\Sigma^+$ ($v = 0$) fragments is probed by inducing downward OH $A \rightarrow X$ transitions (arrow 2), thus depleting the spontaneous fluorescence from OH via stimulated emission.

probed on transitions with the largest line strength factors, $S_{j'j''}$. In most cases, only the higher energy spin-rotation component (j) for a given rotational level, n, was probed due to weak Hönl–London factors for downward transitions from the lower energy component. In those instances when both spin-rotation components could be probed, equal populations were found in each of the two components. Based on this finding, equal population of the two spin-rotation components has been assumed in all cases. By analyzing the fluorescence dip data in this way, *quantitative* OH rotational state distributions have been obtained.

Product rotational distributions have been determined by the SEP probe method following preparation of OH–Ar in either the $(K, s, b) = (0, 5, 0)$ or (1, 3, 0) intermolecular level with a binding energy of -175 cm^{-1} or -157 cm^{-1} with respect to the OH $A\ ^2\Sigma^+$ $(v = 1)$ + Ar dissociation limit [95]. Vibrational predissociation of OH–Ar from these levels releases more than 2800 cm^{-1} of excess energy to products, permitting OH product rotational levels as high as n = 12 to be populated. The product rotational distributions are found to be highly nonstatistical and strongly dependent on the initially prepared level.

When OH–Ar vibrationally predissociates from a pure stretching level (0, 5, 0), OH $A\ ^2\Sigma^+$ $(v = 0)$ fragments are produced in $n = 4$ through 8. Most of the population (72%) is found in $n = 6$, 7, and 8. The relative population of the photofragments in all the energetically allowable OH $A\ ^2\Sigma^+$ $(v = 0)$ rotational levels n are depicted in the upper panel of Fig. 14. The OH product rotational distribution following predissociation from the OH–Ar (1, 3, 0) level, presented pictorially in the upper panel of Fig. 15, differs substantially from that obtained for predissociation from (0, 5, 0). The distribution for predissociation from the (1, 3, 0) level extends from $n = 3$ to 9, and is rather flat from $n = 3$ to 6 with non-negligible population. Also, in contrast with the distribution obtained for dissociation from (0, 5, 0), this distribution is very sharply peaked at $n = 8$ (1221 cm^{-1}), which contains 30% of the OH fragments. At the peak of the product rotational distribution, approximately 43% of the available energy appears as OH rotational excitation.

The experimental results can be compared with the final state distributions predicted theoretically. Before comparison with the full dynamical calculation, however, it is useful to examine the product distribution that is computed for the Franck–Condon limit where only the initial angular motion of the complex is transformed into product rotation. The decompositions of the initial wave functions for OH–Ar in the (0, 5, 0) and (1, 3, 0) intermolecular levels in terms of the free rotor functions of OH $A\ ^2\Sigma^+$ are shown with solid squares and dotted lines in the bottom panels of Figs. 14 and 15, respectively [95]. The wavefunctions for the initially prepared OH–Ar levels have been computed based on an adjusted semiempirical potential energy surface for

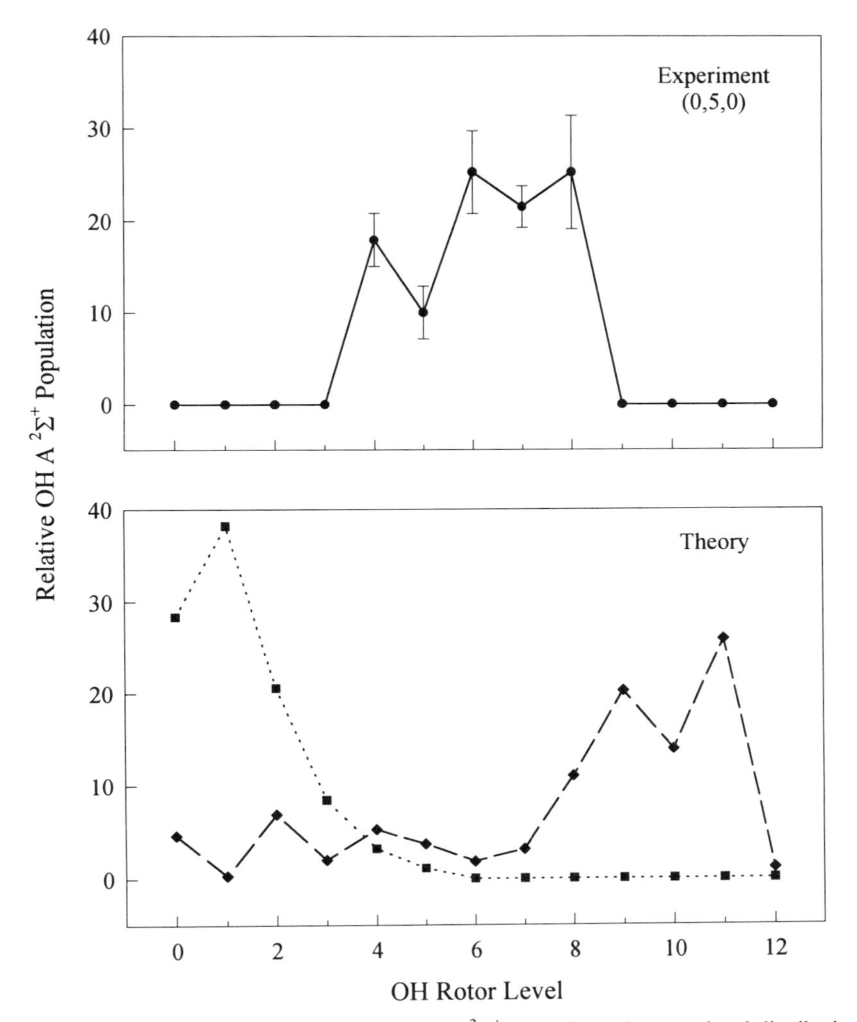

Figure 14. Experimentally determined OH $A\ ^2\Sigma^+$ ($v = 0$) product rotational distribution [95] following vibrational predissociation of OH–Ar from a pure intermolecular stretching level, $(K, s, b) = (0, 5, 0)$, correlating with OH $A\ ^2\Sigma^+$ ($v = 1$) + Ar (top panel). For comparison, the decomposition of the initial OH–Ar wavefunction in terms of OH free rotor states (Franck–Condon limit) is displayed with a dotted line [95] and the final state distribution computed theoretically based on an *ab initio* potential energy surface [96] is shown with a dashed line (bottom panel).

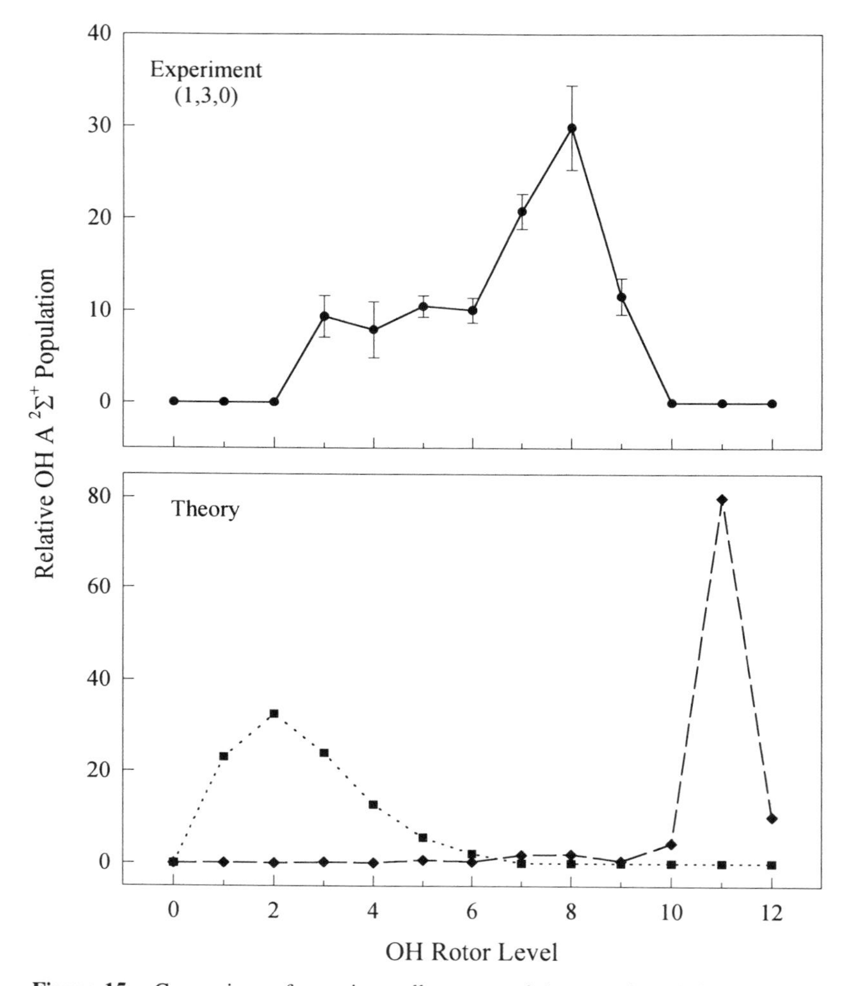

Figure 15. Comparison of experimentally measured (top panel) and theoretically computed (bottom panel) final state distributions following vibrational predissociation of OH–Ar ($A\ ^2\Sigma^+$, $v = 1$) from an intermolecular level with $K = 1$ [95, 96]. Both the decomposition of the initial OH–Ar wavefunction (dotted line) and full dynamics (dashed line) calculations are shown in the lower panel.

OH $A\ ^2\Sigma^+$ ($v = 0$) + Ar [97, 104]. These distributions peak at low n, fall off with increasing OH rotor level, and have negligible population for $n \geq 5$. The distribution for the (1, s, 0) level has a minimum at $n = 0$ as compared to the (0, s, 0) distribution which has a large contribution from $n = 0$. Comparison with the experimental final state distributions from corresponding levels (upper panels of Figs. 14 and 15) shows that the product rotational excitation differs substantially from the prediction of this limiting case and, therefore, is *not* simply a consequence of the initial bending wave function in OH–Ar.

Theoretical calculations performed by Chakravarty et al. [96] have examined the full vibrational predissociation dynamics of OH–Ar $A\,^2\Sigma^+$ (v = 1). Their calculations employ a Golden Rule approach to model vibrational predissociation dynamics occurring on the *ab initio* potential energy surface for OH $A\,^2\Sigma^+$ (v) + Ar. In general, they find that the OH $A\,^2\Sigma^+$ rotational distributions following OH–Ar prepared in both (0, s, 0) and (1, s, 0) levels peak near the energetic limit. For example, when OH–Ar vibrationally predissociates from the (0, 5, 0) intermolecular level at -209.2 cm^{-1} with respect to the OH $A\,^2\Sigma^+$ (v = 1) + Ar asymptote, the population of OH $A\,^2\Sigma^+$ (v = 0) fragments in n = 0 to 7 is predicted to be small and relatively flat; the calculated product distribution reaches maxima at n = 9 and 11. The n = 0 to 6 levels of OH $A\,^2\Sigma^+$ (v = 0) have negligible population when OH–Ar predissociates from (1, 2, 0) at -171.8 cm^{-1}; in contrast, n = 11 and 12 contain 90% of the OH fragments. These results are also presented in the bottom panels of Figs. 14 and 15 with solid squares and dashed lines.

The experimentally determined OH population distribution does not extend as high as that predicted theoretically (n = 9 to 11) when OH–Ar predissociates from the (0, 5, 0) level. When the complex predissociates from a (1, s, 0) level, the calculations again predict that most of the fragments populate the highest OH rotational levels energetically accessible (n = 11, 12). In contrast with these results, no significant population in levels higher than n = 9 has been observed for predissociation from the OH–Ar (1, 3, 0) level. As Chakravarty *et al.* [96] point out, population of high OH rotational levels in the vibrational predissociation process can be produced by the strong anisotropy of the OH $A\,^2\Sigma^+$ + Ar potential energy surface; therefore, since the predicted product state distributions differ markedly from the experimentally observed ones, one can conclude that the anisotropy of the *ab initio* potential energy surface is too high in the regions sampled during the predissociation process. Moreover, spectroscopic studies on excited angular states of the complex provide further evidence that the angular dependence of the *ab initio* intermolecular potential is too strong: the theoretically computed intermolecular vibrational frequencies for these angular levels are much higher than those measured experimentally [97, 104].

The full dynamics calculations [96] also yield vibrational predissociation lifetimes that can be compared with experimental observations. The calculated line widths follow a trend of decreasing line width with increasing intermolecular stretching excitation for (0, s, 0) levels. The quasibound levels with (1, s, 0) show the same trend with increasing energy but have significantly larger line widths than levels with (0, s, 0). The calculated line widths, however, are much smaller than the experimental line widths by at least a factor of 30, suggesting that the coupling between the initial and final OH vibrational states is too weak in the *ab initio* potential [95].

Since the predissociation rate has been shown to be strongly correlated with the square of the spectral shift in many weakly bound systems [7, 107] comparison of the experimentally measured and theoretically calculated spectral shifts for intermolecular levels correlating with Ar + OH $A\ ^2\Sigma^+$ $(v = 0)$ and $(v = 1)$ is also warranted. Experimentally, the binding energy increases by 32.5 cm^{-1} upon OH vibrational excitation [94]; theoretical calculations predict that the OH $(v = 1)$ + Ar binding energy is over 200 cm^{-1} larger than OH $(v = 0)$ [96]. The increase in the binding energies is reflected in the change in the spectral shifts of OH–Ar $(v = 0, 1)$ features from the OH monomer A–X 0–0 and 1–0 transitions, respectively: the theoretically predicted spectral shift is greater than the shift measured experimentally by a factor of 6. Based on the spectral shifts, one would expect the measured predissociation lifetimes to differ from the computed ones by a factor of approximately 30; this agrees well with the fact that the calculated line widths are smaller than the experimental line widths by at least this amount.

Vibrational predissociation of OH–Ar ($A\ ^2\Sigma^+$, $v = 1$) yields OH $A\ ^2\Sigma^+$ $(v = 0)$ fragments with substantial rotational excitation, albeit somewhat less than that predicted theoretically. The OH fragment rotational excitation originates from final state interactions as the OH and Ar fragments separate on a highly anisotropic PES and *not* from the initial bending motion of the complex. The vibrational coupling term controls the rate of vibrational predissociation and may also influence the product rotational distribution. Comparison between experiment and theory suggests that the anisotropy of the *ab initio* potential is too strong, resulting in rotational excitation of the OH fragments greater than that observed experimentally. In addition, the coupling between the interaction potentials for the initial and final vibrational states may be underestimated by *ab initio* potential, such that the calculated vibrational predissociation lifetimes are substantially longer than those observed experimentally.

VI. CONCLUSIONS

The vibrational predissociation dynamics of weakly bound complexes provide detailed information on the intermolecular potential energy surface. Each of the three main experimental observables, (1) rotationally resolved excitation spectra of the complex, (2) the lifetime of the vibrationally excited complex, and (3) the final state distribution of the product molecules, probes different aspects of the PES. Spectroscopic studies primarily yield information on the potential in the vicinity of the minimum; when excited intermolecular vibrations can be accessed, broader regions of the potential can be characterized [97, 104, 105]. Vibrational predissociation lifetimes provide a measure of the coupling between the initially prepared monomer vibration and the intermolecular stretching vibration (dissociation coordinate). In some systems, this coupling includes a vibration-to-vibration energy transfer step, either within one monomer subunit (Ar–CO_2 [29, 30]) or from one monomer subunit to the other across the weak intermolecular bond (D_2–HF [26], N_2–HF [25, 27, 28], CO_2–HF [28], NO–HF [44]). Final state distributions of the products are dependent on these two factors, as well as being sensitive to the regions of the potential sampled by the recoiling fragments [5].

In this review, we have focused on product rotational distributions following vibrational predissociation of weakly bound complexes. The *energy gap* model predicts that translational energy release should be minimized to increase the rate of vibrational predissociation [14, 15]. Since internal excitation of the fragments decreases the energy available to translation, the energy gap model would predict that product channels with vibrational and/or rotational excitation should be favored. As illustrated in this review, the energy gap model provides a useful guide for what final states *may* be important in the dissociation dynamics, The highest energetically accessible product rotational channels, however, do not necessarily have the largest probability even though these channels minimize the energy gap. Rather, dynamical factors appear to control the product rotational distributions. In many cases, the final state distribution provides a direct reflection of the details of the underlying potential energy surface.

The primary origin of the rotational excitation of the fragments has been assessed in several systems. Initial state preparation clearly influences the rotational distribution of the products; the dissociation dynamics are dependent on the bound state wavefunction of the vibrationally excited complex. Excitation of an intermolecular bending level or an excited angular state has been shown to have the greatest effect, as seen in the HF dimer [24, 39] and OH–Ar [95, 96]. In halogen–rare gas systems, the product rotational distributions change with the degree of halogen vibrational excitation [77, 81, 83, 84]. This can be primarily attributed to an *energy gap* criterion. In $NeCl_2$, however, the initial bending wavefunction of the complex may actually control the product Cl_2 rotational distribution (Franck–Condon limit) [72].

Product state distributions are also affected by the coupling term, particularly when V–V energy transfer occurs. In D_2–HF, for example, the dynamics are dominated by this V–V coupling term, resulting in production of the two fragments in low rotational levels [40]. It is possible that this coupling may also be the origin of the highly nonstatistical HF product distributions observed for N_2–HF [25, 27, 28], CO_2–HF [28], and NO–HF [44] where a single HF rotor level is populated for each monomer vibrational channel. Theoretical calculations are needed to test this hypothesis. In NeICl, theoretical calculations have shown that the angular dependence of the coupling term affects the product rotational distribution but is not the most important factor in determining the product rotational distribution [87].

Final state interactions between the separating fragments in the exit channel are the dominant factor in determining product rotational distributions in several systems, such as $(HF)_2$ [21–25], N_2–HF [27, 28], NeICl [75, 84], and OH–Ar [95, 96]. In these cases, the product rotational distributions are principally governed by the forces (torques) on the molecules as they separate. The product rotational distributions are therefore a probe of the regions of the potential sampled in the dissociation process. Comparison of experimentally measured and theoretically computed final state distributions are a sensitive test of the anisotropy of the underlying PES. Such a comparison for the HF dimer suggests that both the semiempirical and *ab initio* PESs underestimate the anisotropy of the potential sampled during the fragmentation process [22, 38]. By contrast, the theoretical calculations for OH–Ar overestimate the degree of rotational excitation of the products, indicating that the anisotropy of the *ab initio* PES is too strong [95, 96].

As seen in this review, product rotational distributions are yielding considerable new insight into the dissociation dynamics of weakly bound complexes. The final state distributions following vibrational predissociation of vdW complexes contain dynamical information similar to that found in the photodissociation of stable molecules, even though vibrational predissociation takes place on a much slower time scale than direct dissociation events [5]. In the future, one can expect developments that parallel those in photodissociation studies of molecules, specifically measurements of vector correlations, theoretical calculations of the experimental observables, as well as experimental and theoretical investigations of more complex systems.

ACKNOWLEDGMENTS

The experimental and theoretical research on NeICl and OH–Ar conducted in the author's group at the University of Pennsylvania has been sponsored by the National Science Foundation and the Division of Chemical Sciences, Office of Basic Energy Sciences of the Department of Energy. Special thanks are due to the present and past members of the Lester group who participated in these projects: John Skene, Janet Drobits, Robert Waterland, Tom Stephenson, Mary Berry, Mitch Brustein, Rob Randall, Leanna Giancarlo, and Susan Choi.

Collaborations and discussions on the theoretical aspects of these projects with Reinhardt Schinke, Nadine Halberstadt, Alberto Beswick, Charusita Chakravarty, and David Clary are gratefully acknowledged.

REFERENCES

1. *Faraday Disc.* **97** (1994).
2. N. Halberstadt and K. C. Janda, Eds., *Dynamics of Polyatomic Van der Waals Complexes*, NATO Advanced Science Institutes Series, Plenum Press, New York, 1990.
3. M. N. R. Ashfold and J. E. Baggott, Eds., *Molecular Photodissociation Dynamics, Advances in Gas-Phase Photochemistry and Kinetics*, The Royal Society of Chemistry, London, 1987.
4. *Photodissociation and Photoionization*, K. P. Lawley, Ed., *Adv. Chem. Phys.* **60** (1985).
5. R. Schinke, *Photodissociation Dynamics*, Cambridge University Press, Cambridge, 1993.
6. J. M. Hutson, *Ann. Rev. Phys. Chem.* **41,** 123 (1990).
7. R. E. Miller, *Science* **240,** 447 (1988).
8. D. J. Nesbitt, *Chem. Rev.* **88,** 843 (1988).
9. M. P. Casassa, A. M. Woodward, J. C. Stephenson, and D. S. King, *J. Chem. Phys.* **85,** 6235 (1986).
10. J. C. Drobits, J. M. Skene, and M. I. Lester, *J. Chem. Phys.* **84,** 2896 (1986).
11. D. M. Willberg, M. Gutmann, J. J. Breen, and A. H. Zewail, *J. Chem. Phys.* **96,** 198 (1992).
12. A. S. Pine and W. J. Lafferty, *J. Chem. Phys.* **78,** 2154 (1983); A. S. Pine, W. J. Lafferty, and B. J. Howard, *J. Chem. Phys.* **81,** 2939 (1984); A. S. Pine and B. J. Howard, *J. Chem. Phys.* **84,** 590 (1986); A. S. Pine and G. T. Fraser, *J. Chem. Phys.* **89,** 6636 (1988).
13. Z. S. Huang, K. W. Jucks, and R. E. Miller, *J. Chem. Phys.* **85,** 3338 (1986).
14. J. A. Beswick and J. Jortner, *Adv. Chem. Phys.* **47,** 363 (1981); *Chem. Phys. Lett.* **49,** 13 (1977).
15. G. E. Ewing, *J. Phys. Chem.* **91,** 4662 (1987); *Faraday Discuss. Chem. Soc.* **73,** 325 (1982); *J. Chem. Phys.* **71,** 3143 (1979).
16. R. L. Waterland, M. I. Lester, and N. Halberstadt, *J. Chem. Phys.* **92,** 4261 (1990).
17. M. F. Vernon, D. J. Krajnovich, H. S. Kwok, J. M. Lisy, Y. R. Shen, and Y. T. Lee, *J. Chem. Phys.* **77,** 47 (1982).
18. M. F. Vernon, J. M. Lisy, D. J. Krajnovich, A. Tramer, H.-S. Kwok, Y. R. Shen, and Y. T. Lee, *Faraday Discuss. Chem. Soc.* **73,** 387 (1982).
19. M. A. Hoffbauer, K. Liu, C. F. Giese, and W. R. Gentry, *J. Chem. Phys.* **78,** 5567 (1983).
20. T. E. Gough, D. G. Knight, P. A. Rowntree, and G. Scoles, *J. Phys. Chem.* **90,** 4026 (1986).
21. D. C. Dayton, K. W. Jucks, and R. E. Miller, *J. Chem. Phys.* **90,** 2631 (1989).
22. E. J. Bohac, M. D. Marshall, and R. E. Miller, *J. Chem. Phys.* **96,** 6681 (1992).
23. M. D. Marshall, E. J. Bohac, and R. E. Miller, *J. Chem. Phys.* **97,** 3307 (1992).
24. E. J. Bohac and R. E. Miller, *J. Chem. Phys.* **99,** 1537 (1993).
25. R. J. Bemish, M. Wu, and R. E. Miller, *Faraday Discuss.* **97,** 57 (1994).
26. E. J. Bohac and R. E. Miller, *J. Chem. Phys.* **98,** 2604 (1993).

27. R. E. Bemish, E. J. Bohac, M. Wu, and R. E. Miller, *J. Chem. Phys.* **101,** 9457 (1994).
28. E. J. Bohac and R. E. Miller, *Phys. Rev. Lett.* **71,** 54 (1994).
29. E. J. Bohac, M. D. Marshall, and R. E. Miller, *J. Chem. Phys.* **97,** 4890 (1992).
30. E. J. Bohac, M. D. Marshall, and R. E. Miller, *J. Chem. Phys.* **97,** 4901 (1992).
31. T. R. Dyke, B. J. Howard, and W. Klemperer, *J. Chem. Phys.* **56,** 2442 (1972).
32. B. J. Howard, T. R. Dyke, and W. Klemperer, *J. Chem. Phys.* **81,** 5417 (1984).
33. H. S. Gutowsky, C. Chuang, J. D. Keen, T. D. Klots, and T. Emilsson, *J. Chem. Phys.* **83,** 2070 (1985).
34. W. J. Lafferty, R. D. Suenram, and F. J. Lovas, *J. Mol. Spectrosc.* **123,** 434 (1987).
35. M. Kofranek, H. Lischka, and A. Karpfen, *Chem. Phys.* **121,** 137 (1988); P. R. Bunker, M. Kofranek, H. Lischka and A. Karpfen, *J. Chem. Phys.* **89,** 3002 (1988); P. R. Bunker, P. Jensen, A. Karpfen, M. Kofranek, and H. Lischka, *J. Chem. Phys.* **92,** 7432 (1990).
36. M. Quack and M. A. Suhm, *Mol. Phys.* **69,** 791 (1990); M. Quack and M. A. Suhm, *J. Chem. Phys.* **95,** 28 (1991).
37. N. Halberstadt, P. Brechignac, J. A. Beswick, and M. Shapiro, *J. Chem. Phys.* **84,** 170 (1986).
38. D. H. Zhang and J. Z. H. Zhang, *J. Chem. Phys.* **98,** 5978; **99,** 6624 (1993).
39. M. von Dirke, Z. Bacic, D. H. Zhang, and J. Z. H. Zhang, *J. Chem. Phys.* **102,** 4382 (1995).
40. D. C. Clary, *J. Chem. Phys.* **96,** 90 (1992).
41. C. M. Lovejoy, D. D. Nelson, Jr., and D. J. Nesbitt, *J. Chem. Phys.* **89,** 7180 (1988); **87,** 5621 (1987).
42. K. W. Jucks and R. E. Miller, *J. Chem. Phys.* **87,** 5629 (1987).
43. D. H. Zhang, J. Z. H. Zhang, and Z. Bacic, *J. Chem. Phys.* **97,** 927; 3149 (1992); *Chem. Phys. Lett.* **194,** 313 (1992).
44. J. H. Shorter, M. P. Casassa, and D. S. King, *J. Chem. Phys.* **97,** 1824 (1992).
45. J. M. Steed, T. A. Dixon, and W. Klemperer, *J. Chem. Phys.* **70,** 4095 (1979).
46. A. M. Hough and B. J. Howard, *J. Chem. Soc., Faraday Trans. 2* **83,** 173 (1987).
47. G. T. Fraser, A. S. Pine, and R. D. Suenram, *J. Chem. Phys.* **88,** 6157 (1988).
48. M. H. Alexander and D. C. Clary, *Chem. Phys. Lett.* **98,** 319 (1983); D. C. Clary, *J. Chem. Phys.* **78,** 4915 (1983).
49. T. G. Kreutz, F. A. Khan, and G. W. Flynn, *J. Chem. Phys.* **92,** 347 (1990).
50. M. P. Casassa, J. C. Stephenson, and D. S. King, *Faraday Discuss. Chem. Soc.* **82,** 251 (1986); *J. Chem. Phys.* **85,** 2333 (1986).
51. M. P. Casassa, J. C. Stephenson, and D. S. King, *J. Chem. Phys.* **89,** 1966 (1988).
52. D. S. King and J. C. Stephenson, *J. Chem. Phys.* **82,** 5286 (1985).
53. Y. Rudich and R. Naaman, *J. Chem. Phys.* **96,** 8616 (1992).
54. J. Serafin, H. Ni, and J. J. Valentini, *J. Chem. Phys.* **100,** 2385 (1994).
55. R. N. Dixon, *J. Chem. Phys.* **85,** 1866 (1986); K.-H. Gericke, S. Klee, F. J. Comes, and R. N. Dixon, *J. Chem. Phys.* **85,** 4463 (1986).
56. C. M. Western, P. R. R. Langridge-Smith, and B. J. Howard, *Mol. Phys.* **44,** 145 (1981).

57. Ph. Brechignac, N. Halberstadt, B. J. Whitaker, and B. Coutant, *Chem. Phys. Lett.* **142,** 125 (1987); Ph. Brechignac, S. De Benedictis, N. Halberstadt, B. J. Whitaker, and S. Avrillier, *J. Chem. Phys.* **83,** 2064 (1985).
58. Y. Matsumoto, Y. Ohshima, and M. Takami, *J. Chem. Phys.* **92,** 937 (1990).
59. J. R. Hetzler, M. P. Casassa, and D. S. King, *J. Phys. Chem.* **95,** 8086 (1991).
60. W. M. Fawzy, G. T. Fraser, J. T. Hougen, and A. S. Pine, *J. Chem. Phys.* **95,** 7086 (1991).
61. T. A. Spiglanin, D. W. Chandler, and D. H. Parker, *Chem. Phys. Lett.* **137,** 414 (1987).
62. R. Callaghan, S. Arepalli, and R. J. Gordon, *J. Chem. Phys.* **86,** 5273 (1987).
63. Y. Rudich, R. J. Gordon, E. E. Nikitin, and R. Naaman, *J. Chem. Phys.* **96,** 4423 (1992).
64. D. C. Dayton, P. A. Block, and R. E. Miller, *J. Phys. Chem.* **95,** 2881 (1991).
65. A. C. Legon, P. D. Aldrich, and W. H. Flygare, *J. Chem. Phys.* **75,** 625 (1981).
66. N. Ohashi and A. S. Pine, *J. Chem. Phys.* **81,** 73 (1984).
67. D. H. Levy, *Adv. Chem. Phys.* **47,** 323 (1981); *Ann. Rev. Phys. Chem.* **31,** 197 (1980).
68. K. C. Janda, *Adv. Chem. Phys.* **60,** 201 (1985).
69. F. M. Tao and W. Klemperer, *J. Chem. Phys.* **97,** 440 (1992).
70. S. S. Huang, C. R. Bieler, K. C. Janda, F. M. Tao, W. Klemperer, P. Casavecchia, G. G. Volpi, and N. Halberstadt, *J. Chem. Phys.* **102,** 8846 (1995).
71. J. A. Beswick and G. Delgado-Barrio, *J. Chem. Phys.* **73,** 3653 (1980).
72. N. Halberstadt, J. A. Beswick, and K. C. Janda, *J. Chem. Phys.* **87,** 3966 (1987).
73. F. G. Celii and K. C. Janda, *Chem. Rev.* **86,** 507 (1986).
74. K. C. Janda, C. R. Bieler, and E. R. Bernstein, Eds., *Atomic and Molecular Clusters*, Elsevier, Amsterdam, 1990, p. 455.
75. J. M. Skene, J. C. Drobits, and M. I. Lester, *J. Chem. Phys.* **85,** 2329 (1986).
76. J. I. Cline, N. Sivakumar, D. D. Evard, and K. C. Janda, *J. Chem. Phys.* **86,** 1636 (1987).
77. J. I. Cline, N. Sivakumar, D. D. Evard, C. R. Bieler, B. P. Reid, N. Halberstadt, S. R. Hair, and K. C. Janada, *J. Chem. Phys.* **90,** 2605 (1989).
78. J. I. Cline, N. Sivakumar, D. D. Evard, and K. C. Janda, *Phys. Rev. A* **36,** 1944 (1987).
79. J. I. Cline, B. P. Reid, D. D. Evard, N. Sivakumar, N. Halberstadt, and K. C. Janda, *J. Chem. Phys.* **89,** 3535 (1988).
80. S. A. Walter and T. A. Stephenson, *J. Chem. Phys.* **96,** 3536 (1992).
81. T. Stephenson, *J. Chem. Phys.* **97,** 6262 (1992).
82. J. C. Drobits and M. I. Lester, *J. Chem. Phys.* **86,** 1662 (1987).
83. J. C. Drobits and M. I. Lester, *J. Chem. Phys.* **89,** 4716 (1988).
84. J. C. Drobits and M. I. Lester, *J. Chem. Phys.* **88,** 120 (1988).
85. J. M. Skene, Ph.D. Thesis, University of Pennsylvania, Philadelphia, 1988.
86. R. L. Waterland, J. M. Skene, and M. I. Lester, *J. Chem. Phys.* **89,** 7277 (1988).
87. O. Roncero, J. A. Beswick, N. Halberstadt, P. Villarreal, and G. Delgado-Barrio, *J. Chem. Phys.* **92,** 3348 (1990).
88. S. K. Gray and C. E. Wonzy, *Ber. Bunsenges. Phys. Chem.* **92,** 236 (1988). *J. Chem. Phys.* **94,** 2817 (1991).

89. M. Zhao and S. A. Rice, *J. Chem. Phys.* **96,** 7483 (1992).
90. N. Lipkin, N. Moiseyev, and C. Leforestier, *J. Chem. Phys.* **98,** 1888 (1993).
91. D. W. Brinza, C. M. Western, D. D. Evard, F. Thommen, B. A. Swartz, and K. C. Janda, *J. Phys. Chem.* **88,** 2004 (1984); D. D. Evard, F. Thommen, and K. C. Janda, *J. Chem. Phys.* **84,** 3630 (1986); D. D. Evard, F. Thommen, J. I. Cline, and K. C. Janda, *J. Phys. Chem.* **91,** 2508 (1987).
92. B. P. Reid, K. C. Janda, and N. Halberstadt, *J. Phys. Chem.* **92,** 587 (1988).
93. M. T. Berry, M. R. Brustein, and M. I. Lester, *J. Chem. Phys.* **90,** 5878 (1989).
94. M. T. Berry, M. R. Brustein, and M. I. Lester, *J. Chem. Phys.* **92,** 6469 (1990).
95. M. I. Lester, S. E. Choi, L. C. Giancarlo, and R. W. Randall, *Faraday Discuss.* **97,** 365 (1994).
96. C. Chakravarty, D. C. Clary, A. Degli Esposti, and H.-J. Werner, *J. Chem. Phys.* **95,** 8149 (1991).
97. M. I. Lester, R. A. Loomis, L. C. Giancarlo, M. T. Berry, C. Chakravarty, and D. C. Clary, *J. Chem. Phys.* **98,** 9320 (1993).
98. M. T. Berry, M. R. Brustein, J. R. Adamo, and M. I. Lester, *J. Phys. Chem.* **92,** 5551 (1988); M. T. Berry, M. R. Brustein, and M. I. Lester, *Chem. Phys. Lett.* **153,** 17 (1988).
99. W. M. Fawzy and M. C. Heaven, *J. Chem. Phys.* **89,** 7030 (1988); W. M. Fawzy and M. C. Heaven, *J. Chem. Phys.* **92,** 909 (1990).
100. J. Schleipen, L. Nemes, J. Heinze, and J. J. ter Meulen, *Chem. Phys. Lett.* **175,** 561 (1990).
101. B.-C. Chang, L. Yu, D. Cullin, B. Rehfuss, J. Williamson, T. A. Miller, W. M. Fawzy, X. Zheng, S. Fei, and M. Heaven, *J. Chem. Phys.* **95,** 7086 (1991).
102. A. Degli Esposti and H.-J. Werner, *J. Chem. Phys.* **93,** 3351 (1990).
103. C. Chakravarty, D. C. Clary, A. Degli Esposti and H.-J. Werner, *J. Chem. Phys.* **93,** 3367 (1990).
104. J. M. Bowman, B. Gazdy, P. Schafer, and M. C. Heaven, *J. Phys. Chem.* **94,** 2226 (1990); *J. Phys. Chem.* **94,** 8858 (1990); Y. Guan and J. T. Muckerman, *J. Phys. Chem.* **95,** 8293 (1991); U. Schnupf, J. M. Bowman, and M. C. Heaven, *Chem. Phys. Lett.* **189,** 487 (1992).
105. T.-S. Ho, H. Rabitz, S. E. Choi, and M. I. Lester, *J. Chem. Phys.* **102,** 2282 (1995).
106. L. C. Giancarlo, R. W. Randall, S. E. Choi, and M. I. Lester, *J. Chem. Phys.* **101,** 2914 (1994).
107. R. J. LeRoy, M. R. Davies, and M. E. Lam, *J. Phys. Chem.* **95,** 2167 (1991).

ELECTRON SCATTERING BY SMALL MOLECULES

CARL WINSTEAD AND VINCENT McKOY

A. A. Noyes Laboratory of Chemical Physics, California Institute of Technology, Pasadena, CA 91125

CONTENTS

Advances in Chemical Physics, Volume XCVI, Edited by I. Prigogine and Stuart A. Rice.
ISBN 0-471-15652-3

I. INTRODUCTION

Theoretical studies of electron–molecule collisions have progressed significantly since the classic review of Lane [1]. Some of these developments have involved the detailed study of processes occurring at very low impact energies, i.e., on the order of 1 electron volt (eV) or less. Much of that work is treated in the review of Morrison [2]. Morrison's article is also an excellent source of references to earlier reviews and books on electron–molecule scattering in general, among which we mention, in particular, the volumes edited by Shimamura and Takayanagi [3] and by Gianturco [4], which cover both experiment and theory. Progress has also been great in extending the scope of electron–molecule collision calculations to encompass collisions with nonlinear polyatomic molecules and electronic excitation by electron impact, areas of study that were in their infancy in 1980. Developments in computational methodology that have facilitated the study of collisions with polyatomic molecules have been the subject of chapters by Rescigno et al. and by the present authors in a recent volume [5], while the monograph of Burke and Berrington [6] is primarily concerned with diatomic molecules and with atoms. Theoretical and computational progress in the closely related area of electron–atom scattering has been reviewed by Henry [7] and by Burke [8]. For recent accounts of experimental progress in electron–molecule scattering, see the reviews of Allan [9] and of Kauppila and Stein [10] (the latter is concerned with both electron and positron scattering). Dunning [11] has recently reviewed experiments at extremely low impact energies. Finally, we mention the recent volume edited by Ehrhardt and Morgan [12], which provides a concise but broad overview of the field.

In the present work, we wish to focus, to some extent, on methods of calculation that are applicable to low-energy electron collisions with polyatomic molecules, but also on the *results* of recent studies of such collisions. Our goal is to describe the current state of the field, including areas where further progress is needed, while illustrating some recent successes. As there are presently few groups engaged in theoretical studies of this kind, much of the theoretical work that we will review is due to the Livermore group and to our own group. Where possible, however, we will make reference to other calculations, and in numerous cases it will be possible to compare calculations with measured cross sections.

In studying electron collisions with polyatomic molecules, it might seem natural to pursue those aspects of the problem that are unique, such as the possibility of exciting vibrational or rotational combination levels. Yet almost all *ab initio* calculations on polyatomics have been conducted in a fixed-nuclei approximation, with an implicit average over the omitted nuclear degrees of freedom. The neglect of rotational motion actually is not a significant limitation, at least in making contact with experiment, since experiments have achieved rotational resolution only for molecules with large rotational constants, such as H_2. Electron spectrometers are commonly able to resolve vibrational structure, however, and correspondingly detailed calculations would obviously be useful. That they are not yet common reflects practical considerations. Because of the low symmetry of the problem, the computational study of low-energy electron–molecule collisions is challenging, even when employing rather more severe approximations than do the best current electron–atom collision studies. These approximations (to be discussed more fully in a later section) often preclude a meaningful study of vibrational excitation. Moreover, even where such a study would be possible, it is usually considered too laborious. With limited exceptions, the *ab initio* study of collisions between electrons and polyatomics is confined to the semiquantitative, and at times qualitative, calculation of cross sections for electronically elastic or inelastic scattering.

Fortunately, with regard to applications, electronic excitation is often the most interesting process, since many applications are concerned with fluorescence (as in gas lasers) or electron-impact dissociation (as in low-temperature plasma processing). Moreover, in polyatomic molecules, the study of electronically inelastic scattering is complicated by the profusion of overlapping vibronic bands, which must be deconvoluted and then integrated in order to obtain cross sections for the electronic states. Consequently, measured electronically inelastic cross sections, already scarce for diatomic molecules, are largely unavailable in the case of polyatomics. Fixed-nuclei calculations thus can be extremely valuable as a source of data, provided they are reasonably reliable and can be carried out for a sufficient number of inelastic channels and a sufficient variety of molecules.

Indeed, in recent years a major goal of our own work has been to contribute to the developing database of collision cross sections relevant to low-temperature plasma processing of materials, especially semiconductor microelectronics. Such plasma-processing applications typically employ polyatomic feed gases whose electron-impact dissociation within the plasma is a source of reactive fragments, including atoms, radicals, and ions, that drive the chemical and physical changes occurring at the semiconductor surface. Increasingly sophisticated numerical models of plasmas and of surface processes, together with increasingly powerful computers on which to run those models, will soon permit robust three-dimensional simulations of plasma reactors; however, the reliability of such simulations will depend in part on the extent and quality of the physical and chemical data employed, including surface reaction rates and collision cross sections for plasma species. As already mentioned, electronically inelastic electron–molecule collisions are particularly important as a dissociation mechanism. Elastic (and, at the lowest energies, vibrationally inelastic) cross sections for electron scattering by heavy particles are also essential in modelling electron transport.

As the capability of modeling low-temperature plasma systems evolves, the demand for cross sections will intensify. Meeting that demand requires not only computational methods of sufficient reliability, but implementations of those methods that can produce timely and cost-effective results. Such implementations, we believe, will rely on massively parallel computers, which can provide the absolute performance required by the scale of *ab initio* calculations on large polyatomics at a price that is not prohibitive. Indeed, many of the calculations we have undertaken in recent years, and that we will report below, would not have been feasible on conventional vector supercomputers.

The following section presents the theoretical background necessary to understand how electron–molecule computations are carried out, followed by an overview of methods of approximation used in such computations. We assume a familiarity with elementary quantum mechanics, but a prior acquaintance with scattering theory is not necessary. Section III is devoted to cross sections for various molecules that have been the subject of recent theoretical studies. An effort will be made in the presentation of these cross sections to illuminate features that are common across a class of related species, as well as interesting variations. We will close with some remarks on the state of the field, summarizing the preceding sections, and its prospects in the near future.

II. THEORETICAL

A. Approximations

It is useful, for the sake of orientation, to review the various approximations that are commonly made in solving the electron–molecule scattering problem. Neglecting relativistic effects, the full problem may be expressed by the Schrödinger equation

$$\mathcal{H}^{\text{tot}}\Psi = \mathcal{E}\Psi \tag{2.1}$$

where $\mathcal{E}$ is the total energy of the molecule in its initial state plus the asymptotic kinetic energy of the scattering electron; the wave function $\Psi(\{\mathbf{R}_j, j = 1, \ldots, K\}, \{\mathbf{r}_i, i = 1, \ldots, N + 1\})$ is a function of the coordinates $\{\mathbf{R}_j\}$ of the K nuclei and $\{\mathbf{r}_i\}$ of the $N + 1$ electrons; and $\mathcal{H}^{\text{tot}}$ is the Hamiltonian

$$\mathcal{H}^{\text{tot}} = -\frac{1}{2M_j}\sum_{j=1}^{K}\nabla^2_{\mathbf{R}_j} - \frac{1}{2}\sum_{i=1}^{N+1}\nabla^2_{\mathbf{r}_i} + \sum_{j=1}^{K}\sum_{\substack{j'=1\\ j'>j}}^{K}\frac{Z_jZ_{j'}}{|\mathbf{R}_j - \mathbf{R}_{j'}|} + \sum_{i=1}^{N+1}\sum_{\substack{i'=1\\ i'>i}}^{N+1}\frac{1}{|\mathbf{r}_i - r_{i'}|} - \sum_{i=1}^{N+1}\sum_{j=1}^{K}\frac{Z_j}{|\mathbf{r}_i - \mathbf{R}_j|} \tag{2.2}$$

In Eq. (2.2), ∇_r^2 is the Laplacian in the variable **r**, and we have adopted Hartree atomic units, $\hbar = e = m_e = 1$, where e is the electron charge and m_e the electron mass; thus, the nuclear charge Z_j and mass M_j are taken to be specified in multiples of e and m_e. We shall use these units throughout, except that, in discussing electron energies, we shall frequently use the conventional units of electron volts (eV; 27.212 eV $\approx$ 1 hartree), while for cross sections conventional units are Å^2 (10^{-16} cm^2).

The first approximation we make is to treat the molecule as infinitely massive compared to the electron, so that the molecule is at rest in the center-of-mass coordinate system of the electron–molecule complex. This approximation is justified by the large factor by which the mass of even the lightest molecule exceeds that of the electron, or, more precisely, by the insignificant difference between the momentum of a low-energy electron in the center-of-mass frame and its momentum in the molecular rest frame.

The next approximation, related to the first, is the adiabatic or Born–Oppenheimer approximation [13] by which nuclear and electronic motion are separated, so that Ψ factors into a nuclear rovibrational wave function $\Gamma(\{\mathbf{R}_i\})$ and an electronic wave function $\psi(\{\mathbf{r}_i\})$. Here ψ depends parametrically on the nuclear coordinates, while the vibrational part $u(\{\mathbf{R}_i\})$ of Γ describes the adiabatic motion of the nuclei in a potential determined by the variation of the electronic energy with nuclear geometry. As with purely bound states, the application of the adiabatic approximation to a scattering situation is based, roughly speaking, on the assumption that the electrons' motion is much more rapid than that of the nuclei. More generally, non-Born–Oppenheimer effects may be expected when the duration of the collision is on the order of a rotational or a vibrational period of the molecule [13]. In nonresonant or direct scattering, the duration of the collision may be taken to be the time required for a classical particle moving with the asymptotic velocity of the scattering electron to traverse a distance equal to a representative molecular dimension. For example, at the velocity corresponding to 1 eV kinetic energy, the time to traverse a 10-Å distance is about 2×10^{-15} s, whereas the period of a vibration with an energy of 3,000 cm^{-1} is about 10^{-14} s. Thus, above 1 eV, electron–molecule collisions can in general be treated as vibrationally (and rotationally) sudden, describable by an electronic scattering wave function determined in the field of the fixed nuclei. A major exception occurs in the case of resonances, or temporary complexes of the electron and the molecule. From the uncertainty relation $\Delta E \Delta t \geq \frac{1}{2}\hbar$ and the preceding discussion, we can see that resonances whose width is less than 1 eV may exist for times comparable to a vibrational period. So-called shape resonances in several molecules are known to fall into this category, and the accurate treatment of these resonances requires an accounting for non-Born–Oppenheimer effects. The classic case is N_2 [14]. Within extremely narrow resonances, such as Feshbach resonances, we might likewise expect a breakdown of the adiabatic approximation in rotational excitation; however, as discussed in the introduction, the resolution necessary to observe such effects is rarely achieved in electron scattering experiments. For most shape resonances and for nonresonant scattering, the Born–Oppenheimer approximation is adequate.

A second major exception to the rule of thumb developed above, i.e., that collisions above 1 eV impact energy are rotationally and vibrationally sudden, occurs when the molecule possesses a permanent electrostatic moment. (Of course, this ''exception'' is in fact the usual case). The reason for this is that the long-range potential created by the permanent moment is felt by the electron at large distances, invalidating the assumption that the interaction time is the time to traverse the immediate vicinity of the molecule [15]. On the other hand, the effects of such multipoles are rarely a serious complication. First of all, the potentials involved are weak: the strongest, that arising from a permanent dipole, decays as r^{-2}. Indeed, effects arising from multipoles higher than the dipole are hardly observed. Electrons that penetrate to regions where the molecular charge density is significant experience much stronger forces than those arising from the long-range field, so that the influence of the multipole field on the scattering is largely obscured. Only those electrons that are scattered weakly—those, classically speaking, whose impact parameter $b = r_{\text{min}}$ is large—evince the effects of the multipole field. These electrons are necessarily scattered in the near-forward direction, and are consequently more difficult to observe experimentally, though effects on the cross section arising from a strong permanent dipole are often seen. As one might anticipate, the fact that the long-range electron–molecule interaction is weak implies that a perturbative treatment will be successful; we will describe in Section II.C.5 approaches that combine a perturbative treatment of the long-range scattering with a more exact treatment of short-range effects.

By means of the above approximations, we have reduced the ''essential'' part of the problem to that of determining an electronic wave function for the N electrons of the target molecule and the scattering electron. This wave function is to satisfy the electronic Schrödinger equation

$$H\psi(\mathbf{r}_i, i = 1, \ldots, N + 1\}) = E\psi(\{\mathbf{r}_i, i = 1, \ldots, N + 1\}) \tag{2.3}$$

with electronic Hamiltonian

$$H = -\frac{1}{2}\sum_{i=1}^{N+1} \nabla_{\mathbf{r}_i}^2 + \sum_{i=1}^{N+1}\sum_{\substack{i'=1 \\ i'>i}}^{N+1} \frac{1}{|\mathbf{r}_i - r_{i'}|} - \sum_{i=1}^{N+1}\sum_{j=1}^{K} \frac{Z_j}{|\mathbf{r}_i - \mathbf{R}_j|} \tag{2.4}$$

In Eq. (2.4), the nuclear coordinates $\{\mathbf{R}_j\}$ are taken to be fixed parameters. Formally, there is no difference between Eq. (2.3) and the bound-state eigenvalue problem that is the subject of molecular electronic structure calculations; however, we will see in the next section that different boundary conditions are involved, and that very different methods of solution are, in consequence, required. For now it is enough to observe that, in the scattering case, Eq. (2.3) does not represent an eigenvalue problem, E being rather a parameter specified prior to solution.

It should be noted that we are not debarred by the approximations made from computing cross sections for rotational or vibrational excitation by electron impact; methods analogous to those used for optical transitions between bound levels can be applied. Vibrational excitation, for example, can be accounted for by solving Eq. (2.3) at multiple geometries and integrating the resulting electronic transition operator T_e (whose form we will later derive) over the vibrational wave functions u_v:

$$T(v \longrightarrow v') \approx \langle u_{v'}(\{\mathbf{R}_j\})|T_e(\{\mathbf{R}_j\})|u_v(\{\mathbf{R}_j\})\rangle_{\{\mathbf{R}_j\}} \tag{2.5}$$

Nevertheless, the situation is slightly more complicated than that of optical excitation. Conservation of energy requires that the electronic energy be different after the collision than it is before; strictly, therefore, the transition operator $T_e(\{\mathbf{R}_j\})$ cannot be determined by solving the electronic scattering problem of Eq. (2.3) at a single energy E. In practice, the percentage change in the electronic energy is small enough to justify neglecting the inelasticity except at collision energies that are close to the vibrational threshold energy. Techniques for treating the scattering at these energies are reviewed by Morrison [2].

Before concluding this discussion of approximations, we would like to point out certain important features that remain, *without* approximation, in Eqs. (2.3) and (2.4). Most salient is the many-electron character of the scattering problem; the relevant Schrödinger equation is not that of one particle moving in a potential field, but that of $N + 1$ indistinguishable, interacting particles. At high impact energies, the necessity of treating the electrons on an equal footing diminishes, and a one-particle description becomes acceptable; however, we are in this review specifically interested in low energies. It is nonetheless sometimes useful to take the high-energy limit as the ''zeroth-order'' approximation, at least for purposes of discussion. Thus, in considering the interaction among electrons, one often thinks in terms of a scattering electron that feels the net effect of, first, a *static* potential arising from the nuclei and an N-electron charge distribution; second, an *exchange* interaction, arising from the Pauli principle; and third, a *polarization* interaction that results from distortion of the molecular electron density in response to the charge of the impinging electron. None of these ''components'' of the electron–molecule interaction is, of course, independently visible in Eq. (2.4). However, besides a certain utility as a guide to intuition, this decomposition of the electron–molecule interaction is the basis of further approximations employed in some methods of computing cross sections; see Section II.C below.

The development we have described also retains a proper accounting for electron spin, at least where spin–orbit coupling can be neglected; however, insofar as most electron–molecule scattering experiments are conducted with unpolarized electrons and molecules (mostly, in fact, with closed-shell singlet molecules), spin will not figure largely in the subsequent discussion. Under these conditions, calculations may be carried out by assuming a fixed value for S_{N+1}, the $N + 1$-electron total spin, and for its projection in the molecular frame, $M_{S,N+1}$; where the target spin S_N is nonzero, it will be necessary to compute cross sections for both $S_{N+1} = S_N + \frac{1}{2}$ and $S_{N+1} = S_N - \frac{1}{2}$. Below, spin will primarily manifest itself in the Pauli exclusion principle and in the selection rule $\Delta S_N = 0, \pm 1$, which follows from conservation of S_{N+1}.

Since we are focusing on polyatomic molecules, it is, finally, germane to emphasize that, except in the special case of linear molecules, the nuclear potential in Eq. (2.4) does not permit any separation of variables to be made. Special techniques developed for electron–atom and electron–diatomic scattering that take advantage of such separations are thus inapplicable to general polyatomics, though they are sometimes applied to quasiatomic molecules like methane. In most cases, however, the scattering problem must be solved in the full $3(N + 1)$-dimensional electronic phase space.

B. Scattering Theory

Quantum mechanical scattering theory is a large and well-developed field in its own right, and is treated in standard textbooks (e.g., [16]) and in several monographs; that of Newton [17] is especially comprehensive, while that of Taylor [18] is especially concise. The classic reference specific to atomic scattering is Mott and Massey [19]; much of the development there is also applicable to molecular problems. Review articles mentioned in Section I provide additional background. Here we will develop notions of scattering theory only so far as is necessary to understand the formulation of methods for solving Eq. (2.3) subject to appropriate boundary conditions, and to introduce basic terminology.

1. Boundary Conditions

As discussed above, Eq. (2.3) is to be solved at an energy specified in advance. At arbitrarily large times before the collision, we may assume that the molecule is in an electronic eigenstate $\phi_0(\{\mathbf{r}_i, i = 1, \ldots, N\})$ with definite energy E_0, while the electron moves freely with a definite velocity $\mathbf{k}_0$, and hence kinetic energy $\frac{1}{2}k_0^2$; thus, $E = E_0 + \frac{1}{2}k_0^2$. Since energy is conserved, we also have $E = E_n + \frac{1}{2}k_n^2$, where E_n is the energy of the eigenstate ϕ_n in which the molecule is left after the collision, and k_n is the asymptotic velocity of the departing electron. Electronically elastic scattering is characterized by $E_n = E_0$ and $\phi_n = \phi_0$; all other processes constitute inelastic scattering. Thresholds at which inelastic processes become energetically possible are determined by the conservation requirement and the physical requirements $k_0 > 0$, $k_n > 0$.

The elementary considerations of the preceding paragraphs express the boundary conditions subject to which Eq. (2.3) is solved. We require, specifically, that ψ have a form that incorporates the behavior of the system both long before and long after the collision. In algebraic form,

$$\psi(r_i \to \infty) = \frac{1}{\sqrt{N+1}} \Bigg[\phi_0(\{\mathbf{r}_{i'}, i' = 1, \ldots, N+1, i' \neq i\}) \cdot \exp(i\mathbf{k}_0 \cdot \mathbf{r}_i) + \sum_{n \in \text{open}} f_{0n}(\hat{\boldsymbol{k}}_0, \hat{\boldsymbol{k}}_n) \cdot \phi_n(\{\mathbf{r}_{i'}, i' = 1, \ldots, N+1, i' \neq i\}) \frac{\exp(ik_n r_i)}{r_i} \Bigg] \tag{2.6}$$

That is, as any one electron's coordinates are taken to infinity, ψ becomes a sum of a term, $\phi_0 \exp(i\mathbf{k}_0 \cdot \mathbf{r}_i)$, representing the initial electronic state of the molecule and a free electron with velocity $\mathbf{k}_0$, and of terms of the form $\phi_n \exp(ik_n r_i) r_i^{-1}$ that are the product of a final molecular state ϕ_n and an outgoing spherical wave representing the scattered electron. There is one term of the latter form for each *open channel,* i.e., for each electronic state that is energetically accessible at energy E, signified by the notation $n \in$ open in Eq. (2.6). The overall factor $(N + 1)^{-1/2}$ on the right-hand side arises because we have chosen the usual $[(N + 1)!]^{-1/2}$ normalization for the antisymmetric function ψ. The crucial role played by the complex weighting factors $f_{0n}(\hat{\boldsymbol{k}}_0, \hat{\boldsymbol{k}}_n)$ should be noted: these *scattering amplitudes* modify the intensity of the outgoing spherical waves as a function of the directions $\hat{\boldsymbol{k}}_0$ and $\hat{\boldsymbol{k}}_n$ of the electron's incidence and departure.

We should point out that, in specifying the asymptotic form of ψ by Eq. (2.6), we have taken the N-electron states ϕ_n to be bound states, and thus implicitly excluded electron-impact ionization, which would lead, for single ionization, to terms with two electrons present in the asymptotic region, and to correspondingly more complicated forms for double and higher levels of ionization. For simplicity in the discussion, we prefer to neglect such terms; moreover, almost all calculations on low-energy electron–molecule scattering have also neglected ionization processes as being too difficult to incorporate, even though cross sections for single ionization tend not to be particularly small in comparison to those for bound excitations.

2. *Cross Sections*

The scattering amplitudes $f_{0n}(\hat{\boldsymbol{k}}_0, \hat{\boldsymbol{k}}_n)$ in Eq. (2.6) contain the measurable information about the scattering process, namely the likelihood that a specified initial configuration develops into a specific final configuration. To see this, recall the expression for the flux density, which in atomic units reads

$$\mathbf{j} = \frac{i}{2} (\psi \nabla \psi^* - \psi^* \nabla \psi) \tag{2.7}$$

Applying Eq. (2.7) to individual terms of Eq. (2.6), we see that the incident flux density is given by $\mathbf{k}_0$ [the factor $(N + 1)^{-1}$ disappears because there is a contribution for each $\mathbf{r}_i$]. To obtain the scattered flux, consider a detector placed at the point $\mathbf{R}$ with respect to the molecule, where the distance R is macroscopic; let the detector's aperture subtend a solid angle $d\Omega$. Under these conditions, only the radial component $\mathbf{j} \cdot \hat{\boldsymbol{R}}$ of the flux is measured, so that we may replace ∇ with $(\partial/\partial r)\hat{\boldsymbol{R}}$ in Eq. (2.7). Then the detected flux F_n due to a single term $f_{0n} \exp(ik_n r)/r$ in Eq. (2.6) is the product of the flux density $k_n|f_{0n}|^2/R^2$ obtained using Eq. (2.7) and the aperture area $R^2 d\Omega$, or

$$F_n = k_n |f_{0n}(\hat{\boldsymbol{k}}_0, \hat{\boldsymbol{k}}_n)|^2 \, d\Omega \tag{2.8}$$

The intrinsic property of the molecule that characterizes the scattering is the coefficient of $d\Omega$ in the *ratio* of F_n to the magnitude k_0 of the incident flux; this ratio, which has the dimensions of area per solid angle, we call the *differential cross section, $d\sigma/d\Omega$*:

$$\begin{aligned} \frac{d\sigma_{0n}}{d\Omega} (\hat{\boldsymbol{k}}_0, \hat{\boldsymbol{k}}_n) &= \frac{F_n}{k_0} \\ &= \frac{k_n}{k_0} |f_{0n}(\hat{\boldsymbol{k}}_0, \hat{\boldsymbol{k}}_n)|^2 \end{aligned} \tag{2.9}$$

In practice Eq. (2.9) is rarely applicable, since most scattering experiments are done on dilute gases or molecular beams, in which vast numbers of randomly oriented molecules are present, and the measured flux is the incoherent superposition of that resulting from many independent collisions. In effect, what is measured is the average of Eq. (2.9) over all possible directions $\hat{k}_0$ and $\hat{k}_n$, with the relative angle Θ, defined by $\cos\Theta = \hat{\boldsymbol{k}}_0 \cdot \hat{\boldsymbol{k}}_n$, held fixed. We then have

$$\frac{d\sigma_{0n}}{d\Omega}(\Theta) = \frac{1}{8\pi^2}\frac{k_n}{k_0}\int d\hat{\boldsymbol{k}}_0 \int d\hat{\boldsymbol{k}}_n \, |f_{0n}(\hat{\boldsymbol{k}}_0, \hat{\boldsymbol{k}}_n : \hat{\boldsymbol{k}}_0 \cdot \hat{\boldsymbol{k}}_n = \cos\Theta)|^2 \quad (2.10)$$

as the relevant differential cross section.

Integral cross sections σ_{0n} are defined in the obvious way, namely as the integral over $d\Omega$ of the corresponding differential cross section $d\sigma_{0n}/d\Omega$:

$$\begin{aligned} \sigma_{0n} &= \int d\Omega \, \frac{d\sigma_{0n}}{d\Omega} \\ &= 2\pi \int d\Theta \, \frac{d\sigma_{0n}}{d\Omega}(\Theta) \end{aligned} \quad (2.11)$$

The *total cross section*, σ_{tot}, is the sum of the individual channel cross sections: $\sigma_{\text{tot}} = \Sigma_{n \in \text{open}} \, \sigma_{0n}$. To simplify the notation we shall often omit the subscript "0"; it will, however, then be understood that a specification of the initial state (which, in practice, is almost always the ground electronic state), as well as the final state, is included in the definition of the cross section and related quantities.

It may be remarked that we have been somewhat loose in deriving expressions for the differential cross sections, in that we have neglected interference terms in the flux associated with the asymptotic form of Eq. (6). The physical justification for neglecting the interference of the incident and scattered waves is discussed, e.g., in Schiff [20]. The essential point is that the incident electron beam may be collimated so as not to reach the detector by passing it through a slit of macroscopic size, which will perturb its plane-wave character (as perceived by the molecule) only negligibly. Scattered waves with different momenta k_n do not interfere because k_n is an observable in the asymptotic region; detection of the electron at $\mathbf{R}$ with momentum $k_n\hat{\boldsymbol{R}}$ "collapses" the right-hand side of Eq. (2.6) to a single term [21]. An alternative point of view is that electrons with different k_n are distinguishable because they reach the detector at different times.

3. *Lippmann–Schwinger Equation*

One way to develop solutions of the Schrödinger equation [Eq. (2.3)] subject to the boundary conditions of Eq. (2.6) is to replace Eq. (2.3) with an equivalent integral equation. We begin that process by rewriting Eqs. (2.3) and (2.4), separating the Hamiltonian of Eq. (2.4) into a zeroth-order Hamiltonian H_0 representing a free electron and an undisturbed molecule,

$$H_{0,N+1} = -\frac{1}{2}\sum_{i=1}^{N+1} \nabla_{r_i}^2 + \sum_{i=1}^{N}\sum_{\substack{i'=1\\ i'>i}}^{N} \frac{1}{|\mathbf{r}_i - \mathbf{r}_{i'}|} - \sum_{i=1}^{N}\sum_{j=1}^{K} \frac{Z_j}{|\mathbf{r}_i - \mathbf{R}_j|} \tag{2.12}$$

and an interaction potential V_{N+1},

$$V_{N+1} = \sum_{i=1}^{N} \frac{1}{|\mathbf{r}_{N+1} - \mathbf{r}_i|} - \sum_{j=1}^{K} \frac{Z_j}{|\mathbf{r}_{N+1} - \mathbf{R}_j|} \tag{2.13}$$

In this partitioning, electron $N + 1$ has a unique status as the "scattering" electron, hence the subscript $N + 1$ on H_0 and V (which we shall omit when no specific partitioning is intended); however, the choice of scattering electron is arbitrary, and final expressions will be fully antisymmetric. Eq. (2.3) may now be rearranged to

$$(E - H_0)\psi = V\psi \tag{2.14}$$

and solved formally to obtain the *Lippmann–Schwinger equation* [22]

$$\psi = \Phi + \frac{1}{E - H_0} V\psi \tag{2.15}$$

In this equation, Φ is a particular solution to the homogeneous version of Eq. (2.14), i.e.,

$$H_0\Phi = E\Phi \tag{2.16}$$

It will be noted that, owing to the asymmetrical partitioning of the Hamiltonian in Eqs. (2.12) and (2.13), not all solutions to Eq. (2.15) are properly antisymmetric. However, we can always obtain an antisymmetric solution by first solving Eq. (2.15) without enforcing the equivalence of the scattering electron with the target electrons, and then and applying the antisymmetrization operator

$$\mathcal{A}_i = \frac{1}{\sqrt{N+1}}\left(1 - \sum_{\substack{j=1\\ j\neq i}}^{N+1} \hat{P}_{ij}\right) \tag{2.17}$$

to the result. Here i is the electron chosen to be the "scattering" electron and the permutation operator $\hat{P}_{ij}$ interchanges electrons i and j.

The desired boundary conditions are incorporated in the Lippmann–Schwinger equation through the choice of Φ and through the limiting process used to give meaning to the singular operator $G_0(E) = (E - H_0)^{-1}$ (known as a Green's operator, or in the coordinate representation as a Green's function). We let $\Phi = \Phi_0$, the initial configuration of the system represented by the first term on the right-hand side of Eq. (2.6):

$$\Phi_0 = \phi_0(\{\mathbf{r}_i, i = 1, \ldots, N\}) \exp(i\mathbf{k}_0 \cdot \mathbf{r}_{N+1}) \tag{2.18}$$

We add the subscript 0 to $\psi^{(+)}$ to reflect this boundary condition. The specification of the Green's operator may be made most clearly in terms of its effect on the complete set of eigenfunctions of H_0, the functions

$$\Phi_n(E') = \phi_n(\{\mathbf{r}_i, i = 1, \ldots, N\}) \exp(i\mathbf{k}'_n \cdot \mathbf{r}_{N+1}) \qquad E' = E_n + \tfrac{1}{2}k_n'^2 \tag{2.19}$$

with $\mathbf{k}'_n$ taking on all possible directions. The resolution of the identity in terms of the $\Phi_n(E')$ is

$$\prod_{i=1}^{N+1} \delta(\mathbf{r}_i - \mathbf{r}'_i) = \frac{1}{(2\pi)^3} \sum_{n=0}^{\infty} \int dE' \int d\mathbf{k}'_n |\Phi_n(E')\rangle\langle\Phi_n(E')| \tag{2.20}$$

Note that the sum on n is *not* restricted to open channels and includes, in particular, ionized states of the molecule.

We require of the Green's operator acting on Eq. (2.20) that

$$\begin{aligned}
G_0(E) \prod_{i=1}^{N+1} \delta(\mathbf{r}_i - \mathbf{r}'_i) &= \frac{1}{(2\pi)^3} \lim_{\epsilon \to 0^+} \sum_{n=0}^{\infty} \int dE' \int d\hat{k}'_n \frac{|\Phi_n(E')\rangle\, \langle\Phi_n(E')|}{E - E' + i\epsilon} \\
&= \lim_{\epsilon \to 0^+} \sum_{n=0}^{\infty} \int dE' \int d\hat{k}'_n \\
&\quad \cdot \exp(i\hat{k}'_n \cdot r_{N+1}) |\phi_n(\{r_i, i = 1, \ldots, N\})\rangle \\
&\quad \times \frac{1}{E - E_n - \frac{1}{2}k_n'^2 + i\epsilon} \exp(-i\mathbf{k}'_n \cdot \mathbf{r}'_{N+1}) \\
&\quad \cdot \langle\phi_n(\{\mathbf{r}'_i, i = 1, \ldots, N\})|
\end{aligned} \tag{2.21}$$

where $\epsilon \to 0^+$ means that ϵ approaches 0 from above. The superscript (+) is used to indicate this limiting process; thus we write the Green's operator defined by Eq. (2.21) as $G_0^{(+)}(E)$ and the corresponding solution to the

Lippmann–Schwinger equation as $\psi_0^{(+)}(E)$:

$$\psi_0^{(+)}(E) = \Phi_0(E) + G_0^{(+)}(E)V\psi_0^{(+)}(E) \tag{2.22}$$

or in slightly more detail,

$$\psi_0^{(+)}(E) = \Phi_0(\mathbf{k}_0) + \frac{1}{(2\pi)^3} \lim_{\epsilon\to 0^+} \sum_{n=0}^{\infty} \int dE' \int d\hat{k}'_n \frac{\Phi_n(\mathbf{k}'_n)T_{0n}(\mathbf{k}_0, \mathbf{k}'_n)}{E - E_n - \frac{1}{2}k_n'^2 + i\epsilon} \tag{2.23}$$

where the quantity T_{0n}, referred to as an element of the transition matrix, or simply the T-matrix, is given by

$$T_{0n}(\mathbf{k}_0, \mathbf{k}_n) = \langle \Phi_n(\mathbf{k}_n)|V|\, \psi_0^{(+)}(\mathbf{k}_0)\rangle \tag{2.24}$$

That $\psi_0^{(+)}(E)$ defined in this manner does indeed have the correct asymptotic form, Eq. (2.6), may be seen by taking Eq. (2.23) into the asymptotic region and comparing the sum involving the Green's function in Eq. (2.23) with the sum over scattered waves in Eq. (2.6). This development is straightforward but lengthy, and is relegated to Appendix A. It should be noted that terms in Eq. (2.23) corresponding to closed channels do not contribute asymptotically, as we would expect on physical grounds. In Eq. (2.25) we have the relation between T-matrix elements and scattering amplitudes [see Eq. (A.6) in Appendix A],

$$f_n(\mathbf{k}_0, \mathbf{k}_n) = -\frac{1}{2\pi} T_{0n}(\mathbf{k}_0, \mathbf{k}_n) \tag{2.25}$$

For future reference, we note that replacing $\epsilon \to 0^+$ with $\epsilon \to 0^-$ in Eq. (2.21) leads to a Green's function $G_0^{(-)}$ that is the complex conjugate of $G_0^{(+)}$. Replacing $G_0^{(+)}$ with $G_0^{(-)}$ and Φ_0 with Φ_n in Eq. (2.21) defines wave functions $\psi_n^{(-)}$,

$$\psi_n^{(-)}(E) = \Phi_n(E) + G_0^{(-)}(E)V\psi_n^{(-)}(E) \tag{2.26}$$

that asymptotically have *incoming* spherical wave behavior. We can rearrange the adjoint of Eq. (2.26) into an expression for Φ_n^*; multiplying that expression from the right by V and substituting into the definition of the T-matrix, Eq. (2.24), gives the alternative form

$$T_{0n}(\mathbf{k}_0, \mathbf{k}_n) = \langle \psi_n^{(-)}(\mathbf{k}_n)|V - VG_0^{(+)}V|\, \psi_0^{(+)}(\mathbf{k}_0)\rangle \tag{2.27}$$

where we have used $G_0^{(-)*} = G_0^{(+)}$. By means of Eq. (2.22), we may derive from Eq. (2.27) a third form for the T-matrix,

$$T_{0n}(\mathbf{k}_0, \mathbf{k}_n) = \langle \psi_n^{(-)}(\mathbf{k}_n) | V | \Phi_0(\mathbf{k}_0) \rangle \tag{2.28}$$

These expressions will be used together with Eq. (2.24) in formulating the Schwinger variational procedure for computing the T-matrix.

C. Methods

We discuss here the principal numerical methods currently used for low-energy electron collisions with polyatomic molecules. Such methods may be classified in a number of ways. One possible distinction is to separate *ab initio* from semiempirical methods, as in bound-state quantum chemistry. *Ab initio* methods aim to compute results from first principles, ideally using no more input than the atomic numbers of the atoms in the molecule (and a specification of the isomer desired, where more than one exists), while semiempirical methods rely to a greater or lesser extent on experimental information. In electron scattering, the most common use of empirical data is the approximation of the polarization effect through a local polarization potential, whose long-range behavior is matched to the measured polarizability. Of course, if a computed polarizability were used, such a method would become "*ab initio*"; conversely, "*ab initio*" calculations frequently make use of empirical excitation thresholds and equilibrium geometries.

A more useful classification separates variational from direct methods, and model-potential methods from all-electron methods. Direct methods are those that attempt to compute the scattering wave function, usually by numerical integration of the appropriate Schrödinger equation, and from that wave function determine scattering information. In practice, however, direct solution of the full electron–molecule scattering problem is too difficult, and direct methods are usually formulated in the context of model potentials, described immediately below. Variational methods, on the other hand, typically employ a finite space of $(N + 1)$-particle functions to represent the scattering wave function, thereby making the problem more tractable, and then seek the stationary value of the T-matrix, or some closely related quantity, within that space. An approximate wave function may be generated by a variational approach, but it is often disregarded. We will describe two variational methods, the Schwinger method [23] and the Kohn method [24], in some detail below. The R-matrix method [25, 26] is something of a hybrid between variational and direct methods: Within a sphere of finite radius, a type of variational problem is constructed and solved, but the final scattering information is extracted by numerically propagating the scattering wave function from the surface of that sphere into the asymptotic region. We shall have less to say about the R-matrix method since its application to polyatomic molecules has so far been very limited. The explicit appearance of a sphere is undoubtedly an obstacle to its application to general molecules, since a partial-wave expansion, i.e., an expansion in eigenfunctions of angular momentum, is then almost unavoidable, and such expansions may converge slowly for asymmetric molecules.

Model-potential methods, are broadly speaking, those that replace the full $(N + 1)$-electron scattering problem of Eqs. (2.4) and (2.6) with the much more tractable problem of a single electron moving in a potential field. A complete model-potential approach might, for example, begin with the electron–nucleus Coulomb potential and the static portion of the electron–electron potential, that is, the Coulombic interaction between an electron and the electron density derived from a bound-state calculation for the N molecular electrons. To this are typically added potential terms intended to represent the effects of electron exchange and of polarization. One drawback of the model-potential approaches is that neither exchange nor polarization is in fact a local effect; hence, no local potential, however carefully designed, is capable of representing these effects with complete fidelity. Moreover, if a local approximation fails to give adequate results, there is usually no systematic way to improve the model towards convergence. Local approximations to exchange, in particular, have proven unreliable at low energies, and in recent years some workers have adopted mixed approximations, in which exchange is treated exactly but polarization is still represented by a local potential. A further drawback of replacing the molecular electrons in whole or in part by effective potentials is that the treatment of electronic excitation by electron impact becomes problematic.

Lima and coworkers [27] have implemented a slightly different form of model potential, using effective core potentials derived from atomic structure calculations that replace only the inner-shell electrons with a local model potential. Such potentials, familiar in bound-state studies involving heavy atoms and especially in solid-state electronic-structure calculations, permit significant reductions in the number of electrons treated explicitly (and thus the number of basis functions) while retaining the full complexity of the many-body problem for the interaction of the scattering electron with the valence shell. The reduced problem is then solved by the Schwinger variational method. Such an approach has considerable advantages over pure model-potential methods but, like them, may not be suitable for electronically inelastic scattering.

1. Kohn Variational Method

The Kohn variational method was originally formulated [24] as an approximation to the *phase shift*, η. In solving one-dimensional potential-scattering problems, it is common to take the "incident" wave to be the real function

$A \sin(kx)$, which in fact represents a standing wave. The asymptotic form, including the scattered wave, may then be written as $A \sin(kx) + B \cos(kx)$, or equivalently as $A[\sin(kx) + \tan \eta \cos(kx)]$, which defines η. Spherically symmetric problems may be separated into radial equations for each angular momentum ℓ and solved analogously, with the scattering characterized by the set of phase shifts η_ℓ. In three dimensions, and when there is more than one channel, the generalization of the phase shift is an infinite-dimensional matrix **K** known as the reactance matrix or simply the K-matrix. The K-matrix is the analogue, for standing-wave boundary conditions, of the T-matrix, to which it is related by

$$\pi \mathbf{T} = -\mathbf{K}(\mathbf{1} - i\mathbf{K})^{-1} \tag{2.29}$$

where **1** is the unit matrix (see for example [28]).

Kohn began with the quantity $I = \langle \psi | E - H | \psi \rangle$. Obviously $I = 0$ when ψ is exact. The first-order variation of I arising from a variation $\delta\psi$ about the exact ψ is

$$\begin{aligned} \delta I &= \langle \psi | E - H | \delta\psi \rangle \\ &= \frac{1}{2} \lim_{x \to \infty} \delta\psi \frac{d\psi}{dx} - \frac{d(\delta\psi)}{dx} \psi \end{aligned} \tag{2.30}$$

as may be shown by integrating the kinetic energy term of the Hamiltonian twice by parts; here $\lim_{x \to \infty}$ does not indicate a proper limit but rather the asymptotic functional form. We note that, as usual in variational problems, the variation $\delta\psi$ is required to satisfy the boundary conditions appropriate to ψ, one of which is $\psi(0) = 0$; this condition accounts for the lack of a contribution from $x = 0$ in Eq. (2.30). If x is large, the same assumption implies that $\delta\psi$ has the form $\delta B \cos(kx)$, from which we can evaluate Eq. (2.30), giving

$$\delta I = -\tfrac{1}{2} kA\delta B \tag{2.31}$$

Thus $I + kAB/2$ is a stationary quantity: first-order errors in B will be offset by the first-order variation of I. Letting B_0 be the value of B obtained from an approximate wave function $\tilde{\psi}$ and recalling that $B = A \tan \eta$, we therefore have

$$kA^2 \tan \eta = I^{(0)} + \tfrac{1}{2} kAB_0 \tag{2.32}$$

as a variational approximation to the tangent of the phase shift, where $I^{(0)}$ means I evaluated for the trial function $\tilde{\psi}$.

To implement the Kohn method, it is usual to introduce a linear approximation to the trial function, i.e., to set

$$\tilde{\psi} = \sum_{j=1}^{M} c_j \chi_j \tag{2.33}$$

where the χ_j are a set of known functions and the coefficients c_j are to be determined. Eq. (2.32) then becomes

$$A^2 k \tan \eta = \sum_{j=1}^{M} \sum_{j'=1}^{M} c_j c_j' I_{jj'}^{(0)} + \tfrac{1}{2} k A B_0 \tag{2.34}$$

with

$$I_{jj'}^{(0)} = \langle \chi_j | E - H | \chi_j' \rangle \tag{2.35}$$

For simplicity we set $A = 1$ and let χ_1 be $\sin(kx)$, χ_2 be $\zeta(x) \cos(kx)$, and let all of the remaining χ_j vanish asymptotically. The function $\zeta(x)$ is a regularization factor with the properties $\zeta(0) = 0$, $\lim_{x \to \infty} \zeta(x) = 1$, whose purpose is to enforce the boundary condition at $x = 0$. With these restrictions, we have $c_1 = 1$ and $c_2 = B_0$, and by imposing the stability requirement

$$\frac{\partial \tan \eta}{\partial c_j} = 0 \qquad \text{for all } j \tag{2.36}$$

we arrive at a set of linear equations determining the c_j. In matrix-vector form these read

$$\mathbf{Ic} = -\mathbf{s} \tag{2.37}$$

where $\mathbf{I}$ is the square matrix with elements $I_{jj'}^{(0)}$, $\mathbf{c}$ is a column vector with elements c_j, and $\mathbf{s}$ is a column vector with elements $s_j = \langle \chi_j | E - H | \chi_1 \rangle = \langle \chi_j | E - H | \sin(kx) \rangle$. In these definitions, j and j' run from 2 to M. Inverting $\mathbf{I}$ gives $\mathbf{c} = -\mathbf{I}^{-1}\mathbf{s}$, which, when substituted back into Eq. (2.34), yields

$$k \tan \eta = c_2 - \frac{2}{k} (\langle \sin(kx) | E - H | \sin(kx) \rangle + \mathbf{c}^T \cdot \mathbf{s}) \tag{2.38}$$

We have developed the Kohn principle so far as a principle for tan η not because that development is particularly appropriate for applications such as those we have in mind (as Kohn already observed [24]), but because, historically, the Kohn method has almost come to be identified with this formulation, which has a serious deficiency. It was early observed [29] that phase shifts calculated according to Eq. (2.38) possessed, as functions of E, not only the physical singularities expected when $\eta(E)$ passes through a multiple of $\pi/2$, but also *spurious* singularities that grew in number as the basis set was extended. These spurious singularities are simply the zeroes of the characteristic polynomial det $\mathbf{I}(E)$, which lie on the real axis because $\mathbf{I}$ is Hermitian. Effort has been devoted (see, for example, [30]) to avoiding or minimizing the effect of these singularities—for example, one can switch back and forth between the Kohn principle for tan η and the analogous principle for cot η as spurious singularities are encountered—but their existence has until recently impaired the popularity of the Kohn method.

In retrospect, it is clear that, since the singularities arise from the Hermiticity of $\mathbf{I}$, which is itself a consequence of the choice of standing-wave boundary conditions, formulations that instead use traveling waves may not suffer from the same defect. Consider, for instance, the variational principle Kohn presented [24] for the scattering amplitude (or, equivalently, the T-matrix),

$$f(\mathbf{k}_1, \mathbf{k}_2) = f^{(0)}(\mathbf{k}_1, \mathbf{k}_2) + \frac{1}{2\pi} \int d^3r \psi_2^{(+)} (\mathbf{r}; -\mathbf{k}_2) (E - H)\psi_1^{(+)}(\mathbf{r}; \mathbf{k}_1) \tag{2.39}$$

where $f^{(0)}$ is the trial scattering amplitude. Observe that $\psi_2^{(+)}$, and not its complex conjugate, occurs in Eq. (2.39), and that $\psi_1^{(+)}$ and $\psi_2^{(+)}$ must be complex to satisfy the boundary conditions of Eq. (2.6). Consequently, when a basis set is introduced, Eq. (2.39) leads, by a development analogous to Eqs. (2.33–2.37), to linear equations requiring the inverse of a complex-symmetric, rather than Hermitian, matrix. Except accidentally, the poles of such a matrix will not lie on the real k-axis, and we can expect the T-matrix Kohn method to be free, or nearly free [31], of spurious singularities. Only quite recently [32] has this property of Kohn-type formulations with traveling-wave boundary conditions been generally appreciated (see, however, [33]).

Rescigno, McCurdy, and coworkers have implemented the Kohn principle for the T-matrix and applied it to electron–molecule collision problems [34]. This requires, first of all, a straightforward generalization of the Kohn principle to a many-particle formulation. The localized one-electron functions χ_j are replaced with properly antisymmetrized $(N + 1)$-electron functions, while $\sin(kr)$, $\exp(i\mathbf{k}\cdot\mathbf{r})$, and so on, are antisymmetrized with localized N-electron functions that represent the appropriate bound state of the molecule. The Hamiltonian H becomes the $(N + 1)$-electron Hamiltonian of Eq. (2.4). Rescigno et al. [34] have chosen not to work with Eq. (2.39), but instead to represent the scattering portion of the wave function in a basis set of angular-momentum eigenfunctions $j_\ell(kr)Y_\ell^m(\theta, \phi)$ and $\zeta(r)h_\ell^{(+)}(kr)Y_\ell^m(\theta, \phi)$, where the term in j_ℓ, a regular spherical Bessel function, arises from the partial-wave expansion of $\exp(i\mathbf{k}\cdot\mathbf{r})$ and the term in $h_\ell^{(+)}$, a spherical Hankel function with outgoing-wave asymptotic behavior, from the expansion of the scattered wave; Y_ℓ^m is a spherical harmonic. The function $\zeta(r)$ is, as in the one-dimensional case, a regularization factor, suppressing the singularity of the Hankel function at $r = 0$. In practice, of course, it is necessary to truncate the expansion in ℓ and m, giving rise, through the variational expression, to a finite-matrix equation for the partial-wave representation of the T matrix, $T_{\ell m0,\ell' m' n}$.

The principal computational advantage of the Kohn method is that it requires only matrix elements of the Hamiltonian. With a proper choice of the asymptotically vanishing basis functions χ_j, the evaluation of almost all of the requisite matrix elements can be made easy. For molecular calculations, these "bound" χ_j can, for example, be Slater determinants of molecular orbitals expanded in a Gaussian basis, in which case standard electronic-structure programs developed for bound-state quantum chemistry can be used to evaluate the Hamiltonian matrix elements that involve only bound functions. The only complication arises from integrals in which the "free" functions $j_\ell(kr)Y_\ell^m(\theta, \phi)$ and $\zeta(r)h_\ell^{(+)}(kr)Y_\ell^m(\theta, \phi)$ occur. Integrals arising from the electron–electron repulsion terms in the Hamiltonian are the most difficult, since these involve the coordinates of two electrons. These integrals include so-called bound-free integrals such as

$$\left\langle a(\mathbf{r}_1)b(\mathbf{r}_2) \left| \frac{1}{|\mathbf{r}_1 - \mathbf{r}_2|} \right| c(\mathbf{r}_1)j_\ell(kr_2)Y_\ell^m(\theta_2, \phi_2) \right\rangle \tag{2.40}$$

free-free Coulomb integrals such as

$$\left\langle a(\mathbf{r}_1)j_\ell(kr_2)Y_\ell^m(\theta_2, \phi_2) \left| \frac{1}{|\mathbf{r}_1 - r_2|} \right| b(\mathbf{r}_1)j_{\ell'}(k'r_2)Y_{\ell'}^{m'}(\theta_2, \phi_2) \right\rangle \tag{2.41}$$

and free-free exchange integrals such as

$$\left\langle a(\mathbf{r}_1) j_\ell(kr_2) Y_\ell^m(\theta_2, \phi_2) \left| \frac{1}{|\mathbf{r}_1 - \mathbf{r}_2|} \right| j_{\ell'}(k'r_1) Y_{\ell'}^{m'}(\theta_1, \phi_1) b(\mathbf{r}_2) \right\rangle \tag{2.42}$$

In Eqs. (2.40–2.42), *a, b,* and *c* represent bound one-electron functions, typically Cartesian Gaussians. However, for Cartesian Gaussians, the integrals in Eqs. (2.40–2.42) are not available in closed form. McCurdy, Rescigno, and coworkers [34] evaluate the bound-free and Coulomb-type free-free integrals by numerical quadrature; after performing the $\mathbf{r}_1$ integration analytically, only a three-dimensional quadrature in the $\mathbf{r}_2$ variable is required. The free-free exchange integral of Eq. (2.42) would require a six-dimensional quadrature, however, which is considerably more expensive: for example, if 8,000 quadrature points suffice in three dimensions, achieving comparable accuracy in six dimensions requires 6.4×10^6 points. To avoid such a serious computational expense, Rescigno et al. [34] instead assume that the bound $(N + 1)$-electron functions χ_j are locally complete, i.e., that within a certain volume centered on the molecule, any $(N + 1)$-electron function may be represented as a combination of the χ_j. Of course, this assumption cannot be true unless the set of χ_j is infinite, but if the χ_j are well chosen, it may be a good *approximation,* at least for functions within the energy and angular-momentum ranges of interest. Subject to this assumption, the "free" terms in the expansion of the trial function may be orthogonalized to the χ_j without changing the result of the variational calculation, and free-free exchange matrix elements involving these orthogonalized functions may be shown to vanish [34]. In other words, the exchange portion of the Hamiltonian matrix is assumed to be representable entirely by the "bound" portion of the basis set $\{\chi_j\}$.

2. *Schwinger Variational Method*

The Schwinger variational method [23] forms a stationary expression for the *T*-matrix by combining the three expressions given in Eqs. (2.24), (2.27), and (2.28):

$$\begin{aligned} \tilde{T}_{0n}(\mathbf{k}_0, \mathbf{k}_n) &= \langle \psi_n^{(-)}(\mathbf{k}_n) \,|V|\, \Phi_0(\mathbf{k}_0)\rangle + \langle \Phi_n(\mathbf{k}_n) \,|V|\, \psi_0^{(+)}(\mathbf{k}_0)\rangle \\ &\quad - \langle \psi_n^{(-)}(\mathbf{k}_n) \,|V - VG_0^{(+)}V|\, \psi_0^{(+)}(\mathbf{k}_0)\rangle \end{aligned} \tag{2.43}$$

That this expression is correct is obvious, since it simply says that $T = T + T - T$. That it is stationary may be demonstrated by varying $\psi_n^{(-)}$ and $\psi_0^{(+)}$ and looking at the resulting first-order variation of $\tilde{T}_{0n}$. For instance, adding an error term $\delta\psi_n^{(-)}$ to $\psi_n^{(-)}$ in Eq. (2.43) gives

$$\delta\tilde{T}_{0n} = \langle \delta\psi_n^{(-)}(\mathbf{k}_n) \,|V|\, \Phi_0(\mathbf{k}_0)\rangle - \langle \delta\psi_n^{(-)}(\mathbf{k}_n) \,|V - VG_0^{(+)}V|\, \psi_0^{(+)}(\mathbf{k}_0)\rangle \tag{2.44}$$

By means of the Lippmann–Schwinger equation, Eq. (2.22), the second term on the right-hand side of Eq. (2.44) is easily shown to be equal to the first, so that $\delta\tilde{T}_{0n} = 0$. Likewise, by means of the adjoint to the Lippmann–Schwinger equation for $\psi_n^{(-)}$, Eq. (2.26), we may show that the first-order variation in $\tilde{T}_{0n}$ due to varying $\psi_0^{(+)}$ vanishes, establishing that the Eq. (2.43) is variationally stable.

As with the Kohn principle, by far the most frequent use of the Schwinger variational expression in computations relies on a linear approximation to the wave functions. Introducing a basis set of $(N + 1)$-electron functions χ_m, $m = 1, \ldots, M$, we approximate $\psi_0^{(+)}$ and $\psi_n^{(-)}$ by the linear combinations

$$\tilde{\psi}_0^{(+)} = \sum_{m=1}^{M} x_m \chi_m \tag{2.45}$$

$$\tilde{\psi}_n^{(-)} = \sum_{m=1}^{M} y_m \chi_m \tag{2.46}$$

The coefficients x_m and y_m are determined by inserting these expansions in Eq. (2.43) and imposing the stability requirement

$$\frac{\partial \tilde{T}_{0n}}{\partial x_m} = \frac{\partial \tilde{T}_{0n}}{\partial y_m} = 0 \qquad \text{for all } m \tag{2.47}$$

For $\tilde{T}_{0n}$ we obtain

$$\tilde{T}_{0n} = \mathbf{y}^{\dagger}\mathbf{b}^{(0)} + \mathbf{b}^{(n)\dagger}\mathbf{x} - \mathbf{y}^{\dagger}\mathbf{B}\mathbf{x} \tag{2.48}$$

subject to the pair of matrix-vector equations

$$\mathbf{B}\mathbf{x} = \mathbf{b}^{(0)} \tag{2.49}$$

$$\mathbf{y}^{\dagger}\mathbf{B} = \mathbf{b}^{(n)\dagger} \tag{2.50}$$

In Eqs. (2.48–2.50) $\mathbf{B}$ is the $M \times M$ matrix with elements

$$B_{mm'} = \langle \chi_m | V - VG_0^{(+)}V | \chi_{m'} \rangle \tag{2.51}$$

$\mathbf{b}^{(n)}$ is a column vector of length M with elements

$$b_m^{(n)} = \langle \chi_m | V | \Phi_n \rangle \tag{2.52}$$

and **x** and **y** are column vectors of the coefficients x_m and y_m; $\mathbf{y}^\dagger$, the adjoint of **y**, is the row vector $(y_1^*, y_2^*, \ldots, y_M^*)$. In the usual case where we are interested in many different angles $\hat{k}_0$ and $\hat{k}_n$, we simply expand **b**, **x**, and **y** into rectangular matrices by adding a column for each angle. It should be noted that **B**, like the matrix arising from the Kohn principle for the T-matrix, is complex-symmetric, and thus does not in general give rise to spurious singularities (see, however, [35] and references therein).

From Eqs. (2.49) and (2.50), it follows that the three terms of Eq. (2.48) are equal. Thus, to obtain the approximate T-matrix, we need only solve either of Eqs. (2.49) and (2.50). For example, solving Eq. (2.49) for **x**, we obtain $\tilde{T}_{0n}$ as

$$\tilde{T}_{0n} = \mathbf{b}^{(n)\dagger}\mathbf{x} \tag{2.53}$$

while from Eq. (2.50) for $\mathbf{y}^\dagger$, we have

$$\tilde{T}_{0n} = \mathbf{y}^\dagger\mathbf{b}^{(0)} \tag{2.54}$$

The Schwinger method has some notable advantages over the Kohn method. One of these is that it is a higher-order method [36], meaning that, in general, it can be expected to give a better approximation to the T-matrix from a given trial function or from a given basis set $\{\chi_m\}$. To see this, consider an iterative solution to the Lippmann-Schwinger equation for a one-electron problem, beginning with the approximation $\psi_{n,1}^{(\pm)} = \exp(i\mathbf{k}_n \cdot \mathbf{r})$ and proceeding by using $\psi_{n,J}^{(\pm)}$ on the right-hand side of Eq. (2.15) to generate $\psi_{n,J+1}^{(\pm)}$:

$$\begin{aligned}\psi_{n,J+1}^{(\pm)} &= \exp(i\mathbf{k}_n \cdot \mathbf{r}) + G_0^{(\pm)} V \psi_{n,J}^{(\pm)} \\ &= \sum_{K=0}^{J} (G_0^{(\pm)} V)^K \exp(i\mathbf{k}_n \cdot \mathbf{r})\end{aligned} \tag{2.55}$$

This scheme is known as the Born series and $\psi_{n,J}^{(\pm)}$ as the Jth Born approximation to $\psi_n^{(\pm)}$. We can compare the Kohn and Schwinger methods by using $\psi_{n,J}^{(\pm)}$ as trial functions in each variational expression and examining the quality of the resulting T-matrix. For example, with $J = 1$, the Kohn method gives, by Eq. (2.39) with $f^{(0)} = 0$, the first Born amplitude

$$f_{\mathrm{B1}}(\mathbf{k}_0, \mathbf{k}_n) = -\frac{1}{2\pi} \langle \exp(i\mathbf{k}_n \cdot \mathbf{r}) \, |V| \, \exp(i\mathbf{k}_0 \cdot \mathbf{r}) \rangle \tag{2.56}$$

whereas the Schwinger method gives, inserting the same first Born trial functions in Eq. (2.43) and using Eq. (2.25), the second Born amplitude

$$f_{B2}(\mathbf{k}_0, \mathbf{k}_n) = -\frac{1}{2\pi} \langle \exp(i\mathbf{k}_n \cdot \mathbf{r}) \, |V| \, \psi_{0,2}^{(+)}(\mathbf{r}; \mathbf{k}_0) \rangle \qquad (2.57)$$

Similar results may be demonstrated when higher Born approximations are used as trial functions [36].

A second advantage to the Schwinger method is that it is more flexible with regard to the choice of trial function. Indeed, the trial function need not possess the asymptotic form of Eq. (2.6); it can even be asymptotically vanishing. A somewhat abstract justification for this statement is that we have built the boundary conditions into the Green's function, and thus need not incorporate them in the trial function. However, it is also easy to see in Eq. (2.43) that every matrix element contains the potential V, and thus depends on the behavior of the trial functions only within the range of the potential. This property of the Schwinger principle allows us to use ordinary Cartesian Gaussian functions as the one-electron basis set for constructing the $(N + 1)$-electron trial functions, greatly simplifying the evaluation of matrix elements. The occurrence of Φ_0 and Φ_n in the terms $\langle \Phi_n | V | \psi_0^{(+)} \rangle$ and $\langle \psi_n^{(-)} | V | \Phi_0 \rangle$ does give rise to integrals in which plane waves occur. However, these integrals can be evaluated analytically [37] and pose no special difficulty.

The price of the higher order of the Schwinger method is the occurrence of the Green's function. The Green's-function matrix elements $\langle \psi_n^{(-)} | V G_0^{(+)} V | \psi_0^{(+)} \rangle$ have no (known) closed form when Slater determinants of Gaussian one-electron functions are used to expand $\psi_0^{(+)}$ and $\psi_n^{(-)}$. Thus, as in the implementation of the Kohn method, numerical quadrature is involved. However, the quadrature used in the Kohn method was over an electron's spatial coordinates, whereas the Green's function quadrature is carried out over the momentum variable $\mathbf{k}'_n$ introduced by the completeness relation, Eq. (2.20). We will say more about this quadrature when we discuss the implementation of the Schwinger multichannel method in a later section. First, however, we discuss the modification of the canonical Schwinger variational principle of Eq. (2.43) that gives rise to the Schwinger multichannel principle.

3. *Schwinger Multichannel Method*

When applied to problems such as electron–molecule scattering, in which inelastic collision channels exist in addition to the elastic channel, the Schwinger variational method has a further drawback, namely the reliance of the underlying Lippmann–Schwinger equation on closed-channel terms in the Green's function to produce a proper solution of the Schrödinger equation. We showed in Appendix A that closed-channel terms in $G_0^{(+)}$ do not contribute to the direct scattering asymptotically but do contribute to exchange scattering. Thus, when multiple scattering channels exist, we should not expect to obtain reliable results from Schwinger variational calculations unless closed channels are included in the expansion of $G_0^{(+)}$; however, inclusion of closed channels, in particular ionization channels, would greatly complicate the implementation of the method.

To remedy this defect, Takatsuka and McKoy [38] introduced a projection operator into the Lippmann–Schwinger equation, and thence into the Schwinger variational principle. The result is a modified variational method called the Schwinger multichannel or SMC method. The projection operator P selects from an expansion $\Sigma_n c_n \Phi_n$ only those terms in which n is an open channel; it may be written

$$P = \sum_{n \in \text{open}} |\Phi_n(\{\mathbf{r}_i, i = 1, \ldots, N\})\rangle \langle \Phi_n(\{\mathbf{r}'_i, i = 1, \ldots, N\})| \tag{2.58}$$

Although P, when applied to an antisymmetric function, destroys its antisymmetry, this will not prove to be a complication. The projected Lippmann–Schwinger equation for $\psi_0^{(+)}$ is

$$\begin{aligned} P\psi_0^{(+)} &= P\Phi_0 + PG_0^{(+)}V\psi_0^{(+)} \\ &= \Phi_0 + G_P^{(+)}V\psi_0^{(+)} \end{aligned} \tag{2.59}$$

where we can replace $P\Phi_0$ with Φ_0 since the elastic channel is always open, and we have introduced the symbol $G_P^{(+)}$ for the projected Greens function,

$$G_P^{(+)} = \lim_{\epsilon \to 0^+} \sum_{n \in \text{open}} \int d^3k'_n \frac{|\Phi_n(\mathbf{k}'_n)\rangle \langle \Phi_n(\mathbf{k}'_n)|}{E - E_n - \frac{1}{2}k_n'^2 + i\epsilon} \tag{2.60}$$

If we use Eq. (2.59) rather than Eq. (2.22) to formulate an expression for the T-matrix analogous to Eq. (2.27), we obtain

$$T_{0n}(\mathbf{k}_0, \mathbf{k}_n) = \langle \psi_n^{(-)}(\mathbf{k}_n)| VP - VG_P^{(+)}V |\psi_0^{(+)}(\mathbf{k}_0)\rangle \tag{2.61}$$

Eq. (2.61) can be used in place of the third term in Eq. (2.43), the Schwinger variational form for T_{0n}, to obtain an alternate expression for the T-matrix involving only the projected Green's function:

$$\tilde{T}_{0n}(\mathbf{k}_0, \mathbf{k}_n) = \langle \psi_n^{(-)}(\mathbf{k}_n) \,|V|\, \Phi_0(\mathbf{k}_0)\rangle + \langle \Phi_n(\mathbf{k}_n) \,|V|\, \psi_0^{(+)}(\mathbf{k}_0)\rangle$$
$$- \langle \psi_n^{(-)}(\mathbf{k}_n) \,|VP - VG_P^{(+)}V|\, \psi_0^{(+)}(\mathbf{k}_0)\rangle \qquad (2.62)$$

However, there is no reason to expect that this modified expression will be also variationally stable, and indeed it is not. Although the first variation with respect to $\psi_n^{(-)}$ vanishes, variation of $\psi_0^{(+)}$ gives

$$\delta\tilde{T}_{0n} = \langle \Phi_n \,|V|\, \delta\psi_0^{(+)}\rangle - \langle \psi_n^{(-)} \,|VP - VG_P^{(+)}V|\, \delta\psi_0^{(+)}\rangle$$
$$= \langle \psi_n^{(-)} \,|PV - VP|\, \delta\psi_0^{(+)}\rangle \qquad (2.63)$$

where the second line is obtained from the first by adding $PV - PV$ in the second term and employing the Lippmann–Schwinger equation. To recover variational stability while retaining a form that gives the correct T-matrix when $\psi_0^{(+)}$ and $\psi_n^{(-)}$ are exact, we consider adding to the Eq. (2.62) a term of the form $\langle \psi_n^{(-)}|Q(E - H)|\psi_0^{(+)}\rangle$, where Q is to be determined [39]. The desired property is

$$\langle \psi_n^{(-)} \,|Q(E - H)|\delta\psi_0^{(+)}\rangle = -\langle \psi_n^{(-)} \,|PV - VP|\, \delta\psi_0^{(+)}\rangle \qquad (2.64)$$

Such a Q is given by $P - R$, where R is a projection operator that distinguishes one electron, say electron $N + 1$, as the "scattering" electron. The effect of R coincides with that of P when applied to the open-channel space, but R does not annihilate the closed-channel space. With some rearrangement, we can show that this choice of Q satisfies Eq. (2.64), a key point being that $Q\delta\psi_0^{(+)}$ vanishes asymptotically [39]. Thus a stable variational expression is given by

$$\tilde{T}_{0n}(\mathbf{k}_0, \mathbf{k}_n) = \langle \psi_n^{(-)}(\mathbf{k}_n) \,|V|\, \Phi_0(\mathbf{k}_0)\rangle + \langle \Phi_n(\mathbf{k}_n) \,|V|\, \psi_0^{(+)}(\mathbf{k}_0)\rangle$$
$$- \langle \psi_n^{(-)}(\mathbf{k}_n) \,|VP - VG_P^{(+)}V|\, \psi_0^{(+)}(\mathbf{k}_0)\rangle$$
$$+ \langle \psi_n^{(-)}(\mathbf{k}_n) \,|(P - R)\,(E - H)|\, \psi_0^{(+)}(\mathbf{k}_0)\rangle \qquad (2.65)$$

Noting that $\langle \psi_n^{(-)}|R(E - H)|\psi_0^{(+)}\rangle$ may be replaced with $(N + 1)^{-1}$ $\langle \psi_n^{(-)}|E - H|\psi_0^{(+)}\rangle$ because of the antisymmetry of $\psi^{(\pm)}$ and rearranging slightly, we have finally the SMC variational expression,

$$\tilde{T}_{0n}(\mathbf{k}_0, \mathbf{k}_n) = \langle \psi_n^{(-)}(\mathbf{k}_n) \,|V|\, \Phi_0(\mathbf{k}_0)\rangle + \langle \Phi_n(\mathbf{k}_n) \,|V|\, \psi_0^{(+)}(\mathbf{k}_0)\rangle$$
$$- \left\langle \psi_n^{(-)}(\mathbf{k}_n) \,\middle|\, \left(\frac{1}{N + 1} - P\right)(E - H) + VP \right.$$
$$\left. - VG_P^{(+)}V \,\middle|\, \psi_0^{(+)}(\mathbf{k}_0) \right\rangle \qquad (2.66)$$

Introducing the linear trial functions of Eqs. (2.45) and (2.46) into the SMC expression of Eq. (2.66) gives rise to equations for the expansion coefficients completely analogous to those arising from the ordinary Schwinger principle. The only difference is that the matrix **B** of Eqs. (2.49) and (2.50) is replaced with the matrix **A** given by

$$A_{jj'} = \left\langle \chi_j \left| \left(\frac{1}{N+1} - P \right) (E - H) + VP - VG_P^{(+)}V \right| \chi_{j'} \right\rangle \tag{2.67}$$

Thus our linear system is

$$\mathbf{Ax} = \mathbf{b}^{(0)} \tag{2.68}$$

$$\mathbf{y}^{\dagger}\mathbf{A} = \mathbf{b}^{(n)\dagger} \tag{2.69}$$

where $\mathbf{b}^{(n)}$ is defined by Eq. (2.52). T-matrix elements may then be obtained by Eqs. (2.53) and (2.54).

4. *Numerical Implementation of the Schwinger Multichannel Method*

We describe briefly the steps that are involved in a practical implementation of the SMC method for electron–molecule scattering. A fuller account may be found in a recent publication [40]; here we wish to give only a feeling for the considerations that are involved in constructing a workable method, for the scale of the resulting computations, and for the limitations that are imposed in the course of devising a practical strategy.

Our implementation of the SMC method is divided into three major phases. The first of these is the specification of the variational problem, and it includes solving the electronic-structure problem for the target molecule in each electronic state of interest to obtain the N-electron (bound) target states ϕ_n. At present we are restricted to single-configuration descriptions of the states ϕ_n, i.e., we neglect electron correlation. Moreover, our program requires that a single set of molecular orbitals ξ_p be used in expanding all of the ϕ_n, a restriction that greatly simplifies the evaluation of matrix elements between different states. We thus obtain the ground state ϕ_0 as the solution of a self-consistent field (SCF) calculation, while excited states are represented as single-particle excitations away from ϕ_0. The orbital into which excitation occurs is optimized in the field of the $(N - 1)$-electron charge density generated by removing an electron from the appropriate ground-state SCF orbital, with an exchange interaction determined by the overall spin of the state ϕ_n. This "improved virtual orbital" or IVO approximation [41] typically gives excitation thresholds that are in error by one or two electron volts, though errors may be larger for Rydberg excitations, where relaxation of the $(N - 1)$-electron core toward an ionic configuration is important, or where the excited state has no good single-configuration description. For low-lying valence excitations, the IVO description is often quite good.

Having obtained the N-electron states ϕ_n in a common basis of molecular orbitals, we form a space of $(N + 1)$-electron trial functions χ_j by coupling the states ϕ_n to unoccupied or partially-occupied molecular orbitals to generate so-called configuration state functions (CSFs), that is, combinations of Slater determinants that are eigenfunctions of the total electron spin. For a singlet target state ϕ_0, for example, which gives rise to singlet and triplet excited states ϕ_n, the χ_j are constructed to be good doublet eigenfunctions of S^2_{N+1}. A doublet target gives rise to both singlet and triplet $(N + 1)$-particle functions; however, singlet and triplet scattering may be considered separately, since overall spin is conserved, leading to a useful reduction in the dimension of the variational problem. Note that, in using CSFs formed from molecular orbitals as our variational basis set, we are taking advantage of the fact that the SMC expression, Eq. (2.66), like the original Schwinger expression of Eq. (2.43), is insensitive to behavior of the trial functions beyond the range of the potential V [38, 39].

In forming the variational basis set χ_j, we should, in principle, include not only the open channels but also the closed channels among the ϕ_n. Equivalently, we should include not only all χ_j that can be formed by adding a virtual orbital to the open-channel functions ϕ_n, but also all unique CSFs χ_j that can be formed by considering single, double, and higher excitations from ϕ_0 coupled to a virtual orbital. The situation is quite analogous to that arising in bound-state electronic-structure calculations, where a single-configuration description is improved by adding excited configurations to the trial space (the so-called configuration interaction, or CI, procedure). Just as in CI calculations, however, the combinatorial growth with N in the number of possible excited configurations precludes our including anything like the complete set of such functions. In practice, we are able to include up to a few thousand CSFs arising from single excitations away from the ground state ϕ_0 in our basis $\{\chi_j\}$, including both open- and closed-channel configurations. Frequently, however, we restrict the basis to configurations that arise from coupling virtual orbitals with a small number of open channels ϕ_n.

Having chosen a set $\{\chi_j\}$ in the manner just described, the expansions of Eqs. (2.45) and (2.46) are introduced into Eq. (2.66). The many-electron matrix elements may then be reduced to one- and two-electron matrix elements using Slater's rules (e.g., [42]). Since the molecular orbitals ξ_p are given as linear combinations of Cartesian Gaussians ζ_a,

$$\xi_p(\mathbf{r}) = \sum_{a=1}^{N_g} D_{ap}\zeta_a(\mathbf{r}) \tag{2.70}$$

where the N_g Gaussians have the form

$$\zeta_{\alpha_a}(\mathbf{r}) = C_a(x - X_a)^{\ell_a}(y - Y_a)^{m_a}(z - Z_a)^{n_a} \exp(-\alpha_a|\mathbf{r} - \mathbf{R}_a|^2) \quad (2.71)$$

with C_a a normalization constant, $\mathbf{r} = (x, y, z)$, $\mathbf{R}_a = (x_a, y_a, z_a)$, and ℓ_p, m_p, and n_p integers, the integrals ultimately evaluated fall into two classes. The first class involves only Gaussians and includes one-electron integrals of kinetic-energy and nuclear-attraction operators,

$$K_{ab} = \int d^3r \zeta_a(\mathbf{r}) \nabla^2 \zeta_b(\mathbf{r}) \quad (2.72)$$

and

$$U_{ab} = \sum_m \int d^3r \zeta_a(\mathbf{r}) \frac{Z_m}{|\mathbf{r} - \mathbf{R}_m|} \zeta_b(\mathbf{r}) \quad (2.73)$$

In Eq. (2.73), the sum on m runs over all the nuclei. Also included in the first class are electron–electron repulsion integrals

$$V_{abcd} = \int d^3r_1 \int d^3r_2 \zeta_a(\mathbf{r}_1)\zeta_b(\mathbf{r}_2) \frac{1}{|\mathbf{r}_1 - \mathbf{r}_2|} \zeta_c(\mathbf{r}_1)\zeta_d(\mathbf{r}_2) \quad (2.74)$$

involving the coordinates of two electrons. As is well known, the integrals of Eqs. (2.72–2.74) can be evaluated analytically when the ζ_a are functions of Gaussian form. The second class of integrals is similar but has a plane wave in the place of one of the Gaussians. Because the kinetic energy only occurs in Eq. (2.66) in Hamiltonian matrix elements that do not involve plane waves, this second class includes only nuclear potential terms

$$U_{a\mathbf{k}} = \sum_m \int d^3r \zeta_a(\mathbf{r}) \frac{Z_m}{|\mathbf{r} - \mathbf{R}_m|} \exp(i\mathbf{k}\cdot\mathbf{r}) \quad (2.75)$$

and electron–electron potential terms

$$V_{abc\mathbf{k}} = \int d^3r_1 \int d^3r_2 \zeta_a(\mathbf{r}_1)\zeta_b(\mathbf{r}_2) \frac{1}{|\mathbf{r}_1 - \mathbf{r}_2|} \zeta_c(\mathbf{r}_1) \exp(i\mathbf{k}\cdot\mathbf{r}_2) \quad (2.76)$$

These mixed plane-wave–Gaussian integrals can also be evaluated in closed form [37].

The second major phase of the SMC procedure is the actual evaluation of all necessary integrals of the forms given in Eqs. (2.72)–(2.76) and the formation, from appropriate combinations of those integrals, of the matrix elements that arise in Eqs. (2.68) and (2.69). This is the most computationally intensive phase of the calculation because the number of required integrals is vast. In calculations on even quite small molecules, the number of Gaussians N_g is on the order of 100 and, for reasons explained in the next paragraph, the number $N_{\mathbf{k}}$ of wave vectors $\mathbf{k}$ is on the order of 10,000. From Eqs. (2.74) and (2.76), we can see that there will be on the order of 10^8 two-electron integrals V_{abcd} and 10^{10} integrals $V_{abc\mathbf{k}}$ to be evaluated in such a calculation, and that these numbers will increase rapidly with increasing N_g. Moreover, the computational effort involved in forming final matrix elements from the Gaussian integrals through the application of Eq. (2.70) and Slater's rules is comparable to that needed for the actual integral evaluation. Molecular symmetry, if any, can be used to reduce the number of integrals that must be evaluated explicitly, and we can always save a factor of two in evaluating the plane-wave integrals by noting that the complex conjugates of Eqs. (2.75) and (2.76) give the integrals for $-\mathbf{k}$; nonetheless, a very considerable computational effort is clearly required to treat any but the smallest molecules.

The number of plane waves $\exp(i\mathbf{k}\cdot\mathbf{r})$ required is determined in part by the number of energies and scattering angles at which we desire to compute the T-matrix. In practice, a few dozen energies usually suffice to resolve all the major features in the cross section over a range of tens of electron volts, while on the order of 100 directions $\hat{\mathbf{k}}_0$ and $\hat{\mathbf{k}}_n$ will be needed at each energy in order to perform reliably both the numerical integration of the differential cross section (to obtain the integral cross section) and the average of the cross section over molecular orientations. A few thousand different $\mathbf{k}$'s are thus required to obtain a sufficient representation of $T_{0n}(\mathbf{k}_0, \mathbf{k}_n)$. Moreover, as mentioned earlier, the evaluation of the Green's-function matrix elements involves a quadrature in the momentum variable $\mathbf{k}_n'$ of Eq. (2.60); when Eq. (2.60) is inserted in Eq. (2.67), we can see that the quadrature will require matrix elements of the form $\langle \Phi_n(\mathbf{k}_i)|V|\chi_j\rangle$, which give rise to the integrals of Eqs. (2.75) and (2.76). This quadrature, which may be formulated so that the quadrature points $\mathbf{k}_i$ depend on neither the energy E nor the channel label n [43], is performed as a product of a radial integration in the magnitude k_i and an integration over angles $\hat{\mathbf{k}}_i$. Very efficient quadratures for the surface of a sphere, developed primarily by Lebedev [44], may be used to perform the angular integration, but when k_i is large, hundreds of points $\hat{\mathbf{k}}_i$ may still be needed. The radial integration, in our experience, can be done to sufficient accuracy with on the order of 50 radial points; we typically distribute these as Gauss–Legendre points below some value $k_0 \sim 1$ and Gauss–Laguerre points above k_0. Altogether we thus require on the order of

10^4 points $\mathbf{k}_i$. We have found that the actual number of points required to obtain convergence shows a rough dependence on the number of heavy (nonhydrogen) atoms in the molecule.

Since the radial points k_i are independent of the scattering energy E, we compute all of the matrix elements necessary for the Green's function quadrature once and store them on disk; matrix elements of H, which give rise to integrals involving only Gaussians, are likewise precomputed. To complete the second phase of the calculation, for each energy of interest we evaluate those matrix elements that do depend on E and assemble and solve the system of linear equations, Eqs. (2.68) and (2.69). This system of equations is of modest dimension since, as mentioned earlier, our variational basis sets contain at most a few thousand functions χ_j, and its solution is not a major component of the calculation. The T-matrix for each open channel n, $T_{0n}(\mathbf{k}_0, \mathbf{k}_n)$, is then evaluated from Eq. (2.53).

The third and final phase of the calculation is the computation of cross sections. Integral cross sections $\sigma_n(E)$ may be obtained very simply by summing over final angles $\hat{\mathbf{k}}_n$ and averaging over initial angles $\hat{\mathbf{k}}_0$ (with appropriate quadrature weights). Differential cross sections, on the other hand, require an average over molecular orientation if, as is usual, we are interested in scattering by a gas [see Eqs. (2.9) and (2.10) and discussion]. This average may be performed conveniently by first expanding $T_{0n}(\mathbf{k}_0, \mathbf{k}_n)$ in partial waves to obtain its angular-momentum representation $T_{0n}(k_0, \ell_0, m_0, k_n, \ell_n, m_n)Y_{\ell_0}^{m_0 *}(\hat{\mathbf{k}}_0)Y_{\ell_n}^{m_n}(\hat{\mathbf{k}}_n)$. The integral in Eq. (2.10) is then easily evaluated using the properties of spherical harmonics.

5. *Correction for Long-Range Potentials*

When the molecule of interest possesses a permanent electrostatic moment, the assumption that V is a short-range interaction breaks down, as was mentioned earlier. The justification for employing only localized functions in the expansion of $\psi_0^{(+)}$ and $\psi_n^{(-)}$ is thus removed, and we should not expect reliable results from the SMC method under these circumstances. Similarly, the truncation of the partial-wave expansion of the T-matrix at small angular momentum implicit in the implementation of the Kohn method by Rescigno et al. [34] becomes invalid, since long-range scattering involves large angular momenta. Indeed, the situation is even more serious, since the adiabatic approximation that allowed us to separate nuclear from electronic motion itself breaks down in the presence of a long-range potential.

However, as Norcross and Padial pointed out [15], the association of small impact parameter with small angular momentum itself justifies the use of the adiabatic-nuclei approximation, *provided* its use is restricted to the computation of the contribution of low partial waves to the scattering. The contribution from higher angular momenta, on the other hand, being associated with the (weak) long-range potential, can be computed accurately by means of a low-order approximation, such as the first Born approximation. Thus, Norcross and Padial [15] recommended an approximation of the form (see also [45])

$$\frac{d\sigma}{d\Omega}(\Theta) = \frac{d\sigma^{\text{Born}}}{d\Omega}(\Theta) + \Delta\sigma(\Theta) \tag{2.77}$$

Here $d\sigma^{\text{Born}}/d\Omega$ is the first Born differential cross section, and the correction $\Delta\sigma(\Theta)$ replaces the components of the Born cross section for low partial waves with the result of a higher-level calculation done in the adiabatic-nuclei approximation:

$$\Delta\sigma(\Theta) = \sum_{L=0}^{\infty}\left(\frac{d\sigma_L^{\text{AN}}}{d\Omega} - \frac{d\sigma_L^{\text{Born}}}{d\Omega}\right) \tag{2.78}$$

In Eq. (2.78), L labels terms in an expansion of $d\sigma/d\Omega$ in Legendre polynomials $P_L(\Theta)$. Although the Born cross section is to be computed nonadiabatically, it may also be computed using a very simple approximation to the interaction potential V, since only the long-range behavior of V is relevant. In particular, one can treat the molecule as a rotating point dipole [46], ignoring its internal structure altogether. Furthermore, it is not actually necessary to carry out the infinite sum indicated in Eq. (2.78), because the difference between $d\sigma_L^{\text{AN}}/d\Omega$ and $d\sigma_L^{\text{Born}}/d\Omega$ should become negligible beyond some fairly small value L_{max}.

Rescigno and Schneider [47] observed that Eq. (2.77) is not completely satisfactory, in that the low-L terms in the expansion of $d\sigma/d\Omega$ do not arise solely from low values of ℓ and ℓ' in the partial-wave expansion of the scattering amplitude; in fact, the connection between L and ℓ, ℓ' is of the form $|\ell - \ell'| \leq L$, which allows both high and low angular-momentum contributions at small L. Thus Eq. (2.78), restricted to $L < L_{\text{max}}$, does not have the desired effect of removing only low partial-wave contributions to $d\sigma^{\text{Born}}/d\Omega$. Worse yet, in practice the adiabatic-nuclei calculation will be restricted to small values of ℓ and ℓ', so that the components of the Born cross section inadvertently removed cannot be restored by the first term on the right-hand side of Eq. (2.78). This defect manifests itself in an imperfect matching of $d\sigma/d\Omega(\Theta)$ at small Θ (i.e., near-forward scattering), where it is dominated by the Born contribution, with its value at large Θ, where it is dominated by low ℓ and ℓ' [48]. At intermediate angles, unphysical (negative) cross sections may even be observed. Thus it was suggested [47] that the Born correction procedure be applied to the scattering amplitude itself, rather than to the cross section. This simple but important improvement replaces

Eqs. (2.77) and (2.78) with an analogous expression for the corrected scattering amplitude,

$$f(\mathbf{k}, \mathbf{k}') = f^{\text{Born}}(\mathbf{k}, \mathbf{k}') + \sum_{\substack{\ell\ell' \\ mm'}} (f^{\text{AN}}_{\ell m, \ell' m'} - f^{\text{Born}}_{\ell m, l' m'}) Y_{\ell}^{m*}(\hat{\mathbf{k}}) Y_{\ell'}^{m'}(\hat{\mathbf{k}}') \quad (2.79)$$

Inelastic scattering corresponding to an optically forbidden electronic (or vibrational) excitation has no long-range contribution, but long-range terms do arise for optically allowed transitions. However, for dipole-allowed transitions, the effective interaction at long range can be shown to reduce to the *transition dipole* between the two states involved [49]. The correction procedure of Eq. (2.79) may thus be applied, with the static dipole moment being replaced by the transition dipole (obtainable from an electronic structure calculation) in computing the Born cross section [47]. It is also possible to compute the Born approximation to the scattering amplitude for the transition in question without reducing the interaction to a transition dipole, i.e., to evaluate the matrix element $\langle \Phi_n(\mathbf{k}_n) | V | \Phi_0(\mathbf{k}_0) \rangle$. When the target states ϕ_0 and ϕ_n are expanded in CSFs composed from Gaussian orbitals, all necessary integrals can be performed analytically [50]. Such an approach may be useful, for example, where there is evidence that nondipole long-range interactions are important to the scattering. An example is the quadrupole-allowed $X^1\Sigma_g^+ \rightarrow a^1\Pi_g$ transition in N_2 [51].

6. *Parallel Computation*

We have found it essential to our work to make use of the most powerful computers available. Many of the individual studies that we have carried out would be completely impractical on a workstation or even, given the necessity for sharing an expensive resource among many users, on a conventional supercomputer. Massively parallel processors or MPPs, on the other hand, currently provide both the highest absolute performance and the best cost–performance ratio among supercomputers, and they seem likely to continue to do so for the foreseeable future. Since implementing the SMC procedure for MPPs has been as important to our work as any of our recent developments in methodology, we give a brief description of that implementation here. Details may be found elsewhere [40, 52].

The particular class of parallel computer we have used, which is also the class that has come to dominate the high-performance sector of the MPP field, is the distributed-memory, multiple-instruction, multiple-data design. Such a computer consists of a collection of computational *nodes* connected by a high-speed communications network to each other and to peripheral devices such as disk storage and ports for external connections. A node contains its own memory and one or more processing units (CPUs). In current machines such as the Intel Paragon, IBM SP-2, and CRAY T3D, each node contains a RISC (reduced-instruction-set) microprocessor (Paragon nodes actually contain either two or three CPUs) and is more or less the equivalent in computational power of a high-end workstation. The aggregate computational power of several hundred to a thousand or more such nodes assembled into large MPPs is thus quite considerable.

If a single program is to make use of that power, however, it must be capable of running in *parallel,* with different portions of the computation executing on different nodes simultaneously. Moreover, massive parallelism, meaning roughly dozens to hundreds of processors, places special demands on the program design. A computing truism known as Amdahl's law points out that, no matter how many processors are used, the maximum speedup a parallel program can obtain is S^{-1}, where S is the fraction of the total work that is sequential, i.e., that cannot be parallelized. If a program is to make effective use of a massively parallel computer, S must be no larger than a fraction of a percent. Achieving such a high parallel content is in fact possible for many types of computations, including ours, but it requires care in the program design. A key idea, noted by Fox [53], is that often it is possible to identify some component of the calculation that grows more rapidly with some measure of the problem size than any other component. If that rapidly growing component can be parallelized, then the parallel program will automatically be efficient on large problems—which, presumably, are the interesting ones—because the fraction of the total work associated with the parallel component will grow without limit. Conversely, of course, if the fastest-growing portion of the computation is not parallelized, or is parallelized poorly, very little can be expected of the resulting program.

In earlier discussion, we have already identified the two principal computational challenges in the SMC procedure, namely the evaluation of two-electron integrals over three Gaussians and a plane wave, and the subsequent transformation of those integrals into matrix elements between CSFs. Moreover, the labor involved in each of these steps grows rapidly with the size of the molecule: the scaling of the integral evaluation, we saw, was $N_g^3 N_{\mathbf{k}}$, where N_g is the number of Gaussians and $N_{\mathbf{k}}$ the number of plane waves, and we may also show [40] that the work in the transformation step is proportional to $N_g^4 N_{\mathbf{k}}$. Clearly these two operations should be the primary target for parallelization.

In fact, parallelizing the integral evaluation is trivial. Since we typically have billions of integrals to evaluate in the course of a single collision study, we need only devise some rule for allocating the evaluation of different batches of integrals to different processors. We have great flexibility in choosing this division of labor, as long as we obtain good *load balance,* or, in other words, as long as the work is shared nearly equally. Thus we focus instead on the transformation step [52] and seek a partitioning of work (and therefore data) among processors that will facilitate the integral transformation. The distribution we actually use arranges the integrals $V(abc\mathbf{k})$ of Eq. (2.76) into rectangular arrays, with one dimension labelled by the angle $\hat{\mathbf{k}}$ and the other by an index to unique (a, b) pairs; the c and k labels are treated serially. The integral array is distributed over the processors of the MPP by (mentally) arranging the processors into a rectangular grid and cyclically assigning successive rows and columns of the array to rows and columns of the processor grid. Each processor computes the integrals needed to fill in the portion of the array that it owns. It is then possible to construct a coefficient matrix, likewise distributed over processors, that, when multiplied against this integral array, will take us in a single step from the $V(abc\mathbf{k})$ to the desired $\langle \chi_j | V | \Phi_n(\mathbf{k}) \rangle$ matrix elements, bypassing the usual intermediate step of transforming the integrals over Gaussians into integrals over molecular orbitals [52]. This approach has the advantage of encapsulating the communication among processors needed in the transformation within a multiplication of distributed matrices. This not only leads to a straightforward program, but also, and more importantly, ensures high efficiency.

Parallelizing the integral evaluation and transformation is most crucial, but performance has been enhanced by parallelizing several other steps in the SMC procedure as well. In particular, the angular integration over $\hat{\mathbf{k}}$ needed in evaluating the Green's function is performed as a distributed-matrix multiplication, and the program is able to solve the final linear system using a parallel algorithm. Moreover, on machines that support parallel access to disk, we are able to parallelize the principal input/output (I/O) component, the writing and later reading back of the quadrature data used in constructing the Green's function. The resulting program has shown very satisfactory scalability, being able to run large problems efficiently on such machines as a 512-node Intel Paragon and a 256-node CRAY T3D.

III. APPLICATIONS

A. Preliminary

We turn now to a description of calculations and related experiments on the scattering of low-energy electrons by small polyatomic molecules. Related studies of all kinds (rotational and vibrational excitation, electron-impact ionization and dissociation, scattering at intermediate energies, model calculations of various kinds, etc.) comprise a large and active field, to which we cannot do justice at reasonable length and with reasonable coherence. Moreover, earlier work is largely covered in the excellent reviews mentioned in Section I. We will therefore concentrate on recent work on electronically elastic and electronically inelastic scattering by nonlinear polyatomics, for the most part neglecting vibrational and rotational degrees of freedom, and focusing on molecules for which fully *ab initio* calculations have been carried out. We begin with a survey of elastic scattering results, turning later to studies of electronic excitation.

It is appropriate to comment here on the meanings of different cross sections that will be encountered and the relations among them. Theoretical studies, as described in Sections I and II.A, are generally done in the fixed-nuclei approximation and at a single nuclear geometry, with an average over all molecular orientations as in Eq. (2.10). The resulting cross sections correspond to an average over all possible rotationally elastic and inelastic processes; owing to a lack of resolution, so do most measurements. With regard to vibrational excitation, the situation is less clear. For electronically inelastic scattering, a cross section calculated with the nuclei fixed is usually taken to represent a sum over vibrational levels with appropriate Franck–Condon weights, and thus to correspond to the vibrationally unresolved (or vibrationally summed) experimental cross section. In the case of electronically elastic scattering, fixed-nuclei calculations are often identified with the $v = 0 \rightarrow 0$ vibrationally elastic cross section. This is certainly appropriate when the Franck–Condon approximation applies; however, in that case vibrationally inelastic cross sections simply vanish. Turning this statement around, the observation of significant vibrational excitation by electron impact *implies* the breakdown of the Franck–Condon approximation (or of the adiabatic approximation itself, as at very low energies and in narrow resonances). If the vibrationally inelastic scattering cannot be neglected in comparison to the vibrationally elastic scattering, we can calculate vibrational cross sections within the adiabatic approximation by Eq. (2.5). On the basis of this expression, we expect that the vicinity of the equilibrium geometry will make a large contribution not only to $v = 0 \rightarrow 0$ but also to $v = 0 \rightarrow 2n$, $n > 0$; for sufficiently anharmonic potentials, the equilibrium geometry may even contribute to the transition matrix element for odd vibrational levels. We will return to this point in our discussion of electronically elastic scattering by CF_4.

For some molecules, we will discuss, in addition to elastic and inelastic cross sections, a *momentum transfer cross section* σ_{MT}. This quantity, which is important in modeling electron transport in gases, is an integral cross section that is weighted to emphasize large-angle scattering (which, by conservation of momentum, transfers momentum to the molecule):

$$\sigma_{n,\mathrm{MT}} = \int d\Theta \, \frac{d\sigma_n}{d\Omega}(\Theta)\,(1 - \cos\Theta) \tag{3.1}$$

In practice, $\sigma_{n,\mathrm{MT}}$ is most often computed for $n = 0$ (i.e., electronically elastic scattering). The momentum transfer cross section obtained in this way is comparable to the experimental σ_{MT} derived from electron swarm measurements below the first vibrational threshold, and at higher energies as well, if inelastic processes are sufficiently weak.

It is frequently useful, in the absence of experimental measurements of the elastic cross section, to compare calculations with measured *total* cross sections, even above the first electronic threshold. For nonpolar or weakly polar gases, the (integral) total cross section σ_{tot} can be determined relatively easily by measuring the attenuation of an electron beam passed through a gas cell, and it is therefore often available when cross sections for individual channels are not. Measurements on polar gases are complicated by the need to correct for strong forward scattering.

B. Electronically Elastic Scattering

1. *Hydrides XH_2, XH_3, and XH_4*

Electron scattering by hydride molecules has received far more attention than scattering by other polyatomics, particularly among theorists. This attention is certainly justified in part by the importance of molecules such as methane and water in natural science and technology, but it also reflects the relative simplicity of computations on hydrides. In particular, single-center expansions of the target electron density, and hence of the electron–molecule interaction potential, tend to converge rapidly, facilitating the use of techniques based on such expansions. Most model-potential calculations on polyatomics have in fact been carried out on hydrides, and they have also been popular subjects for *ab initio* methods. We begin a survey of some recent results for elastic scattering by hydrides with the highly symmetrical XH_4 species, CH_4, SiH_4, and GeH_4, which as a class have been the subject of the greatest number of studies.

a. Methane. Methane has been the subject of a great number of theoretical and experimental studies, dating back to measurements of the total cross section by Brode [54], by Brüche [55, 56], and by Ramsauer and Kollath [57]. The pace of work has only accelerated in recent years, and it is not possible to do full justice to every topic here. In keeping with our emphasis on *ab initio* theoretical methods, we provide a brief summary of the status of other areas of research before focusing on the vibrationally elastic, rotationally summed scattering that is most comparable to computed fixed-nuclei cross sections.

ROTATIONAL EXCITATION. Methane is one of a very few polyatomic molecules with sufficiently large rotational spacings for rotational excitation cross sections to have been measured [58]. Even for CH_4, however, individual transitions were not fully resolved; rather, $J = 0 \rightarrow 0$, $0 \rightarrow 3$, and $0 \rightarrow 4$ cross sections were obtained indirectly by means of a theoretically based deconvolution procedure [59]. These measurements have been complemented by both model-potential [60–64] and *ab initio* [65] calculations, all within the adiabatic approximation. The theoretical cross sections agree qualitatively, and in some cases quantitatively, with each other, but overall the agreement with experiment is rather poor. Some of the disagreement may be due to uncertainties in the experimental data and/or its analysis, but limitations of the adiabatic approximation may also be involved. Possible effects of gas temperature on the comparison between theory and experiment have been discussed by Jain [66]. Additional experiments would be helpful.

VIBRATIONAL EXCITATION. Cross sections for excitation of the ν_1, ν_2, ν_3, and ν_4 modes of CH_4 have been measured recently by Shyn [67] and by Mapstone and Newell [68]. Also, high-resolution relative measurements (i.e., measurements of the shape, but not the absolute magnitude, of the cross sections) below 2 eV have been reported by Lunt et al. [69]. Schmidt [70] derived cross sections by analysis of electron swarm data. So far, experiments have not resolved ν_2 from ν_4 (the bending modes), nor ν_1 from ν_3 (stretching). Calculations of vibrational excitation cross sections within the adiabatic approximation require the evaluation of fixed-nuclei results at multiple geometries and are therefore considerably more difficult than calculations of rotational excitation. Theoretical results are to our knowledge absent, apart from dipole Born calculations for the optically allowed modes, which show poor agreement with the measured data [69].

TOTAL SCATTERING. The total electron scattering cross section for CH_4 has been the object of much recent experimental study, in part aimed at resolving the disagreements among earlier measurements [54–57, 71–77]. Measurements carried out within the last 5 years [78–80] are in generally good agreement and suggest that the low-energy results of Brode [54], of Barbarito et al. [71], and of Sueoka and Mori [76] are too low, while those of Dababneh et al. [77] are somewhat too high. Recommended total, elastic, and inelastic cross sections have been presented by Kanik et al. [81]. Theoretical calculations of the total cross section [82–85] have employed very simple models that are unreliable at low energies.

VIBRATIONALLY ELASTIC SCATTERING. Differential cross sections for vibrationally elastic scattering have been measured by several groups. Recent experiments include those of Tanaka et al. [86], Curry et al. [87], Sohn et al. [88], Shyn and Cravens [89], and Mapstone and Newell [90]. References to earlier work may be found in the paper by Mapstone and Newell. Schmidt [70] derived integral elastic cross sections below 3 eV by analysis of electron swarm data. Integral elastic cross sections may be computed from the measured differential cross sections, although extrapolation into the experimentally inaccessible near-forward and near-backward directions is required. As pointed out by Kanik et al. [80], the integral elastic cross sections thus obtained are on the whole inconsistent with the total and vibrational cross sections at low energy. Some of the discrepancy may be due to extrapolation error.

Theoretical calculations of elastic scattering by CH_4 have been extensive. Early model-potential calculations (e.g., [60], [62]) have recently been improved upon by replacing local exchange potentials with an exact representation of exchange [64, 91–93], although local-exchange models continue to be employed [94]. Within the past 10 years, methane has been the nonlinear polyatomic target of choice for the initial applications of fully *ab initio* methods. The Schwinger multichannel method was employed first at the static-exchange level [95] and later with a limited representation of polarization effects, through the inclusion of closed-channel terms in the expansion of the wave function [96]. These studies were soon followed by T-matrix Kohn calculations in the static-exchange [97] and static-exchange plus polarization [98] approximations. Nestmann et al. [99] very recently reported the first application of the R-matrix method to a nonlinear polyatomic, a study of CH_4 elastic scattering that included polarization effects.

In Figs. 1 and 2 we show selected results from recent calculations of the integral elastic cross section with selected experimental elastic cross sections and with the total cross sections of Ferch et al. [73] and Kanik et al. [80]. Before comparing the various results, we point out two prominent features of the cross section: a *Ramsauer minimum* between 0 and 1 eV [57] and a broad maximum at approximately 7.5 eV. The minimum, a feature also observed in noble gases and in other molecules, is a purely quantum mechanical effect. It arises, in a quasiatomic view, from the passing of the $\ell = 0$ phase shift η_0 through π at a low energy, where there is very little scattering due to higher partial waves [100]. The broad maximum is generally agreed to be due to a *shape resonance,* in which the scattering electron is temporarily captured by the interaction potential with the target. From the width of the resonance it is obvious that its lifetime is short.

Although much attention has been given to obtaining an accurate description of the Ramsauer minimum, the situation cannot be considered completely satisfactory even today, insofar as there are rather large (percentage) disagreements among the various calculations, and between them and experiment. The Kohn calculation [98] agrees well with the elastic cross section of Sohn et al. [88] above 0.6 eV, but it is significantly smaller than

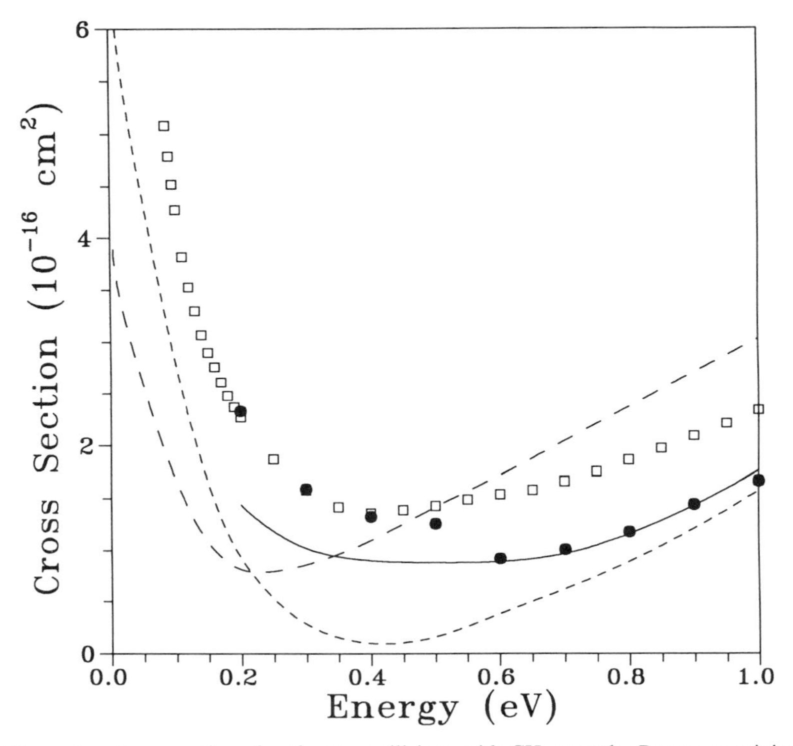

Figure 1. Cross sections for electron collisions with CH_4 near the Ramsauer minimum. Open squares: measured total cross section [73]; filled circles: measured elastic cross section [88]; solid line: Kohn calculation [98]; short dashes: *R*-matrix calculation [99]; long dashes: model-potential calculation [92].

experiment at lower energies. The recent calculations of Gianturco et al. [92], on the other hand, place the minimum at somewhat too low an energy and give too large a cross section above the minimum. These results are, however, quite sensitive to the polarization model employed; with a less sophisticated model, somewhat better results were obtained [92]. Finally, the *R*-matrix calculation gives a minimum too low in energy and a cross section too small in magnitude. All three studies emphasize the delicate balancing of effects that enters into producing the precise form of the Ramsauer minimum.

At higher energies (Fig. 2), theoretical calculations including polarization are seen to agree well with the experimental *total* cross section on the magnitude and position of the shape resonance, but to be considerably larger than the experimental elastic cross sections, with the exception of that of

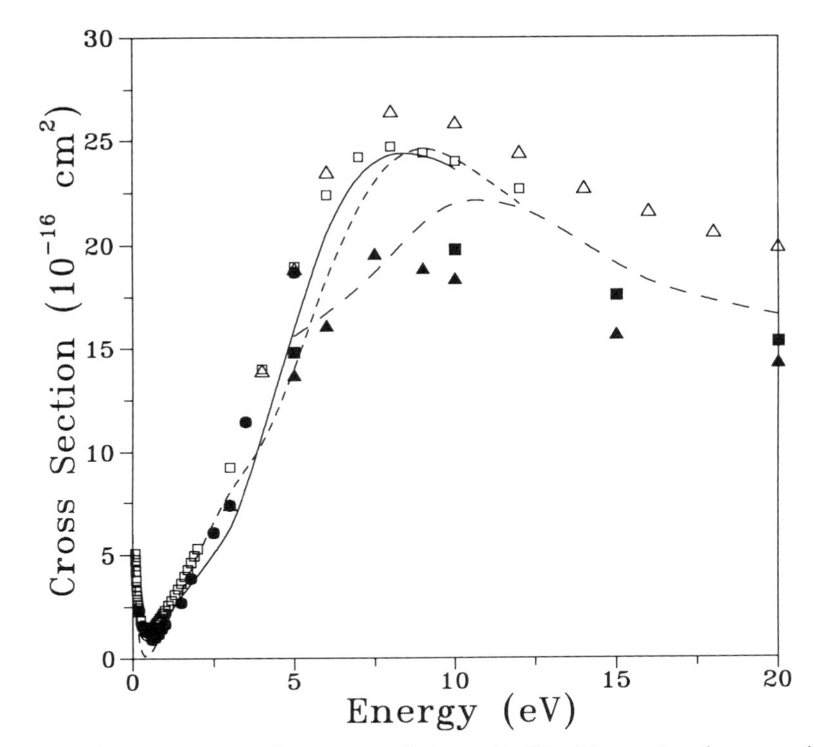

Figure 2. Cross sections for electron collisions with CH_4. Measured total cross sections: open squares, Ferch et al. [73]; open triangles, Kanik et al. [80]. Measured elastic cross sections: filled triangles, Tanaka et al. [86]; filled circles, Sohn et al. [88]; filled squares, Shyn and Cravens [89]. Calculated elastic cross sections: long dashes, Lima et al. [95]; solid line, Lengsfield et al. [98]; short dashes, Nestmann et al. [99].

Sohn et al. [88]. On the other hand, theory and experiment give very similar results for the *differential* elastic cross section at these energies (Fig. 3). These considerations support the conclusion that the discrepancy noted by Kanik et al. [80] between the experimental total and elastic cross sections is due to underestimation of the forward and backward scattering when extrapolating measured differential cross sections.

b. Silane. Silane has been almost as well studied by theorists as methane, though, being a difficult gas to work with, it has been the subject of fewer experiments. Nonetheless, motivated partly by its importance in such applications as plasma deposition of amorphous silicon, a body of cross sections has been accumulated for SiH_4.

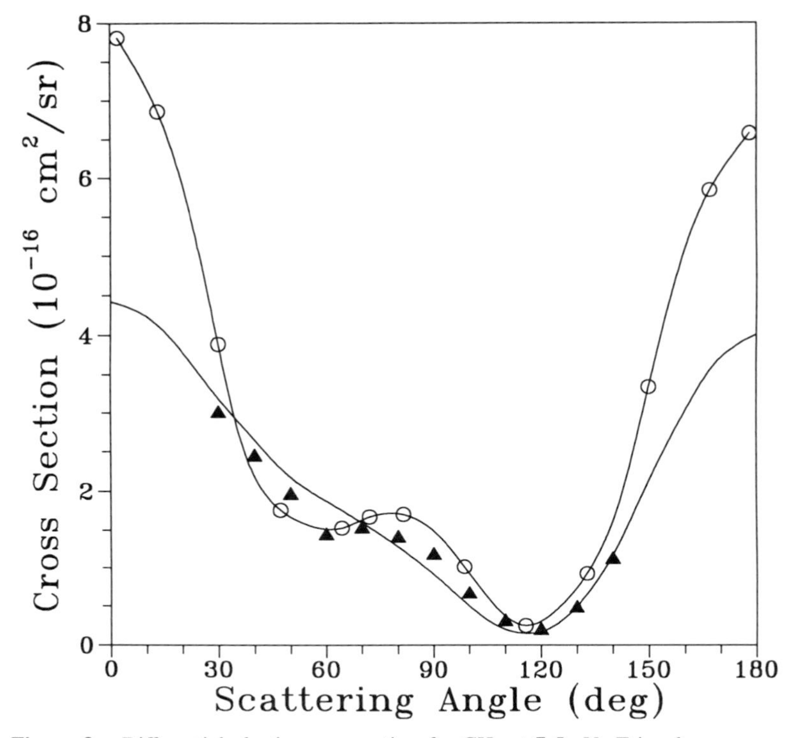

Figure 3. Differential elastic cross section for CH_4 at 7.5 eV. Triangles: measurement of Tanaka et al. [86]; solid line: calculation of Lima et al. [96]; solid line with circles: calculation of Lengsfield et al. [98].

ROTATIONAL EXCITATION. Jain and Thompson [101] have computed rotational-excitation cross sections for SiH_4 using an exact-exchange, model-polarization approximation. There appear to be no experimental results.

VIBRATIONAL EXCITATION. Differential cross sections for vibrational excitation of SiH_4 were measured by Tanaka et al. [102]. As in CH_4, the ν_1 and ν_3 modes were not resolved, nor were ν_2 and ν_4. The same authors found that their results could be reproduced reasonably well by a fairly simple model calculation. Tronc et al. [103] measured an excitation function (i.e., relative cross section) for the $\nu_{1,3}$ excitation at a scattering angle of 30° and found a result in qualitative agreement with the calculation of Tanaka et al. Integral vibrational cross sections have also been obtained by analysis of swarm data [104].

TOTAL SCATTERING. Total electron scattering cross sections for SiH_4 have been measured by Wan et al. [105] (0.2 to 12 eV), by Sueoka et al. [106] (1 to 400 eV), and by Zecca et al. [107] (75 to 4000 eV). As for CH_4, calculated total cross sections [82, 108] employing simple models give reasonably accurate results only at high energies.

ELECTRONICALLY ELASTIC SCATTERING. Differential elastic cross sections for silane were measured by Tanaka et al. [102] from 1.8 to 100 eV and extrapolated into the forward and backward directions to produce integral elastic and momentum-transfer cross sections. Integral elastic cross sections have also been extracted from measurements on electron swarms [104]. At low collision energies, model-potential calculations have been performed on silane using a variety of approximations. The early results of Tossell and Davenport [109] using a muffin-tin–type potential gave cross sections that were qualitatively correct, but much too large; the similar calculations of Tanaka et al. [102] are significantly smaller than the experimental results but likewise have the correct qualitative behavior. Spherical model potentials have been employed, with both approximate [110–112] and exact [113] treatments of exchange, as have nonspherical model potentials, again with both approximate [114–116] and with exact [101, 117] exchange treatments. The Schwinger multichannel method has been applied to SiH_4 at the static-exchange level of approximation [118, 119], while Sun et al. [120] have applied the Kohn method, with polarization effects included.

Integral elastic cross sections for SiH_4 are shown in Fig. 4 along with the total cross sections of Wan et al. [105] and Sueoka et al. [106]. Like CH_4, SiH_4 exhibits both a Ramsauer minimum and a shape resonance. Comparing Figs. 2 and 4, we see that the SiH_4 resonance occurs several electron volts lower in energy. This observation is consistent with the assignment of the resonance to the lowest unoccupied molecular orbital in each molecule, which is a t_2 orbital with some X—H σ^* character; such an orbital would be expected at a lower energy in silane, given its longer bond length (1.48 Å, vs. 1.08 Å in methane). Calculations support this assignment in that the resonance occurs in overall $^2T_2$ symmetry.

The model-potential results shown in Fig. 4, which are representative, are significantly larger than the experimental total cross section from the resonance maximum onward. The Kohn results are closer to the experiment in magnitude. The SMC result also has approximately the correct magnitude above 4 eV, although, since it omits polarization, it places the resonance maximum at too high an energy and is qualitatively wrong at the lowest energies. Both defects are expected, given the omission of the (attractive) polarization effect. Static-exchange results of Jain and Thompson [114] and of Sun et al. [120], not shown in Fig. 4, are very similar to the SMC result.

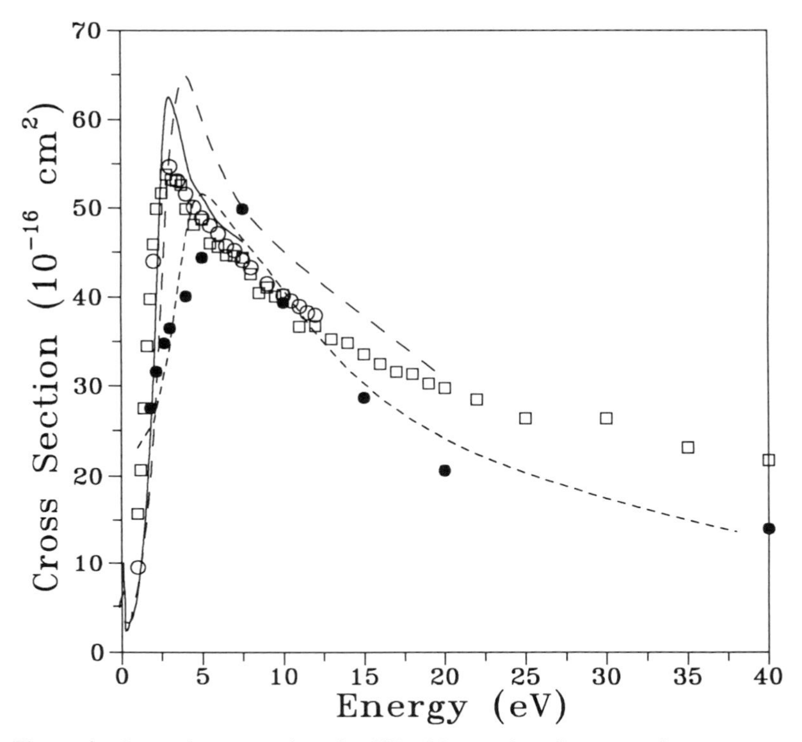

Figure 4. Integral cross sections for SiH_4. Measured total cross sections: open squares, Sueoka et al. [106]; open circles, Wan et al. [105]. Filled circles are the experimental elastic cross section of Tanaka et al. [102]. Calculated elastic cross sections: long dashes, Jain and Thompson [101]; short dashes, Winstead et al. [119]; solid line, Sun et al. [120].

The discrepancy seen in Fig. 4 between the results of Jain and Thompson [101] and those of Sun et al. [120] seems to suggest a deficiency in the polarization potentials for SiH_4 employed in the model-potential calculations [101]. Much better agreement was found in the case of CH_4 (Fig. 2) between the model-potential and Kohn results, perhaps reflecting the smaller polarizability of methane.

c. *Germane.* Germane has been the subject of much less study than either CH_4 or SiH_4. Integral elastic and vibrational-excitation cross sections were extracted from swarm measurements by Soejima and Nakamura [121], and measured directly (including angular dependence) by Dillon et al. [122]. Total cross sections were measured by Karwasz [123], but only at intermediate to high energies (75–4000 eV). On the theoretical side, model-potential calculations using a spherical potential were performed by Jain et al. to obtain elastic differential, integral, and momentum-transfer cross sections from 1 to 100 eV [124] and integral elastic and total cross sections from 10 to 1000 eV [125]. Integral elastic and momentum-transfer cross sections from 0 to 50 eV were recently reported by Kumar et al. [126], also using a spherical model potential. Preliminary results in a nonspherical model (not including cross sections) are found in Gianturco et al. [116]. Differential and integral elastic cross sections have also been calculated by the SMC method [127]. These calculations, which now appear not to have been fully converged, were repeated with a pseudopotential representation of the core electrons by Bettega et al. [27]. Integral elastic cross sections for GeH_4 are shown in Fig. 5. As for SiH_4, the model-potential calculations appear to overestimate the cross section at low energy. The result of Bettega et al. [27] is in better agreement with the experimental cross section than that of Jain et al. [124]. However, the integral elastic cross section of Dillon et al.

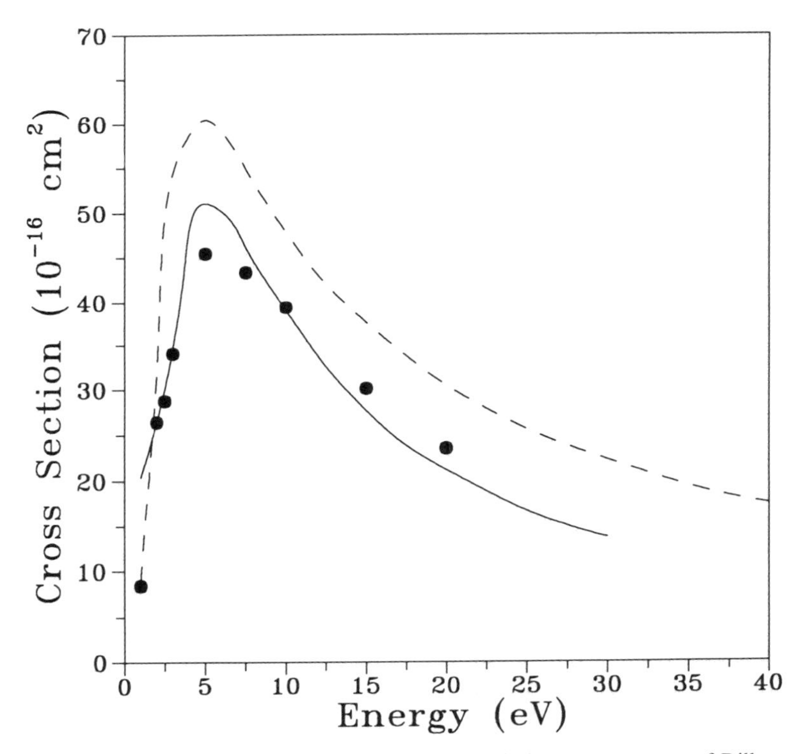

Figure 5. Integral elastic cross sections for GeH_4: circles, measurement of Dillon et al. [122]; dashed line, calculation of Jain et al. [124]; solid line, calculation of Bettega et al. [27].

must be treated with some caution, since, as we have emphasized, such integral cross sections rely on an extrapolation of the actual measurements. Germane and silane have very similar bond lengths, and we might expect their electron-scattering behavior to be similar; indeed, the static-exchange cross sections for the two molecules are very similar (cf. Figs. 4 and 5). It therefore seems likely that the resonance maximum is actually at a somewhat lower energy than 5 eV. Measurements of the total cross section in this energy range would be helpful, as would *ab initio* calculations incorporating polarization.

We turn now to the XH_3 hydrides, ammonia, phosphine, and arsine. The salient feature of these molecules, in comparison to the XH_4 species, is their possession of a permanent dipole moment. As discussed in Section II.C.5, special procedures are necessary to account properly for the long-range scattering by the dipole potential. An alternative, when the dipole moment is small, is to proceed as for a nonpolar molecule, recognizing that the resulting differential cross sections will be incorrect in the extreme forward direction and that the integral cross sections will be affected correspondingly. The momentum-transfer cross section, due to the $(1 - \cos \Theta)$ weighting, will be affected much less.

d. Ammonia. The prototype XH_3 system ammonia has received less attention than has the prototype XH_4 molecule, methane. Total cross section measurements using attenuation methods are complicated by the need to discriminate the unscattered flux from the elastic and rotationally inelastic scattering at very small angles. At low energies, the measured cross sections of Sueoka et al. [128] and of Szmytkowski et al. [129] disagree significantly, while the measurements of Zecca et al. [107] (75 to 4000 eV) have a reported uncertainty of $\pm 30\%$ at 75 eV. Cross sections for rotational excitation at extremely low (~ 1 meV) impact energies have been measured by Ling et al. [130]. Vibrational excitation measurements have been reported by Ben Arfa and Tronc [131, 132], by Furlan et al. [133], and by Gulley et al. [134]. Absolute differential elastic cross sections were measured by Ben Arfa and Tronc [132] at 7.5 eV and at several energies from 22 to 30 eV by Alle et al. [135]. The latter workers also report integral elastic cross and momentum-transfer cross sections, obtained by combining their measurements with the calculated results for forward and backward scattering of Rescigno et al. [136], which are described in the next paragraph.

Calculations on NH_3 have concentrated on electronically elastic scattering, although the total cross section has been obtained within a simple spherical model by Jain [137], who also reports an integral elastic cross section. Earlier model-potential calculations were carried out by Jain and Thompson [138], with subsequent studies by Gianturco [139] and by Yuan and Zhang

[140], who also report rotationally inelastic cross sections. These model-potential studies all used local exchange potentials, except that of Yuan and Zhang, which, however, employed a spherical approximation to the static electronic potential. Schwinger multichannel calculations of electron–NH_3 scattering within the static-exchange approximation were reported by Pritchard et al. [141]. More recently, the Kohn method has been applied to NH_3 by Rescigno et al. [136], with polarization effects included at 7.5 eV and below. The Schwinger calculations did not include a correction for the long-range dipole scattering, and therefore only differential cross sections at angles greater than 30° and momentum-transfer cross sections were reported.

Figure 6 shows calculated and measured differential elastic cross sections for NH_3 at 5 eV. Polarization appears to be relatively unimportant at this energy, given the agreement between the static-exchange result and those calculated with polarization included. The Kohn calculation is in better

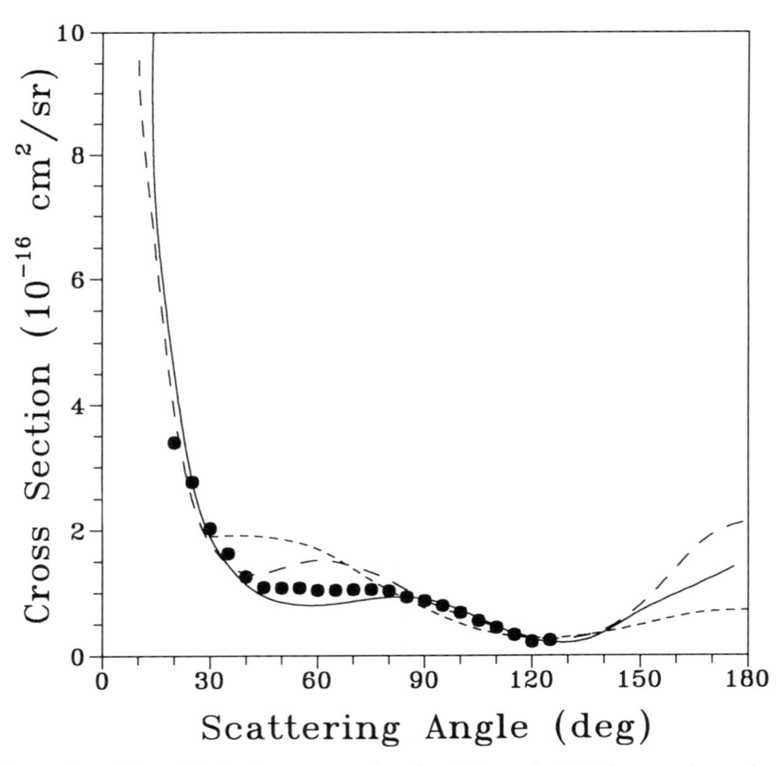

Figure 6. Differential elastic cross section for NH_3 at 5 eV. The experimental cross section of Alle et al. [135] is shown by circles. Calculations are by Gianturco [139], long dashes; Pritchard et al. [141], short dashes; and Rescigno et al. [136], solid line.

agreement with the measurements from 45° to 90°, while below 45° inclusion of the dipolar scattering is essential.

e. Phosphine. Phosphine has a smaller dipole moment than ammonia (0.58 Debye vs. 1.47 D). Since the dipole Born cross section is proportional to the square of the dipole moment, we therefore expect much less dramatic effects on the PH_3 cross sections than are seen for NH_3. Unfortunately, there is very little experimental work available on phosphine, possibly because of its unpleasant properties. Tossell et al. [142] did, however, report a derivative transmission spectrum, which is sensitive to the presence of resonances. Their study indicated the existence of a shape resonance at 1.9 eV, which they assigned to the lowest unoccupied molecular orbital, the antibonding $3e$ orbital. There appear to be no other results relating to elastic scattering, though Ben Arfa and Tronc did report energy-loss spectra showing electronic thresholds [143]. Theoretical total and integral elastic cross sections were reported by Jain and Baluja [82] and by Yuan and Zhang [144]; the latter report also includes differential cross sections. Both calculations employ spherical models of the scattering potential, to which Yuan and Zhang add (incoherently) a dipole-scattering term. At about the same time, the Schwinger multichannel method was used to carry out static-exchange calculations of differential, integral, and momentum-transfer cross sections from 1 to 40 eV [145]. In these calculations, no particular accounting was made for the dipole scattering, since the effect on the cross sections at these energies is expected to be small, except for angles very near forward.

In Fig. 7, we compare integral elastic cross sections obtained with model-potential methods [144] and those obtained with the SMC method [145]. Although the model-potential calculation gives a much larger cross section, the two calculations agree on the position of the resonance, in spite of the fact that the SMC calculation omits polarization. Since we believe the SMC result is accurate, it appears that the calculation of Yuan and Zhang overestimates the cross section. Experimental data will be necessary to resolve the issue.

f. Arsine. Arsine appears so far to have been studied only in our Schwinger calculations [145]. The static-exchange integral cross section is shown in Fig. 8, with that of PH_3 for comparison. As may be seen, there is a great deal of resemblance between the cross sections of the two molecules. This may reflect the similar bond lengths (1.42 Å for PH_3, 1.52 Å for AsH_3) and valence electronic structure. On the whole, the trend in the XH_3 series is very similar to that seen in the XH_4 series, with a smaller cross section and broader resonance in the first-row species and a strong resemblance between the second- and third-row species.

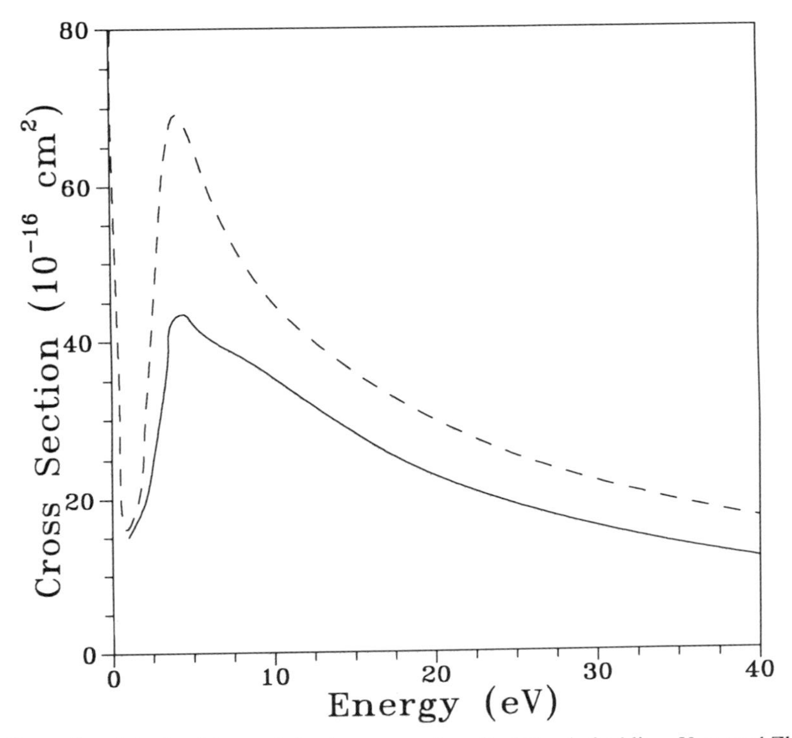

Figure 7. Calculated integral elastic cross sections for PH_3: dashed line, Yuan and Zhang [144]; solid line, Winstead et al. [145].

g. Water. Water has been the subject of numerous recent studies, both experimental and theoretical. For clarity, we organize these according to the type of cross section, as we did for methane and silane.

ROTATIONAL EXCITATION. Differential cross sections for rotationally elastic and inelastic scattering were measured by Jung et al. [146]. Corresponding theoretical studies have been reported by Gianturco [147], using model potentials for exchange and polarization, and by Greer and Thompson [148], with an exact treatment of exchange and a model polarization potential. The two calculations produced similar results, in good agreement with experiment, for rotationally inelastic scattering; however, Greer and Thompson appear to obtain better results for the rotationally elastic cross section.

VIBRATIONAL EXCITATION. Differential cross sections for vibrationally inelastic scattering at low energy were measured by Seng and Linder [149]

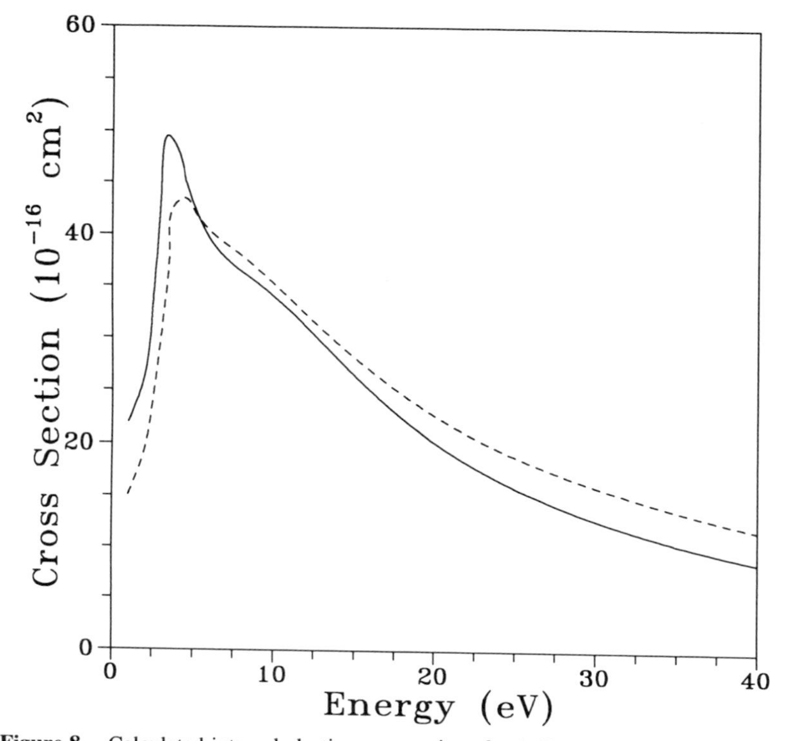

Figure 8. Calculated integral elastic cross sections for AsH_3 (solid line) and PH_3 (dashed line), from Winstead et al. [145].

and more recently by Shyn et al. [150]. Furlan et al. [151] have studied vibrational excitation at higher impact energies (30 and 50 eV). In these measurements, the two stretching modes are not resolved, and only a single vibrational quantum is excited. Agreement between the results of Seng and Linder and those of Shyn et al. is generally good, except in the forward direction. Cvejanović et al. [152] have measured excitation functions for multiple-quantum vibrational excitations. For the single-quantum excitations, model-potential calculations that include a Born correction at high angular momentum [153] agree rather well with the measured low-energy cross section for the bending mode, but not with that for the stretching modes. Similar calculations by Nishimura and Itikawa [154] over a wider energy range appear to give reasonably good results for the bending mode at low energies, but to overestimate the large-angle scattering at higher energies, and also to disagree with the measurements for the stretching-mode excitation.

TOTAL SCATTERING. There are several recent measurements of the total cross section for electron–H_2O scattering. At low energies, results have been reported by Sueoka et al. [128], by Szmytkowski [155], by Nishimura and Yano [156], and by Sağlam and Aktekin [157]. Zecca et al. [158] reported results at intermediate and high energies (81–3000 eV). There is considerable scatter in the experimental results at low energy. In particular, the cross section of [128] is much smaller than the other results [155–158], which cluster around the early measurement of Brüche [159]. The total cross section above 10 eV calculated within a simple model [137] is somewhat larger than any of the experiments at low energies.

ELECTRONICALLY ELASTIC SCATTERING. Elastic differential cross sections have been measured by Danjo and Nishimura [160], by Shyn and Cho [161], and by Johnstone and Newell [162]. These measurements are in generally excellent agreement with each other. The results of Shyn and Cho, which extend over a wider angular range, are notable for displaying unexpectedly strong backward scattering. Various calculations of the elastic cross section based on model potentials have been reported recently [137, 140, 147, 148, 163, 164]. Among these calculations, that of Greer and Thompson [148] is the only one to include exchange effects exactly. Elastic cross sections in the static-exchange approximation were computed by Brescansin et al. [165] using the Schwinger multichannel method and more recently by Machado et al. [166] using an iterative Schwinger procedure [167] and including a dipole Born correction. Calculations based on the Kohn principle were reported by Rescigno and Lengsfield [168] both in the static-exchange approximation and, at 8 eV and below, with polarization effects included.

Selected experimental and theoretical differential cross sections at 6 eV are compared in Fig. 9. As was the case with ammonia, polarization does not appear to be as important as the dipole potential in determining the scattering behavior. None of the calculations reproduces the strong backward scattering observed by Shyn and Cho [161], although the agreement among all of the results shown is quite good at smaller angles. At higher energies, good agreement is obtained at all angles.

h. Hydrogen Sulfide. Hydrogen sulfide has been little studied until quite recently. For the total cross section, the only measurements at low energies appear to be those of Sokolov and Sokolova [169] and of Szmytkowski and Maciąg [170], while Zecca et al. [107] have made measurements above 75 eV. Jain and Baluja [82] have calculated total cross sections at 10 eV and above. Based on subsequent experiments and calculations, the earlier measurements [169] do not appear to be reliable. Measurements of vibrational excitation and vibrationally elastic scattering by Rohr [171] were limited in

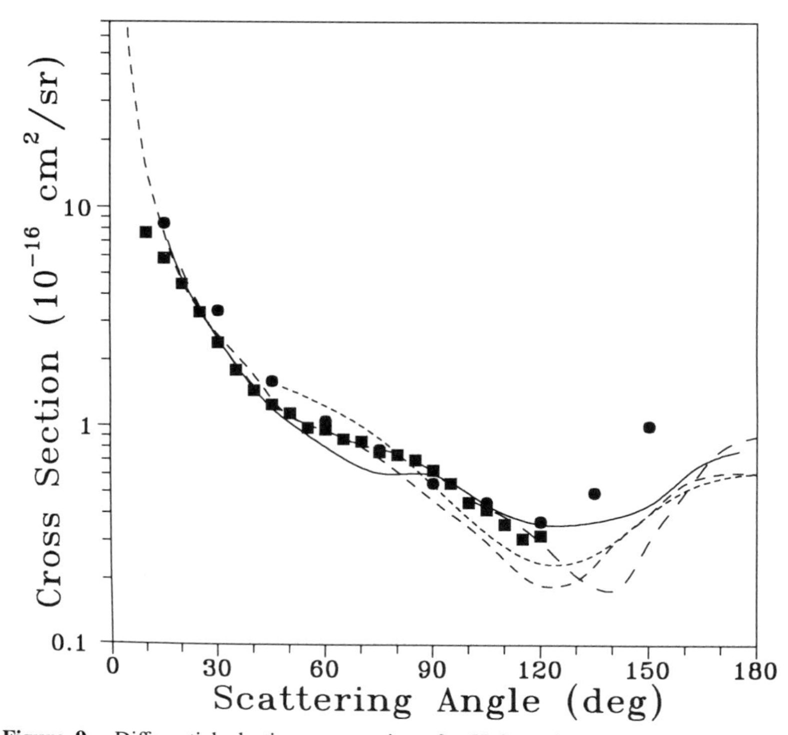

Figure 9. Differential elastic cross sections for H_2O at 6 eV. Measurements: circles, Shyn and Cho [161]; squares, Johnstone and Newell [162]. Calculations: short dashes, Brescansin et al. [165]; medium dashes, Machado et al. [166]; long dashes, Greer and Thompson [148]; solid line, Rescigno and Lengsfield [168].

extent and subject to large uncertainties. The situation has improved with the recent work of Gulley et al. [172], who report differential and integral cross sections for elastic and vibrationally inelastic scattering from 1 to 30 eV. Model-potential calculations for elastic scattering have been performed by Gianturco and Thompson [173], Jain and Thompson [174], Gianturco [139], Jain and Baluja [82], Yuan and Zhang [144], and Greer and Thompson [148]. Fully *ab initio* studies have been conducted using the Kohn method [175] and the iterative Schwinger method [176]. Total and integral elastic cross sections are compared in Fig. 10. There are differences in detail among the calculated cross sections, particularly below 5 eV, where resonances and the dipole potential significantly affect the scattering, but at higher energy all of the theoretical cross sections agree fairly well with each other. They also agree with the data of Gulley et al. [172] within the reported error limits, which are about $\pm 20\%$.

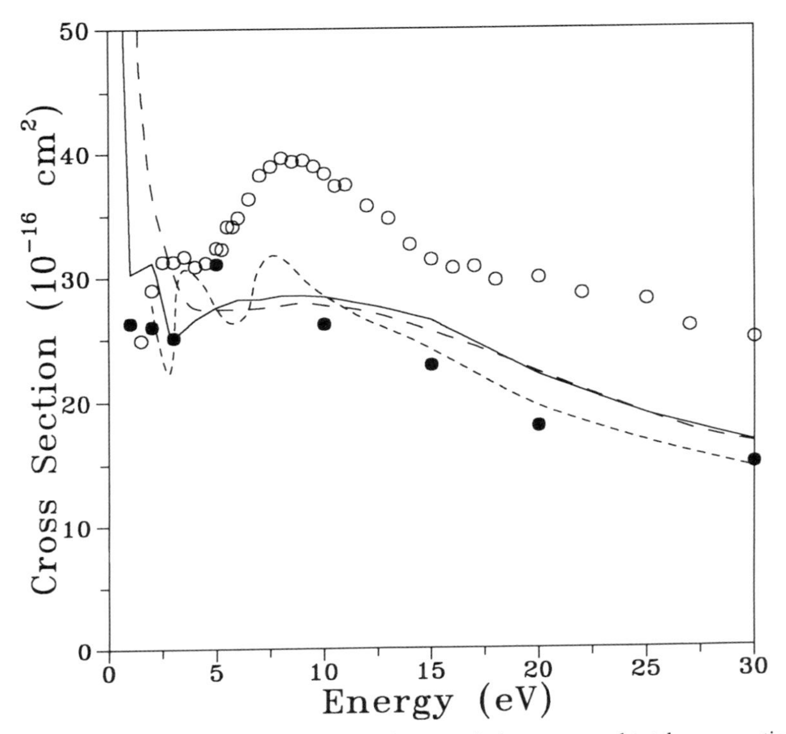

Figure 10. Integral cross sections for H_2S: open circles, measured total cross section of Szmytkowski and Maciąg [170]; filled circles, experimental elastic cross section of Gulley et al. [172]; long dashes, calculation of Gianturco [139]; short dashes, calculation of Machado et al. [176]; solid line, calculation of Lengsfield et al. [175].

2. *Hydrocarbons and Si_2H_6*

Unlike the hydrides, nonlinear hydrocarbons larger than CH_4 have in the past been the subject of very little theoretical research. The difficulty lies in the unsuitability of methods based on partial-wave expansions about a single center for molecules with low symmetry and multiple nonhydrogen atoms, and also in the power-law scaling of the work required by more general methods with the number of target electrons. Recent years have, however, seen the publication of studies of collisions with the two- and three-carbon alkanes and alkenes, and with the smallest cycloalkane, cyclopropane. We have also looked at elastic scattering by disilane, the silicon analog of ethane [127].

a. Ethylene. The prototype alkene, ethylene (or ethene in the IUPAC nomenclature), has been comparatively well studied by low-energy electron scattering experiments. Early measurements did not yield cross sections but were useful in identifying the existence of low-energy resonances [177–179]. Vibrational excitation cross sections have been extracted from electron swarm measurements by Duncan and Walker [180], and measured directly by Walker et al. [181] and by Lunt et al. [69]. Total cross sections were measured by Brüche [182] and have been remeasured by Floeder et al. [75], by Sueoka and Mori [183], and by Nishimura and Tawara [184]. As was observed above for other molecules, the cross section of Sueoka and Mori is somewhat smaller than that measured by other investigators. Calculated total cross sections of Jiang et al. [85, 185] are several times larger than any of the experiments at low energy. Relative measurements of the differential and integral elastic cross sections were reported by Mapstone and Newell [90] and by Lunt et al. [69]. Calculations of the elastic scattering cross section were carried out by us [127], using the SMC method and neglecting polarization, and by Schneider et al. [186], using the Kohn method and including polarization. Schneider et al. report only two symmetry components (^2A_g and $^2B_{2g}$) of the integral cross section. These components exhibit, respectively, a Ramsauer minimum at 0.2 eV and a shape resonance, assigned to the π^* virtual orbital, at 1.8 eV. The positions of both features are in agreement with experimental data [177, 179]. We show in Fig. 11 our calculated differential cross sections at 15 eV, together with the relative measurements of Mapstone and Newell at 15.5 eV, normalized to our calculation. As may be seen, the agreement on the shape of the cross section is quite good.

b. Propylene. Propylene or propene has been the subject of few measurements. Derivative transmission spectra (sensitive to resonances) were reported by Jordan and Burrow [187] and total cross sections by Brüche [56], Floeder et al. [75], and Nishimura and Tawara [184]. We have used the SMC method to conduct a comparative study of elastic scattering by propylene and its isomer cyclopropane [188], results of which are presented in the following paragraph.

c. Cyclopropane. Total cross sections for cyclopropane were measured by Floeder et al. [75] and by Nishimura and Tawara [184]. Howard and Staley [189] reported derivative transmission spectra. Allan has recently measured differential cross sections for both vibrational excitation and vibrationally elastic scattering [190, 191]. Elastic differential and integral cross sections were computed by us using the SMC method [188]. In Fig. 12, we compare our calculated integral elastic cross sections for propylene and cyclopropane with measurements of the total cross sections. There is a noticeable isomer effect on the magnitude of the cross section, which is correctly reproduced

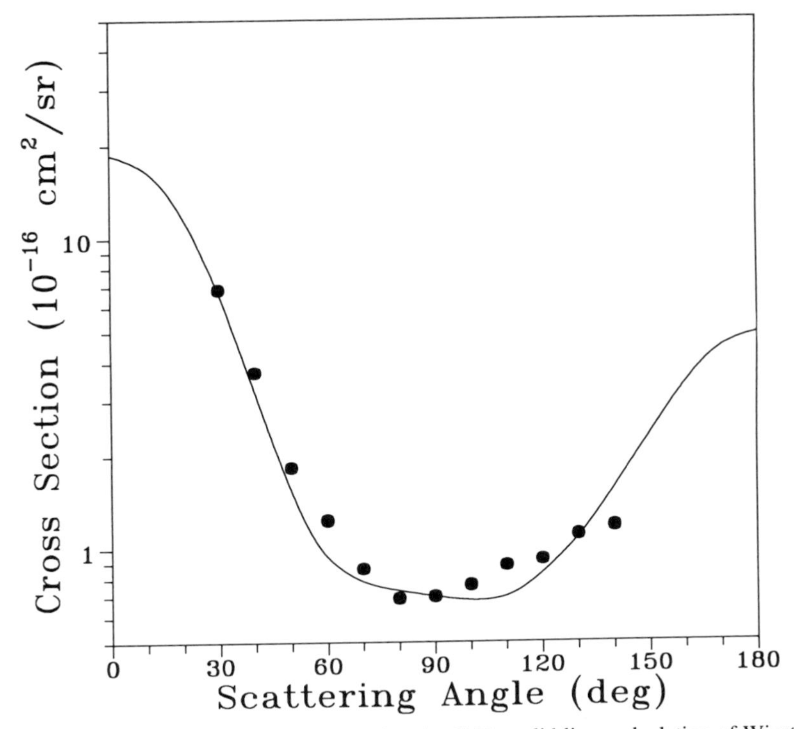

Figure 11. Differential elastic cross section for C_2H_4: solid line, calculation of Winstead et al. [127] at 15 eV; circles, relative cross section of Mapstone and Newell [90] at 15.5 eV, normalized to the calculation at 90°.

in the calculations. Figure 13 shows measured and calculated differential cross sections for cyclopropane at 10 eV, along with the calculated cross section for propylene. The detailed agreement between experiment and theory is very satisfactory (note that the experimental values are *not* normalized to the calculation). At lower energies, where polarization is more important, the agreement is not as good. Static-exchange calculations may still be of value, however, since they err *systematically* in placing resonance energies too high. Thus, we associated a resonance in $^2A_2'$ symmetry with the resonance observed by Howard and Staley [189] at 5.3 eV. As shown by Allan [191], the calculated differential cross section at 7 eV agrees well with the measurement at 5 eV. This resonance appears to be intermediate in both character and energy between the π^* resonances of alkenes and the broad σ^* resonances of alkanes. Its influence on the differential cross section of cyclopropane is still visible at 10 eV (Fig. 13).

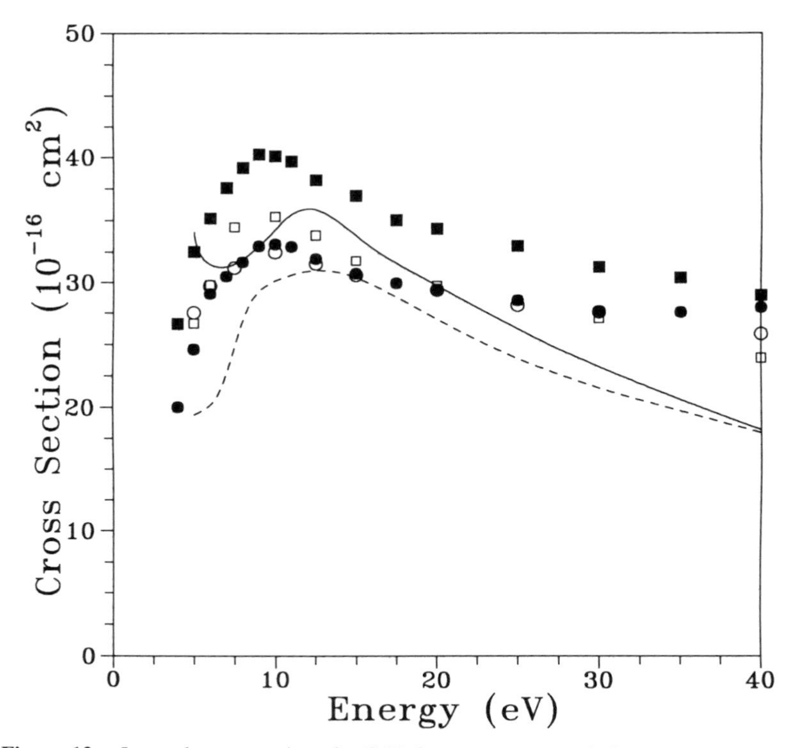

Figure 12. Integral cross sections for C_3H_6 isomers: open symbols, measured total cross sections of Floeder et al. [75]; filled symbols, measured total cross sections of Nishimura and Tawara [184]. Squares are propylene data and circles are cyclopropane. Calculated elastic cross sections from Winstead et al. [188] are shown with a solid line for propylene and dashed line for cyclopropane.

d. Ethane. Ethane has been the subject of total cross section measurements by Brüche [56], Floeder et al. [75], Sueoka and Mori [183], and Nishimura and Tawara [184]. Relative measurements of the total cross section were also reported by Lunt et al. [69]. Differential cross sections for vibrationally inelastic scattering were measured by Curry et al. [87] and by Boesten et al. [192]; the latter study also includes integral cross sections. Low-energy momentum-transfer cross sections were extracted from electron swarm data by Duncan and Walker [180], and by McCorkle et al. [193]. These studies indicate the presence of a Ramsauer minimum. Vibrationally elastic differential and integral cross sections were reported by Curry et al. [87], by Tanaka et al. [194], and by Mapstone and Newell [68]. Theoretical studies of elastic scattering were performed using the SMC method in the static-

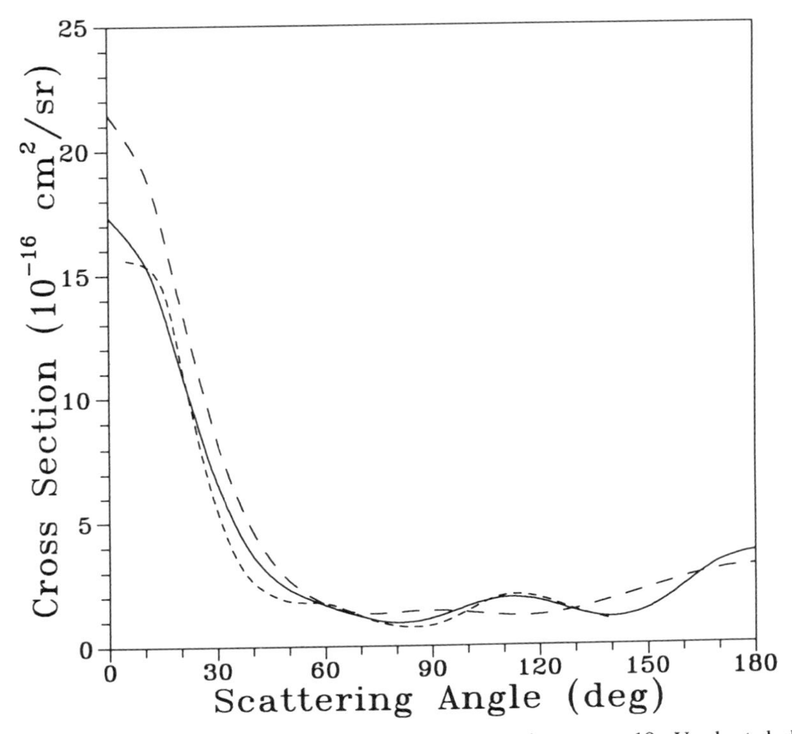

Figure 13. Differential elastic cross sections for C_3H_6 isomers at 10 eV: short dashes, measurements for cyclopropane by Allan [191]; solid line, calculation for cyclopropane by Winstead et al. [188]; long dashes, calculation for propylene.

exchange approximation [127], and using the Kohn method with polarization effects included below 12 eV [195]. Sun et al. [195] considered both eclipsed and staggered conformations of ethane, but concluded that, in the cases where there were substantial differences between the two, the cross sections obtained from the staggered conformation agreed better with the experimental results. A Ramsauer minimum observed by Sun et al. in their momentum-transfer cross section was in reasonably good agreement with the results of Duncan and Walker [180] and of McCorkle et al. [193].

Integral elastic cross sections are compared in Fig. 14. The maximum in the cross section calculated with polarization included is similar in width, position, and magnitude to that observed in the *total* cross section of Nishimura and Tawara [184]. As expected, the static-exchange calculation gives a broadened peak at somewhat higher energy but agrees well with the measured elastic cross section above 15 eV. Differential cross sections near the

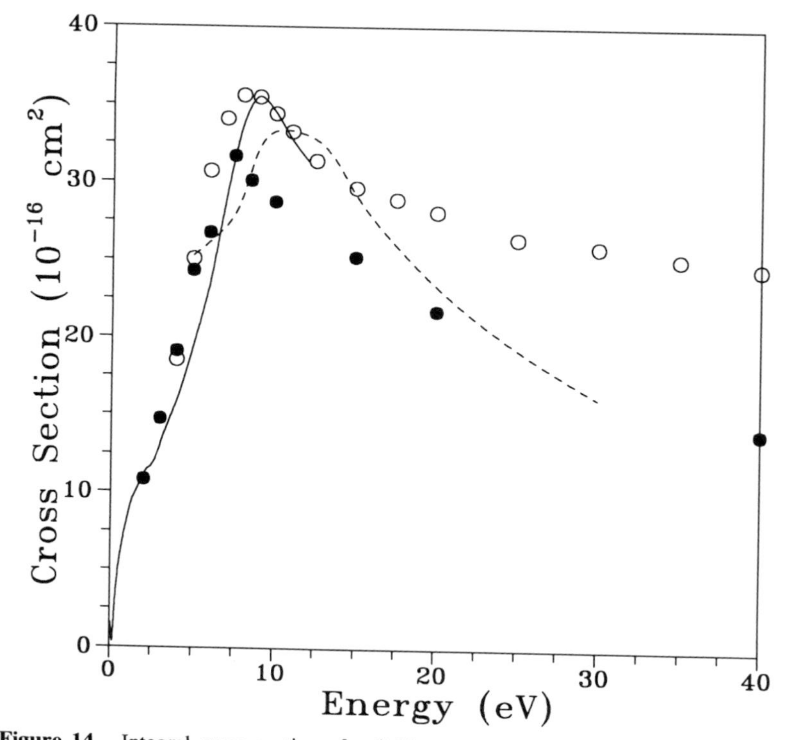

Figure 14. Integral cross sections for C_2H_6: open circles, total cross section measurements of Nishimura and Tawara [184]; filled circles, experimental elastic cross section of Tanaka et al. [194]; dashed line, calculated elastic cross section of Winstead et al. [127]; solid line, calculation of Sun et al. [195].

maximum in the integrated cross sections are shown in Fig. 15. Polarization is clearly important at these energies. At higher energies, however, static-exchange results are in good agreement with experiment [127, 195].

e. Disilane. The silicon analogue of ethane, disilane, has been studied experimentally by Dillon et al. [196], who reported differential and integral elastic cross sections, the momentum-transfer cross section, and a few results for vibrationally inelastic scattering. In addition to their measurements, Dillon et al. reported cross sections obtained from model-potential calculations. We have also studied Si_2H_6, using the SMC method within the static-exchange approximation [127]. Integral cross sections for Si_2H_6 are shown in Fig. 16, and differential cross sections at 5 eV in Fig. 17. In contrast to the case of C_2H_6, polarization does not appear to be necessary to obtain the

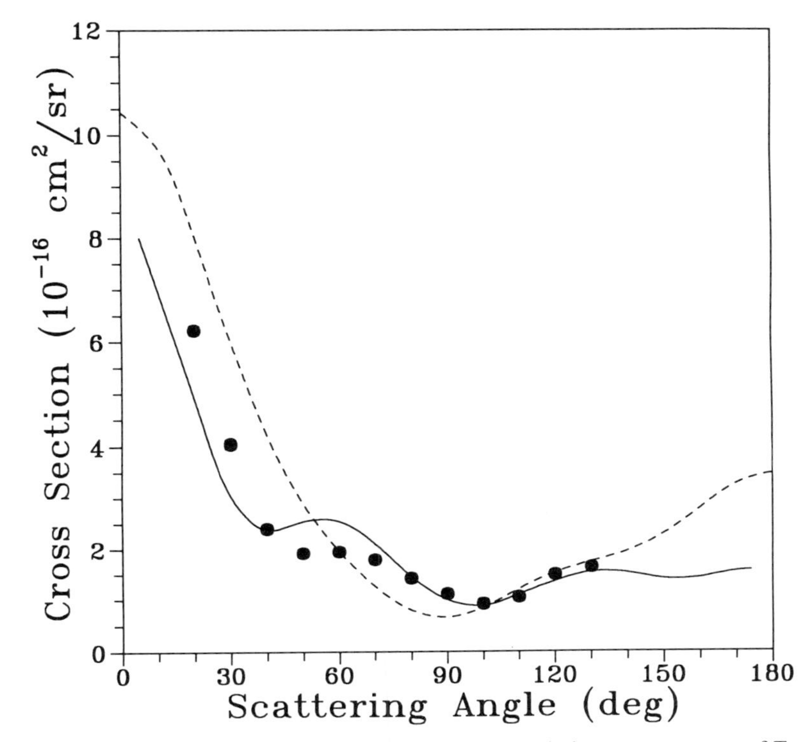

Figure 15. Differential elastic cross sections for C_2H_6: circles, measurements of Tanaka et al. [194] at 6 eV; dashed line, calculation of Winstead et al. [127] at 7.5 eV; solid line, calculation of Sun et al. [195] at 6 eV.

correct qualitative behavior of the cross section even at energies as low as 5 eV. This may be because the shape resonance in Si_2H_6 is much narrower and lies at much lower energy than the resonance in C_2H_6 (cf. Figs. 14 and 16), and is comparatively well described in the static-exchange approximation.

f. Propane. Total cross sections of propane have been measured by Brüche [56], Floeder et al. [75], and Nishimura and Tawara [184]. Momentum-transfer cross sections in the vicinity of the Ramsauer minimum were extracted from swarm data by McCorkle et al. [193]. Boesten et al. [197] recently reported their measurements of differential elastic and vibrational-excitation cross sections, together with integral elastic and momentum-transfer cross sections derived from their measurements, as well as model-potential calculations of the differential and integral cross section. Static-exchange

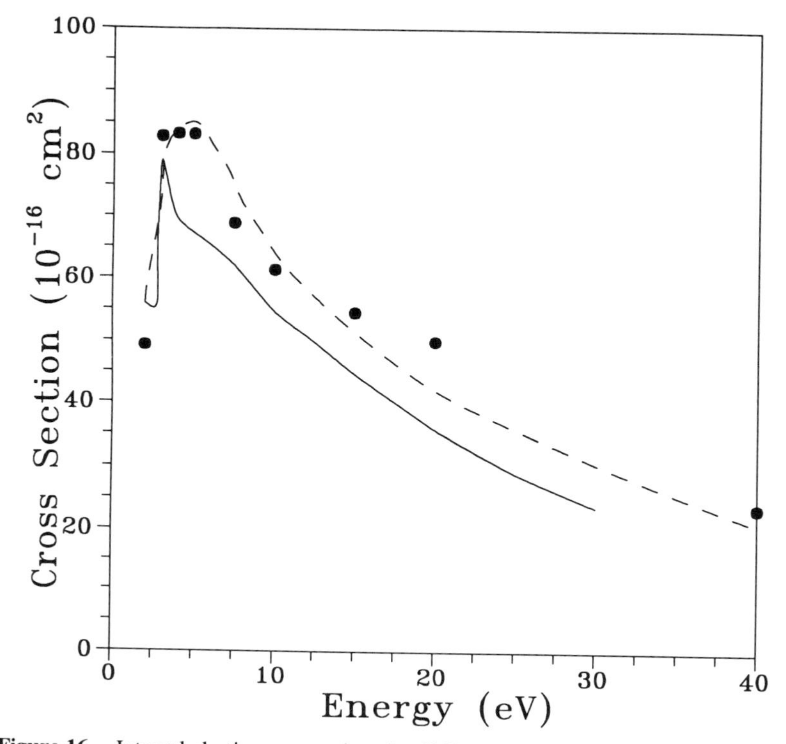

Figure 16. Integral elastic cross sections for Si_2H_6: circles, experimental values of Dillon et al. [196]; dashed line, calculation of Dillon et al. [196]; solid line, calculation of Winstead et al. [127].

cross sections were also calculated using the SMC method [127]. Integral elastic and total cross sections are compared in Fig. 18. The calculations agree quite well with each other, except that Boesten et al. [197], who included a polarization potential, place the resonance maximum somewhat lower in energy, as expected. On the other hand, neither calculation agrees very well with the measured integral elastic cross section, nor does that measurement seem to be completely consistent with the total cross section reported by Nishimura and Tawara [184]. More extensive calculations, as well as additional experiments, seem to be required.

g. Summary of Results for Hydrides and Hydrocarbons. We have necessarily passed rather quickly over a great deal of territory in surveying the hydrides and hydrocarbons. Before continuing, it may be useful to look back

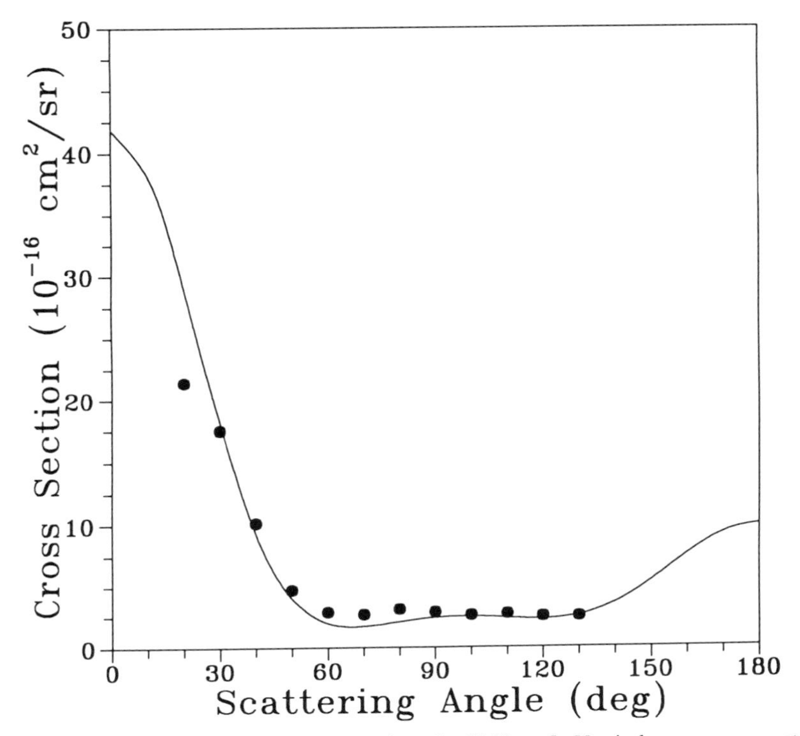

Figure 17. Differential elastic cross sections for Si_2H_6 at 5 eV: circles, measurements of Dillon et al. [196]; solid line, calculation of Winstead et al. [127].

at some of the results in overview. Considering the alkanes (including methane) as a series, their scattering behavior is quite consistent: a Ramsauer minimum at low energy, a broad maximum around 8 eV, and a steady increase in the magnitude of the cross section with increasing molecular size. This trend extends, moreover, at least as far as *n*-butane [75, 193]. In the gross features of their electron-scattering behavior, the alkanes appear to act more or less as aggregations of methyl units. Discussions along similar lines may be found, e.g., in Brüche [56] and Boesten et al. [197]. Comparing silane and disilane to methane and ethane, we see a different sort of trend; the silanes, with greater bond lengths and polarizabilities, are qualitatively similar in their scattering behavior to the alkanes, but with a much narrower resonance maximum occurring at markedly lower energy. Based on the example of GeH_4, we may surmise that the scattering behavior of the higher germanes will strongly resemble that of the corresponding silanes.

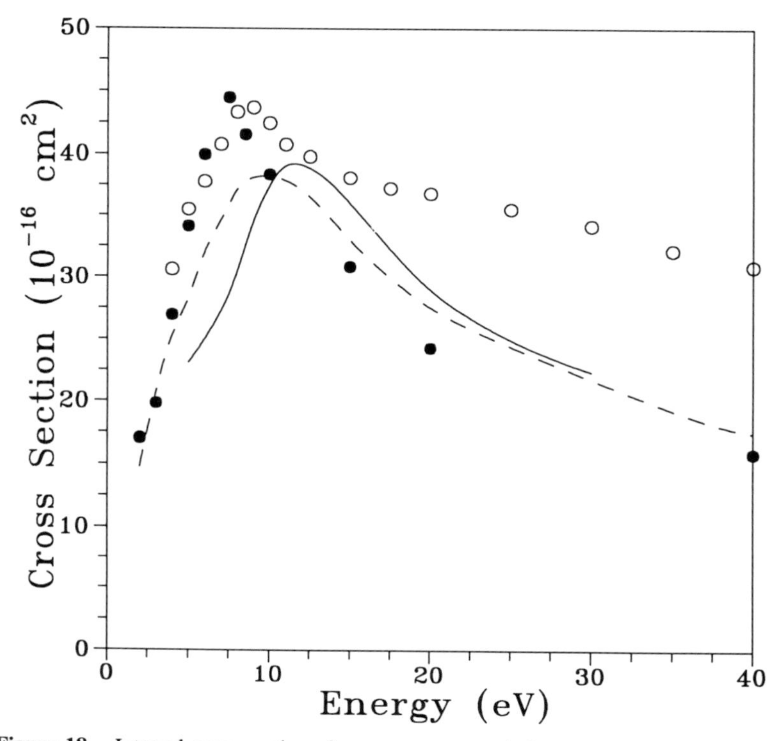

Figure 18. Integral cross sections for propane: open circles, total cross section measurements of Nishimura and Tawara [184]; filled circles, experimental elastic cross section of Boesten et al. [197]; dashed line, calculation of Boesten et al. [197]; solid line, calculation of Winstead et al. [127].

3. *Halogenated Species*

Halogenated gases are important as sources of halogen atoms and ions in plasma etching applications. Cross-section data are also required for etch products, particularly silicon–halogen species, that are found in such plasmas. Nonetheless, electron cross-section data for most halogenated polyatomics are very scarce. The best-studied species is tetrafluoromethane, CF_4. We concentrate here on low-energy elastic scattering. A review of CF_4 cross sections in general has been given by Bonham [198].

Differential cross sections for elastic scattering by CF_4 have been measured by Boesten et al. [199] and by Mann and Linder [200]. Although the differential cross sections obtained in these experiments agree quite well, the integral elastic cross sections reported are markedly different, evidently due to the quite different extrapolation procedures used at high and low angles. Vibrationally inelastic scattering was studied by the same groups [199, 201]. Total electron–CF_4 cross sections have been measured by Jones [202], by Sueoka et al. [106], and by Nishimura [203]. Elastic scattering calculations employing model potentials [109, 125] show large disagreements with the experimental data. Recently there have been several calculations of the elastic cross section employing more sophisticated methods. The first was that of Huo [204] using the SMC method in the static-exchange approximation. Subsequently we re-examined electron–CF_4 scattering using a larger basis set and an improved representation of the Green's function [205]. Natalense et al. have reported SMC results using a pseudopotential representation of the core electrons [206] that are essentially identical to our results [205], except for a slight shift of resonance positions to lower energy, while Gianturco et al. [207] have reported results both in the static-exchange approximation and including a model polarization potential. Thus the past 10 years have seen the accumulation of a fairly substantial body of low-energy scattering data for CF_4.

Figure 19 presents experimental and theoretical integral elastic cross sections along with the total cross sections of Jones [202] and Sueoka et al. [106]. Also shown is the measured *total dissociation* cross section [208, 209]. This makes an interesting point of reference, since all electronically excited and ionized states of CF_4 are thought to be dissociative, and therefore (neglecting vibrational excitation, as seems reasonable above 15 eV) we may identify the total dissociation cross section with the total inelastic cross section. Examining Fig. 19, we observe first of all that the measured total cross sections are in good mutual agreement, while, as mentioned earlier, there is a large difference between the two experimental elastic cross sections. The larger elastic cross section of Boesten et al. [199] is clearly more consistent with the difference between the total cross section and the dissociation cross section. At higher energies, therefore, the recent static-exchange results obtained by the SMC method [205, 206] seem to be closer to the actual elastic cross section than either the static-exchange or the static-exchange plus polarization results of Gianturco et al. [207]. In this connection, we note that, although the magnitude obtained by Gianturco et al. is close to that reported by Huo [204], the latter calculations are now known to be uncoverged [210]. Below 15 eV, the calculations, except the static-exchange result of Gianturco et al., are in better agreement with each other, showing a broad maximum between 10 and 15 eV and becoming more or less flat below 5 eV. The broad peak can be assigned to overlapping resonances of $^2A_1$ and $^2T_2$ symmetry that may be associated with σ^* virtual molecular orbitals (e.g., [205]). Neither the broad peak nor the low-energy plateau is reflected in the experimental elastic cross section of Boesten et al. [199], though there is a broad peak in the *total* cross section. Since this

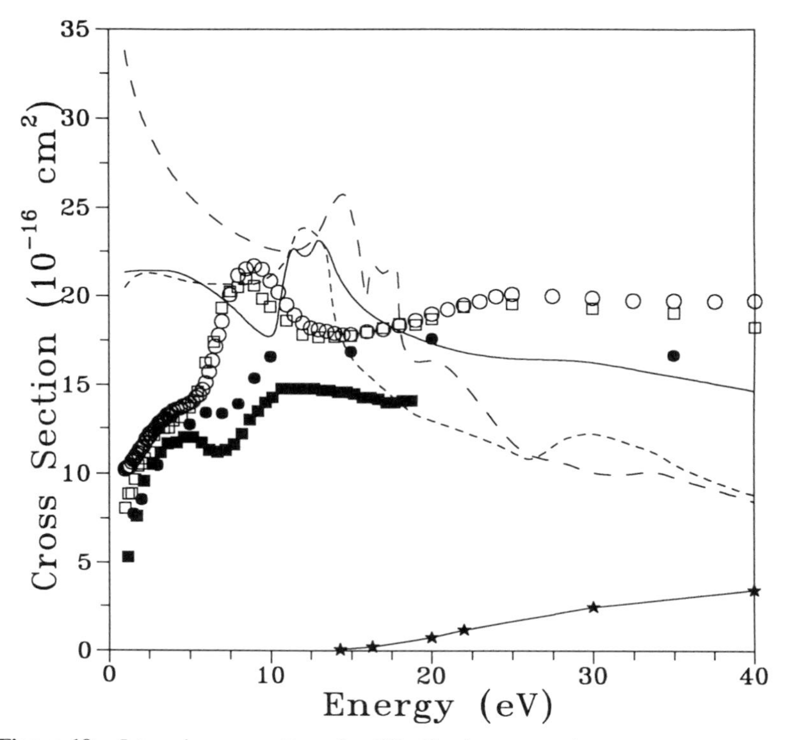

Figure 19. Integral cross sections for CF_4. Total cross section measurements of Jones [202] are shown by open circles and of Sueoka et al. [106] by open squares. Experimental elastic cross sections are from Boesten et al. [199] (filled circles) and Mann and Linder [200] (filled squares). Calculated elastic cross sections are from Gianturco et al. [207] without polarization (long dashes) and with polarization (short dashes), and from Winstead et al. [205] (solid line). The stars connected by lines are the measured dissociation cross section of Winters et al. [208, 209].

peak lies below the first electronic threshold, it can only be due to vibrational excitation, given the absence of a peak in the measured vibrationally elastic cross section. Large vibrational cross sections were indeed reported in this energy range by Mann and Linder [201], though the limitations of their extrapolation procedure cast some doubt on the accuracy of their integrated cross sections, as mentioned earlier. While static-exchange calculations are not expected to give good results below 5 eV, we do expect a qualitative agreement between such calculations and experiment where strong resonances are involved. The most plausible interpretation of the resonances in the fixed-nuclei cross sections is that they correspond to the peak in the total cross section, which is due to vibrational excitation; the shift in energy is not unreasonable for static-exchange results. As discussed in Section III.A, fixed-nuclei calculations cannot be taken as representing the vibrationally elastic scattering alone; however, neither do they represent a proper vibrational sum. A full resolution of the situation at low energies will probably require calculations at multiple nuclear geometries, preferably with polarization effects included.

Calculations on other halogenated gases are sparse. Gianturco et al. [211] have recently calculated elastic cross sections for SF_6 using a static-exchange plus polarization treatment similar to that in their CF_4 study [207]. At higher energies, their cross section is fairly close to the measured total cross section [77, 212], while between 5 and 15 eV they reproduce the presence, but not the position, of two resonant peaks. Rescigno [213] recently published elastic and momentum-transfer cross sections for NF_3 calculated by the Kohn method; corresponding measurements are not available. Natalense et al. [206] reported, in addition to their CF_4 results mentioned above, integral elastic cross sections for CCl_4, $SiCl_4$, $SiBr_4$, and SiI_4. Their cross sections are in fair agreement, as to magnitude, with the total cross sections below 12 eV measured by Wan et al. [105], but there are considerable differences in form. The $SiCl_4$ cross section of Natalense et al. [206] also agrees well with our SMC results [214] from about 7 eV upward, with significant differences at lower energies. We are currently completing a study of elastic and inelastic scattering by $SiCl_4$, BCl_3, and the radicals ($SiCl_3$, $SiCl_2$, SiCl, BCl_2, and BCl) that result from their fragmentation. These results will be published in the near future.

C. Electronic Excitation

Model-potential methods are not, in general, applicable to inelastic scattering, although potentials with an imaginary component may be used to simulate the "absorption" of flux into inelastic channels, as in calculations of total cross sections (e.g., [82]) mentioned in the preceding section. Low-order approximations such as the first Born approximation are computationally feasible, but they are not sufficiently accurate at low impact energies. Somewhat better results are obtained using the distorted-wave Born (DWB) approximation, in which the T-matrix is approximated by a matrix element of the potential between initial- and final-state wave functions obtained by solving separate (single-channel) elastic scattering problems for the target states ϕ_0 and ϕ_n; some recent calculations of this type are mentioned below. The R-matrix method, meanwhile, has proven to be a powerful tool for the study of electronic excitation in atoms and linear molecules, but has not been applied to nonlinear molecules. Progress in computing cross sections for electronic excitation of polyatomics by low-energy electron impact has mainly been associated with implementations of the SMC and Kohn *ab initio* variational procedures. In this section, we describe the calculations that have been undertaken to date, comparing them to experimental results and other calculations where possible.

1. Water

Electron-impact excitation of the dissociative $(3a_1 \rightarrow 3sa_1)^3A_1$ transition in H_2O was studied by Pritchard et al. [215] using the SMC method and a two-channel approximation, i.e., including only the elastic channel and the $^3A_1$ channel in the calculation. Recently Lee et al. [216, 217] studied the same transition using the distorted-wave Born approximation, also reporting results for the $(3a_1 \rightarrow 3pa_1)^3A_1$ channel. The SMC and DWB differential and integral cross sections for the $3a_1 \rightarrow 3sa_1$ channel are qualitatively similar to each other, though the SMC cross section is somewhat larger at most energies. In the integral cross section, a broad maximum is observed at about 17 eV in both calculations. Calculations employing the Kohn method were recently reported [218] for the $(3a_1 \rightarrow 3sa_1)^3A_1$ excitation, for the corresponding singlet transition, and for the $(1b_1 \rightarrow 3sa_1)^{1,3}B_1$ excitations. These results, which were obtained with all five channels (including the elastic channel) coupled, disagree with the SMC (and DWB) results and show a weaker maximum located closer to threshold and a more rapid decrease at higher energies. Gil et al. [218] report that these discrepancies persist in a two-channel Kohn calculation. Excitation cross sections have not been measured for any of the channels considered in the calculations, nor are there obvious reasons for the diagreement, given the similarity between the approximations made in the SMC and Kohn calculations. None of the computed integral excitation cross sections is inconsistent with the indirect estimate of triplet-state excitation based on ·OH fluorescence measurements [219], described in Pritchard et al. [215]. Additional calculations and experiments are clearly called for.

2. Formaldehyde

Calculated cross sections for electron-impact excitation of H_2CO were first reported by Rescigno et al. [220, 221] as the initial application of their implementation of the Kohn method to the excitation of a polyatomic molecule. These calculations, which examined the optically forbidden $\tilde{a}\ ^3A_2$ and $\tilde{A}\ ^1A_2\ (n \rightarrow \pi^*)$ intravalence excitations, were followed by SMC calculations [222]. The Kohn and SMC results are compared in Figs. 20 and 21. Overall, the agreement between the results obtained by the two methods is excellent. Both calculations are fairly consistent with the limited experimental data, a relative cross section for the unresolved singlet and triplet excitations [223], except for a near-threshold peak in the experimental spectrum; however, this peak is now thought to be an artifact [224]. Both channels, and especially

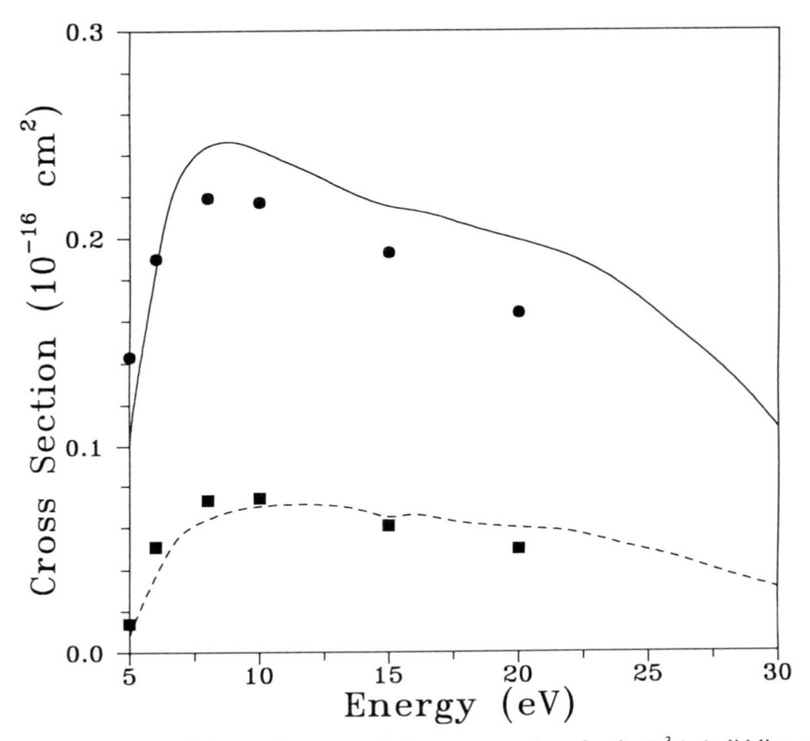

Figure 20. Integral electron-impact excitation cross sections for the $\tilde{a}\ ^3A_2$ (solid line and circles) and $\tilde{A}\ ^1A_2$ (dashed line and squares) $n \rightarrow \pi^*$ transitions in H_2CO. The circles and squares are the calculations of Rescigno et al. [221], and the curves are the calculations of Sun et al. [222].

the triplet, exhibit an unusual angular dependence, with minima in the forward and backward directions. This behavior can be rationalized [221, 222] if we note that the change in the target symmetry from A_1 to A_2 during the collision forces the scattering amplitude to vanish when the plane defined by $\hat{\mathbf{k}}_0$ and $\hat{\mathbf{k}}_n$ coincides with a plane of reflection symmetry. Indeed, for analogous $\Sigma^+ \rightarrow \Sigma^-$ excitations in molecules such as O_2, the forward and backward scattering cross sections vanish rigorously [225].

3. *Ethylene*

The ($\pi \rightarrow \pi^*$) transition in ethylene leading to the $\tilde{a}\ ^3B_{1u}$ and $\tilde{A}\ ^1B_{1u}$ excited states (the *T* and *V* states of Mulliken [226, 227]) has long been a prototype for studies of electronically excited states in polyatomics. Thus the *T* and *V* states are natural subjects for calculations of low-energy electron-impact

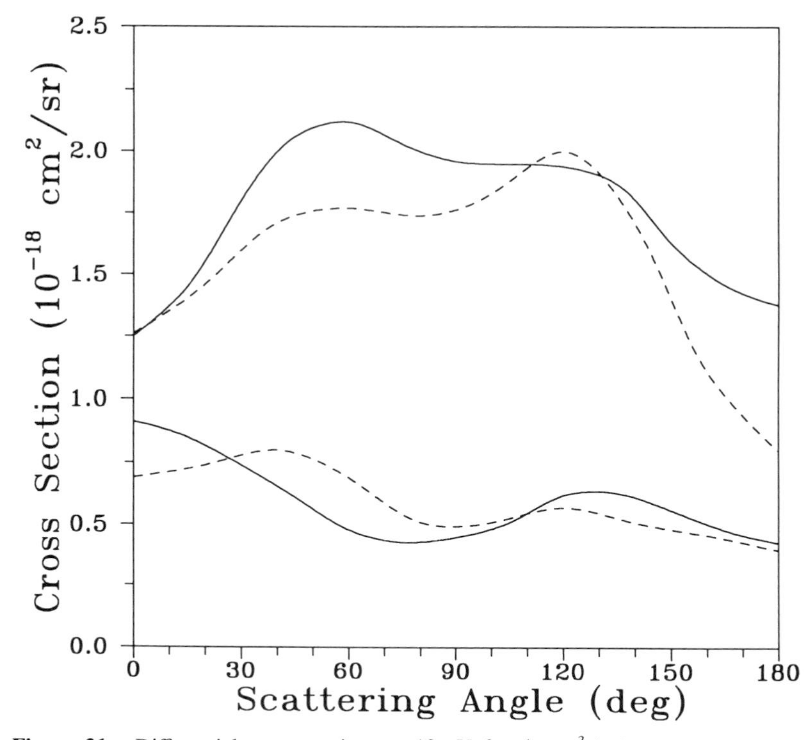

Figure 21. Differential cross sections at 10 eV for the $\tilde{a}\ ^3A_2$ (upper two curves) and $\tilde{A}\ ^1A_2$ (lower two curves) excitations in H_2CO: dashed lines, calculations of Rescigno et al. [221]; solid lines, calculations of Sun et al. [222].

excitation. The SMC method has been used to compute cross sections for the T state in a two-channel approximation [228], while the Kohn method has been applied to both the T and V excitations, in a three-channel approximation above the V threshold and with closed-channel (polarization) effects included below the V threshold [47]. Subsequently, Allan [191, 229] measured absolute values for the differential cross section of the T state, providing a valuable point of reference for comparison of the theoretical cross sections. Love and Jordan [230] have measured relative cross sections for both the T and the V excitations. Their measurements correspond to the summed differential cross sections for $\Theta = 0°$ and $\Theta = 180°$.

Experimental and theoretical cross sections for the T state are compared in Figs. 22 and 23. Considering that the calculations employed different channel-coupling schemes, the agreement between them is very good indeed. Agreement with experiment is less good; the calculated cross sections are

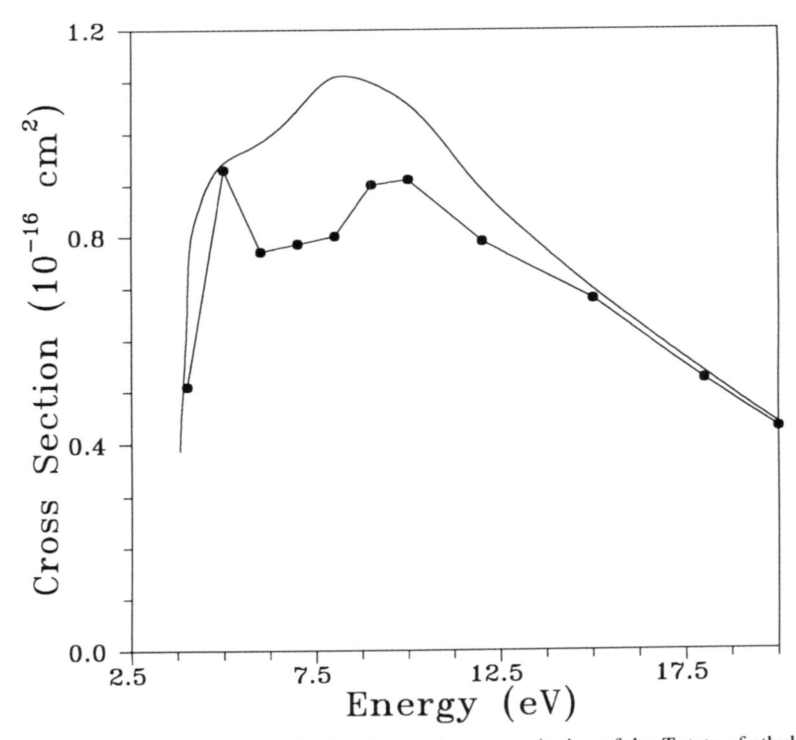

Figure 22. Integral cross section for electron-impact excitation of the T state of ethylene: solid line with circles, calculation of Rescigno and Schneider [47]; solid line, calculation of Sun et al. [228].

approximately twice as large as the measured cross section. Nonetheless, the qualitative features observed in the experiment, such as the strong backward scattering at higher energies, are to be found in the calculated cross sections as well. Both the narrow peak in the Kohn cross section and the rapid onset of the SMC cross section at threshold are associated with strong d-wave scattering ($\ell = 2$) in the entrance channel. Rescigno and Schneider [47] have interpreted this d-wave behavior as being due to the tail of the π^* shape resonance in the elastic channel, which appears to be a reasonable interpretation.

4. *[1.1.1]Propellane*

The [1.1.1]propellane molecule (Fig. 24) is an unusual C_5H_6 isomer of D_{3h} symmetry. Between the two bridgehead carbons is an *inverted* bond, directed 180° away from the direction that would produce a tetrahedral structure

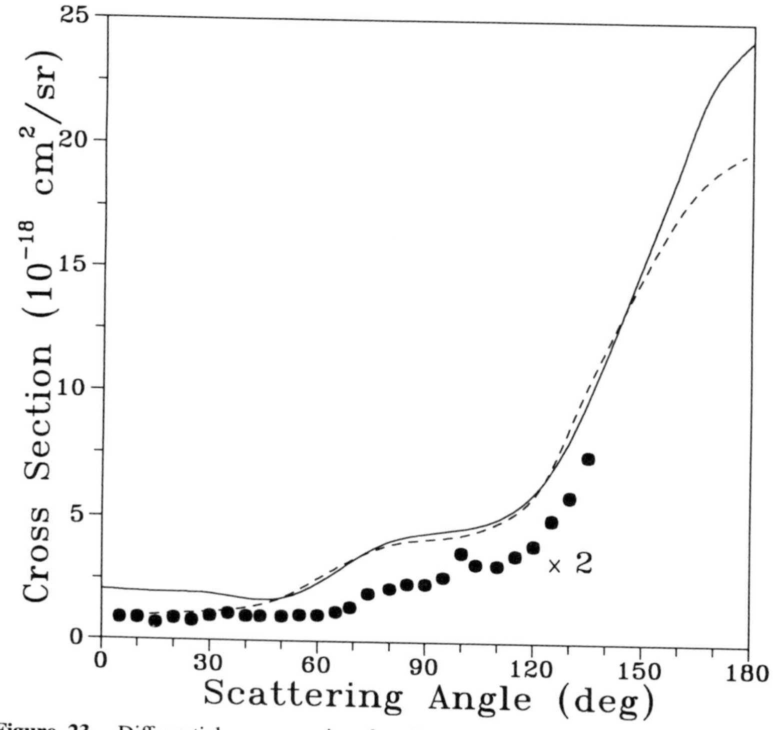

Figure 23. Differential cross section for electron-impact excitation of the *T* state of ethylene: circles, selected values from the measurement of Allan [229] at 14.18 eV; dashed line, calculation of Rescigno and Schneider [47] at 15 eV; solid line, calculation of Sun et al. [228] at 15 eV.

about the bridgehead carbons. In fact, calculations suggest that the very existence of this bond is somewhat problematic (e.g., [231, 232]), and that the ground state has considerable diradical character [232]. With a view to elucidating the structure of the ground and excited states of [1.1.1]propellane, Allan and coworkers [233] measured relative cross sections for elastic, vibrationally inelastic, and electronically inelastic scattering, including excitation of the first triplet and singlet excited states. Measurements of the differential cross section for the $(5a_1' \rightarrow 3a_2'')^3A_2''$ excitation were subsequently extended and placed on an absolute scale [234]. Meanwhile, differential and integral excitation cross sections for the $^3A_2''$ state were calculated using the SMC method [235].

The measured and calculated cross sections for the $(5a_1' \rightarrow 3a_2'')^3A_2''$ excitation of [1.1.1]propellane are compared in Figs. 25 and 26. As was the

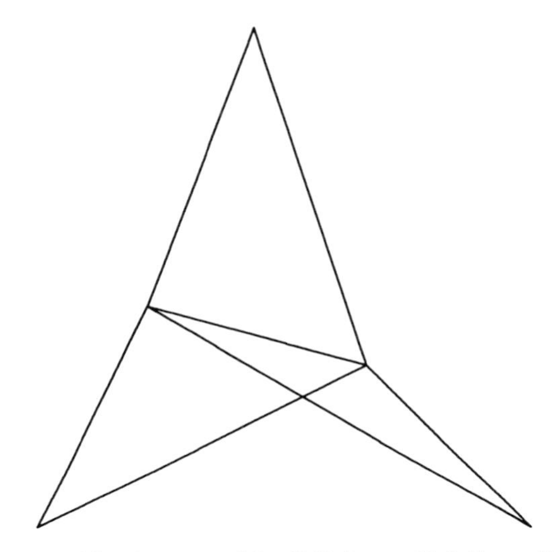

Figure 24. Structure of the C_5H_6 isomer [1.1.1]propellane.

case in ethylene, rather good agreement is observed, except that the theoretical result is about a factor of two larger than the experimental result. Of special note is the resonant feature observed in both cross sections. Symmetry analysis of the calculated cross section confirms the assignment [233] of this feature as a $(5a_1')^1(3a_2'')^2$ core-excited shape resonance. As seen in Fig. 25, the shapes of the calculated and measured integral cross sections are quite similar, except for the location of the resonance. Since the calculation omits polarization, it is not surprising that it places the resonance too high in energy. In Fig. 26, we allow for the shift in the resonance position by comparing differential cross sections at somewhat different energies. There is some suggestion of structure in the experimental cross section, but it is certainly not as pronounced as in the calculation. Given the limitations of the calculation—in particular, a two-channel approximation and a single-configuration description of the ground and excited states—the overall agreement between experiment and theory is perhaps better than could have been anticipated.

5. *Methane and Silane*

We have applied the SMC method to electron-impact excitation of methane [236] and silane [119]. In both molecules, we looked at excitations from the highest occupied orbital, which has t_2 symmetry and σ-bond character, into the lowest empty orbital, which is of a_1 symmetry and mostly *ns* Ryd-

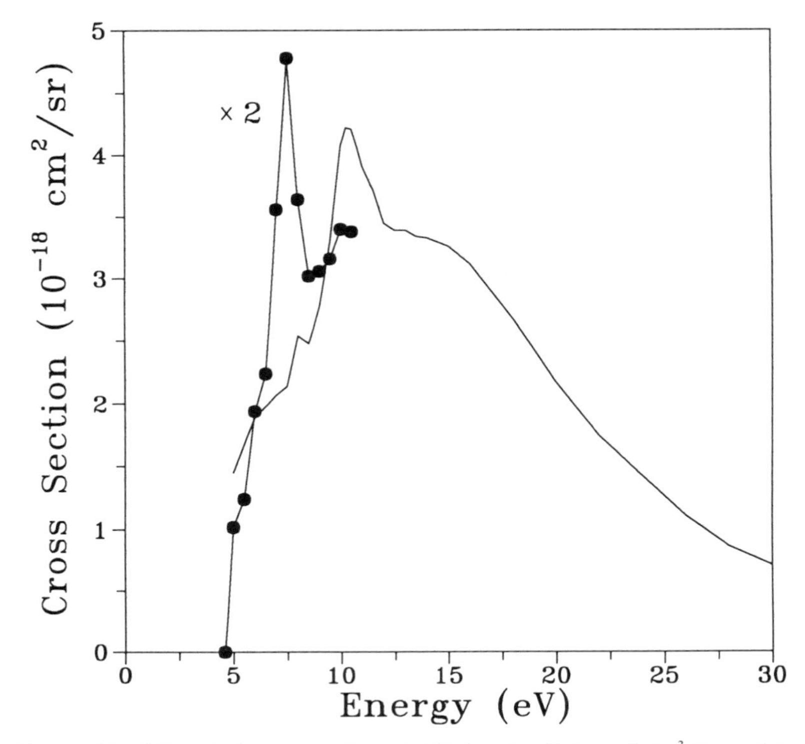

Figure 25. Differential cross section at 90° for the $(5a_1' \rightarrow 3a_2'')\ ^3A_2''$ excitation of [1.1.1]propellane: connected circles, measurement of Allan [234], multiplied by two; solid line, calculation of Winstead et al. [235].

berg character ($n = 3$ for CH_4 and 4 for SiH_4). Both possible spin couplings were considered; thus, including the threefold degeneracy of the $^{1,3}T_2$ excited states, seven open channels were involved in the calculations. Electron-impact excitation of methane has also been studied with the Kohn method [237]. These authors considered singlet and triplet excitations from the $1t_2$ orbital into the $3p$ and $3d$ Rydberg orbitals as well as the $3s$ orbital; excitations into $3s$, $3p$, and $3d$ were treated separately, as were the singlet and triplet manifolds, so that the largest channel-coupling schemes included 16 open channels (the elastic channel and all $1t_2 \rightarrow 3d$ excitations of a given spin).

A comparison of the SMC and Kohn results for the $1t_2 \rightarrow 3sa_1\ ^{1,3}T_2$ excitations in CH_4 is shown in Fig. 27. The $^3T_2$ cross sections are quite similar, except that the Kohn result has a somewhat smaller peak value. Larger differences are observed for the $^1T_2$ channel; however, these are due

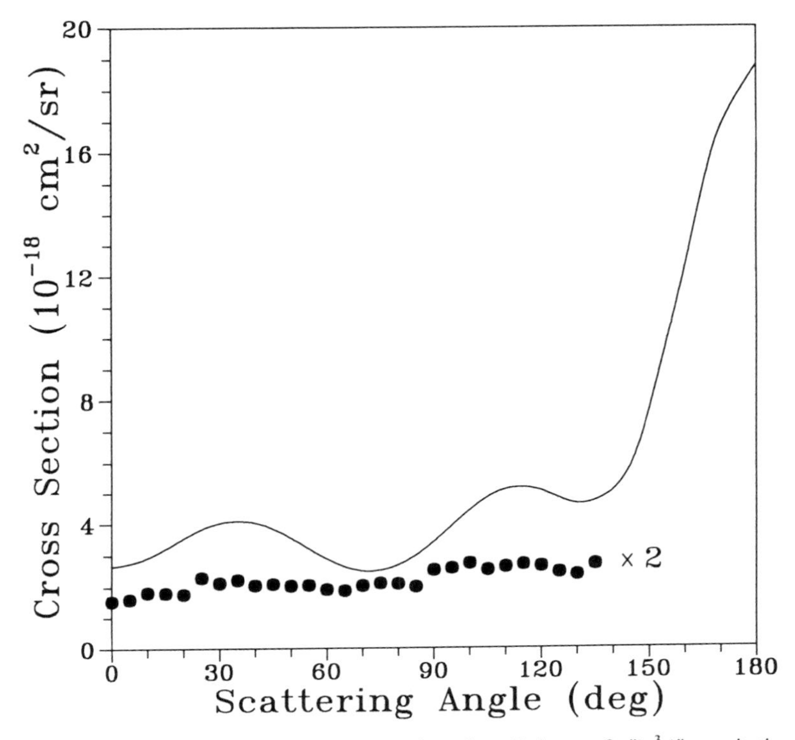

Figure 26. Differential cross sections for the $(5a_1' \rightarrow 3a_2'')$ $^3A_2''$ excitation of [1.1.1]propellane: circles, selected points from the measurements of Allan [234] at 9.68 eV, multiplied by two; solid line, calculation of Winstead et al. [235] at 12 eV.

more to differences in the Born correction procedure than to differences in the underlying *ab initio* calculations. In both calculations, scattering above 20 eV is dominated by the dipole contribution. Since the $1t_2 \rightarrow 3sa_1$ oscillator strength used in the SMC study (0.51), which is obtained using the $3sa_1$ orbital from a $^3T_2$ calculation, is considerably smaller than that used in the Kohn calculation (0.65), the difference in the magnitude of the $^1T_2$ cross sections at higher energies is not surprising. Both values for the oscillator strength are probably too high [238, 239], implying that the $^1T_2$ cross section is actually much smaller at higher energies.

The summed $3s$, $3p$, and $3d$ excitation cross sections of Gil et al. [237] are compared in Fig. 28 to the recommended [81] total electronic excitation cross section for CH_4. Evidently the theoretical cross sections are too large. Gil et al. [237] attribute much of the difference to an overestimation of the $3p$ excitation. Some of the difference may also be due to overestimation of

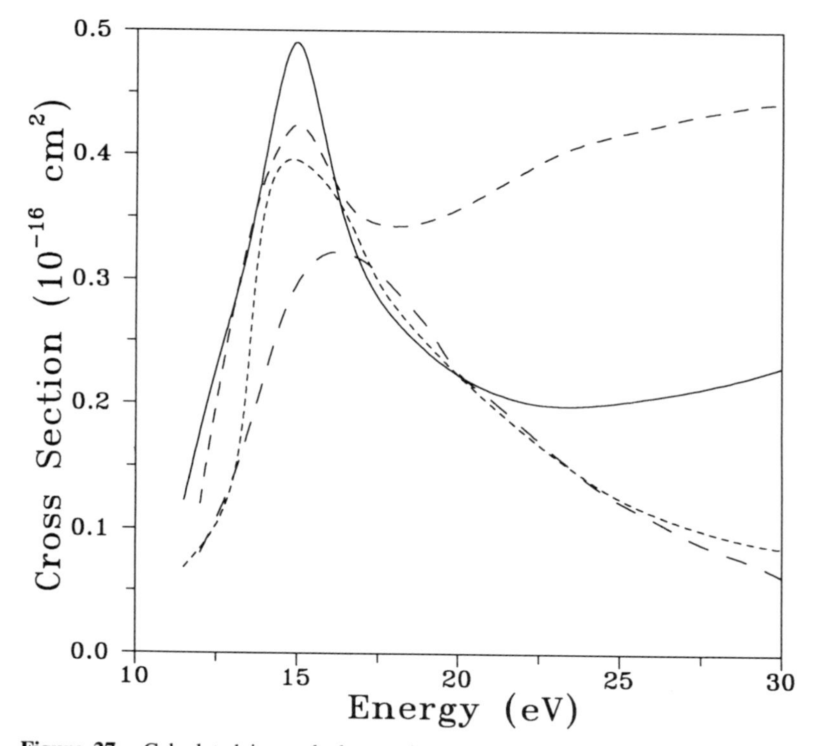

Figure 27. Calculated integral electron-impact excitation cross sections for the $1t_2 \rightarrow 3sa_1$ $^{1,3}T_2$ states of CH_4: solid line, $^1T_2$ cross section of Winstead et al. [236]; medium dashes, $^1T_2$ cross section of Gil et al. [237]; short dashes: $^3T_2$, Winstead et al.; long dashes: $^3T_2$, Gil et al.

the long-range scattering included in the dipole-Born approximation, as discussed above. Further, the use of a fixed-nuclei approximation at a single geometry may be a factor, since the character of the dissociative excited states changes rapidly with nuclear geometry [240–242], possibly leading to large vibrational-averaging effects. Nonetheless, it appears likely that slow convergence of Rydberg excitation cross sections with respect to the size of the coupling scheme employed may be a major factor in the disagreement with experiment [236, 237].

Calculations on SiH_4 [119] tend to support the notion that convergence is difficult to achieve for Rydberg excitations. Indeed, we found, in contrast to the CH_4 case, that singlet-triplet coupling strongly affected the cross sections below 25 eV. The 4*s* excitation cross sections for SiH_4 are on the whole similar to the 3*s* cross sections of CH_4 shown in Fig. 27, except that

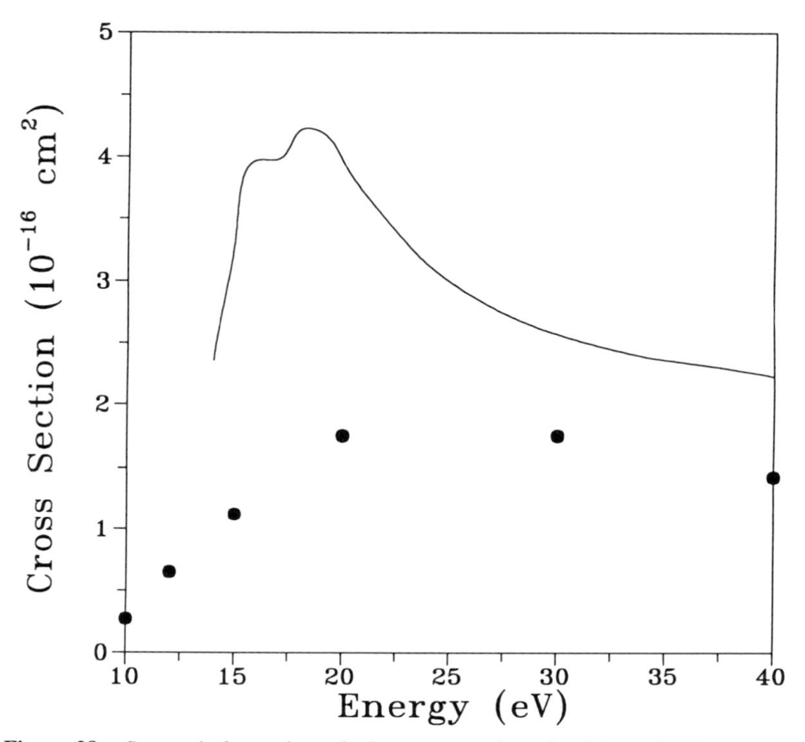

Figure 28. Summed electronic-excitation cross sections for CH_4: solid line, calculated result of Gil et al. [237], including all excitations from the $1t_2$ orbital into $3s$, $3p$, and $3d$ Rydberg orbitals; circles, recommended total electronic-excitation cross section of Kanik et al. [81].

the cross section maxima are shifted closer to threshold. As for CH_4, the dipole-Born correction dominates the summed cross section at higher energies, and, again as for CH_4, the strength of this contribution is probably overestimated. We obtained a $2t_2 \rightarrow 4sa_1$ $^1T_2$ oscillator strength of 0.60, whereas more elaborate calculations on the excited state yielded values of 0.42 [243] and 0.34 [244].

6. *Halogenated Species*

In spite of the importance of halogen-containing molecules in plasma etching applications, there has been little theoretical work on their inelastic electron-scattering behavior. Rescigno [213] recently reported cross sections for the first singlet and triplet excitations of NF_3. As mentioned in Section III.B, we are completing a study of the elastic and inelastic scattering by BCl_3,

$SiCl_4$, and related radicals, which will be published shortly. It is probable that halogenated gases will be the targets of considerable future research.

IV. CONCLUSION

Our hope has been to provide an overview of modern computational methods for studying low-energy collisions between electrons and polyatomic molecules, including their theoretical foundation, and to summarize what has been accomplished so far by applying such methods to specific molecules. In concluding that overview, we evaluate the present state of the field, as inferred from the examples considered in Section III.

Elastic scattering appears, on the whole, to be "under control," in the sense that a systematic improvement of calculations along well-understood lines (increasing the size of the one-electron basis set, improving the representation of polarization) can give satisfactory differential and integral elastic cross sections at low energy, while static-exchange calculations are generally adequate at energies above the location of any pronounced shape resonance. The recent development of methods adapted to general polyatomics, moreover, is beginning to extend the range of systems studied beyond the hydrides, which are still the main topic of studies based on model potentials. Calculations at very low energies, e.g., in the vicinity of the Ramsauer minimum of CH_4, are inherently difficult, however, and success appears to depend on a balancing of several different physical effects. Such a balance may in general be difficult to achieve, since it involves, in part, balancing the description of correlation effects in the N-electron target wave function and in the $(N + 1)$-electron scattering wave function. At sufficiently low energies, the adiabatic-nuclei approximation breaks down, and calculations become much more involved (e.g., [245]). Moreover, even when the adiabatic-nuclei approximation suffices, if strong vibrational excitation is observed or anticipated, as was the case in CF_4, the only alternative for obtaining a detailed theoretical picture appears to be the expensive one of repeating the calculation at enough nuclear geometries to provide an accurate quadrature over the important vibrational modes. Even considering these limitations, however, the situation is much improved from that of 15 or even 5 years ago.

Certainly the situation for inelastic scattering is much improved also, insofar as a reasonably accurate calculation for any nonlinear polyatomic simply did not exist 10 years ago. We have seen above that it is now possible to obtain theoretical results in good quantitative and qualitative agreement from completely independent calculations, as for H_2CO and C_2H_4, and, equally important, to obtain results that compare favorably to measurements, as for ethylene and [1.1.1]propellane. Even in these favorable cases, however, there remain significant differences between calculated and measured cross sections, particularly in magnitude. These differences in magnitude may originate from a variety of causes, and more extensive theoretical and experimental studies, encompassing a wider range of molecules, will be required before any definite conclusions can be drawn. In particular, theoretical studies that include more extensive channel coupling will be informative.

For several molecules, CH_4 being a notable example, excitation cross sections have proven to be sensitive to details of the calculation. Final results in good agreement with experiments do not appear to have been obtained, although comparisons are hindered by the scarcity of experimental data. As discussed, for example, in [237], these convergence difficulties parallel observations of slow convergence in comparable studies on atoms. We have observed similar behavior in studies of the diatomic CO, where studies of valence excitations [246] proved easier to converge and produced more satisfactory results than studies of Rydberg excitations conducted using the same methods [247]. The latter study did suggest that qualitative agreement with experiment can be improved, at least in some cases, by allowing for core relaxation, i.e., by allowing the occupied orbitals in the excited state to relax toward those of the positive ion. Based on current knowledge, it seems likely that truly satisfactory calculations for Rydberg channels will depend on an extensive treatment of open and closed channels (including ionization channels) as well as a provision for core relaxation. In the near future, such calculations are likely to be uncommon due to their demands on computational resources.

To conclude on a more optimistic note, the electronic excitation cross sections for both CH_4 and SiH_4 turned out, except near threshold, to depend mostly on the treatment of the long-range scattering associated with dipole-allowed excitations. There is every reason to suppose that dipole-allowed transitions will dominate in electron collisions with many similar species that lack low-lying virtual valence orbitals. Admittedly, the goal of calculations has been and will remain getting the cross section right near threshold; in the meantime, however, dipole Born results (if based on accurate transition dipoles) may be of practical use in estimating cross sections at higher energies, with optically forbidden excitations being simply ignored. Moreover, a simple estimate of the total cross section for inelastic scattering can be obtained from the difference between a computed elastic cross section and a measured total cross section; though such a subtractive estimate of a small quantity from two large quantities is obviously sensitive to error, it requires only the measurements and calculations that are most reliable and easiest to perform. In CH_4, SiH_4, CF_4, and other molecules where all electronic excitation processes are dissociative, the estimate provides the dis-

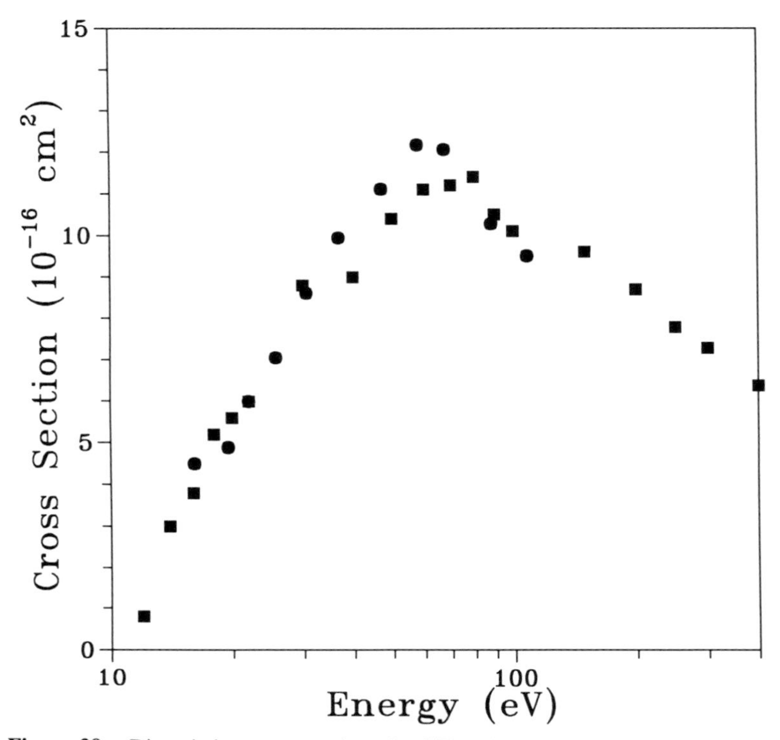

Figure 29. Dissociation cross sections for SiH_4: circles, direct measurements of Perrin et al. [248]; squares, derived values, obtained by subtracting the calculated elastic cross section [119] from the measured total cross section [106].

sociation cross section, which is valuable in modeling. An example is shown in Fig. 29, where we have used our static-exchange cross section for silane [119] together with the most recent total cross section data [106] to produce a dissociation cross section that compares rather well with the direct measurements of Perrin et al. [248]. In cases where direct measurements are unavailable, estimates of this kind may be of practical value.

Although addressing some of the problems that we have identified will likely require great increases in computing power, we are confident that those increases will be forthcoming, given the trends in microprocessors and in parallel computers based on such microprocessors. The challenge in the coming years will be to make intelligent use of that power within well-designed numerical procedures. If this can be achieved, the development of electron–molecule collision calculations, as both a tool for fundamental investigations of molecular structure and dynamics and a tool for the generation of technologically relevant data, is likely to be even more rapid than it has been up to now.

ACKNOWLEDGMENTS

It is a pleasure to thank collaborators who have contributed to the SMC calculations described above, in particular Howard Pritchard, Qiyan Sun, Paul Hipes, and Marco Lima. Our research employed facilities of the Concurrent Supercomputing Consortium, which is supported in part by the National Science Foundation, of the JPL/Caltech Supercomputing Project, and of the Air Force Phillips Laboratory. We thank the staff of the Center for Advanced Computational Research at Caltech and of the JPL Supercomputer Center for continued technical assistance. Research funding was provided by the National Science Foundation, by the Air Force Office of Scientific Research, and by SEMATECH, Inc.

APPENDIX A

We show here that $\psi^{(+)}$ of Eq. (2.23) has the physically appropriate asymptotic form, Eq. (2.6). Eq. (2.18) for Φ_0 obviously reproduces the first term on the right-hand side of Eq. (2.6), so the question is whether the Green's function term in Eq. (2.22), with $G_0^{(+)}$ given by Eq. (2.21), leads to the outgoing spherical wave terms of Eq. (2.6). We begin by converting the integral over E' into an integral over k_n' and writing out in full the T-matrix element, Eq. (2.24). With these changes, the Green's function term of Eq. (2.23) becomes

$$\frac{1}{8\pi^3} \lim_{\epsilon\to 0^+} \sum_{n=0}^{\infty} \int d^3k_n' \, \phi_n(\{\mathbf{r}_i, i = 1, \ldots, N\})$$

$$\cdot \exp(i\mathbf{k}_n' \cdot \mathbf{r}_{N+1}) \frac{\langle \phi_n \exp(i\mathbf{k}_n' \cdot \mathbf{r}_{N+1}') | V | \psi^{(+)}(\mathbf{k}_o) \rangle}{E - E_n - \frac{1}{2}k_n^2 + i\epsilon} \tag{A.1}$$

where the integration signified by the Dirac brackets is over $\{\mathbf{r}_i', i = 1, \ldots, N+1\}$. Performing the angular portion of the integral over $\mathbf{k}_n'$ reduces this to

$$-\frac{1}{\pi^2} \lim_{\epsilon\to 0^+} \sum_{n=0}^{\infty} \phi_n(\{\mathbf{r}_i\}) \int d^3r_1' \ldots \int d^3r_{N+1}' \int_0^{\infty} dk_n'$$

$$\cdot \frac{k_n' \sin(k_n'\rho)/\rho}{k_n'^2 - 2(E - E_n + i\epsilon)} \phi_n^*(\{\mathbf{r}_i'\}) V(\{\mathbf{r}_i'\}) \psi^{(+)}(\{\mathbf{r}_i'\}) \tag{A.2}$$

where $\rho = |\mathbf{r}_{N+1} - \mathbf{r}'_{N+1}|$, and henceforward the upper limit on i in $\{\mathbf{r}_i\}$ (either N or $N + 1$) will be assumed to be understood from context.

The integral over k'_n is evaluated by contour integration in the complex plane. We first expand $\sin(k'_n\rho)$ as $(e^{ik'_n\rho} - e^{-ik'_n\rho})/2i$ and note that the integral over the second term can be replaced with an integral from $k = -\infty$ to 0 over the first term. It is then possible to close the contour with an infinite semicircle in the $\text{Im}(k'_n) > 0$ half-plane, along which the integrand vanishes. The poles of the integrand are at $\pm(k_n + i\epsilon)$, where $k_n^2 = 2(E - E_n)$, and only the pole at $k_n + i\epsilon$ is enclosed by the contour chosen (it is here the $\epsilon \to 0^+$ limit comes into play.) Thus by Cauchy's theorem we have

$$\frac{1}{\rho}\int_0^\infty dk'_n \frac{k'_n \sin(k'_n\rho)}{k'^2_n - 2(E - E_n + i\epsilon)} = \frac{1}{2i\rho}\oint \frac{k'_n \exp(ik'_n\rho)}{k'^2_n - k_n^2 - i\epsilon/2}$$
$$= \frac{\pi}{2}\frac{\exp(ik_n\rho)}{\rho} \tag{A.3}$$

Substituting this result back into Eq. (A.2), we have

$$-\frac{1}{2\pi}\sum_{n=0}^{\infty} \phi_n(\{\mathbf{r}_i\}) \int d^3r'_1 \ldots \int d^3r'_{N+1} \frac{\exp(ik_n\rho)}{\rho} \phi_n^*(\{\mathbf{r}'_i\})V(\{\mathbf{r}'_i\})$$
$$\cdot\, \psi^{(+)}(\{\mathbf{r}'_i\}) \tag{A.4}$$

Now we are ready to evaluate the limit of large r_{N+1}. We perform this operation and the radial integration in r'_{N+1} in such a way that $r_{N+1} >> r'_{N+1}$, and $\hat{\mathbf{k}}_n$ is the direction in which r_{N+1} is taken to infinity. Under these conditions, we may replace $|\mathbf{r}_{N+1} - \mathbf{r}'_{N+1}|$ with $r_{N+1} - \mathbf{k}_n \cdot \mathbf{r}'_{N+1} + O(r_{N+1}^{-1})$; the second term is necessary in the exponential but may be neglected in the denominator. For $E < E_n$, that is, for closed channels, we have $k_n = i\kappa$, with $\kappa > 0$ (because only the pole on the positive imaginary axis is inside the contour). Therefore the closed-channel terms in Eq. (A.4) vanish exponentially as $r_{N+1} \to \infty$ and may be dropped from the sum in Eq. (A.4). We have finally for the Green's function contribution

$$-\frac{1}{2\pi}\sum_{n\in\text{open}} \phi_n(\{\mathbf{r}_i\}) \frac{\exp(ik_n r_{N+1})}{r_{N+1}}$$
$$\cdot \int d^3r'_1 \ldots \int d^3r'_{N+1}\, \phi_n^*(\{\mathbf{r}'_i\}) \exp(-i\mathbf{k}_n \cdot \mathbf{r}'_{N+1})V(\{\mathbf{r}'_i\})$$
$$\cdot\, \psi^{(+)}(\{\mathbf{r}'_i\}) = -\frac{1}{2\pi}\sum_{n\in\text{open}} \phi_n(\{\mathbf{r}_i\}) \frac{\exp(ik_n r_{N+1})}{r_{N+1}} T_{0n}(\mathbf{k}_0, \mathbf{k}_n) \tag{A.5}$$

By comparison with Eq. (2.6), we can see that Eq. (A.5) is indeed the desired asymptotic form. In fact, this comparison establishes the connection between the scattering amplitude $f_n(\mathbf{k}_0, \mathbf{k}_n)$ and the corresponding T-matrix element:

$$f_n(\mathbf{k}_0, \mathbf{k}_n) = -\frac{1}{2\pi} T_{0n}(\mathbf{k}_0, \mathbf{k}_n) \tag{A.6}$$

In electron–molecule scattering, an important qualification on the above demonstration is that it considers only the asymptotic form of $\psi^{(+)}$ in the variable r_{N+1}. Among the excited states $\phi_n(\{\mathbf{r}_i\})$ of the target will be ionized states, in which electrons other than electron $N + 1$ may be found in the asymptotic region. If these are among the open channels, they represent electron-impact ionization. Closed-channel states of this kind, on the other hand, represent *exchange scattering:* electron $N + 1$ becomes bound to the target, while a molecular electron is ejected. Because of the indistinguishability of electrons, the contribution of exchange scattering must be included in elastic and inelastic cross sections.

REFERENCES

1. N. F. Lane, *Rev. Mod. Phys.* **52,** 29 (1980).
2. M. A. Morrison, *Adv. At. Mol. Phys.* **24,** 51 (1988).
3. I. Shimamura and K. Takayanagi, Eds., *Electron-Molecule Collisions,* Plenum, New York, 1984.
4. F. A. Gianturco, Ed., *Collision Theory for Atoms and Molecules (NATO ASI Series B, Vol. 196),* Plenum, New York, 1989.
5. D. F. Yarkony, Ed., *Modern Electronic Structure Theory,* World Scientific, Singapore, 1995.
6. P. G. Burke and K. A. Berrington, *Atomic and Molecular Processes: An R-Matrix Approach,* Institute of Physics, Bristol, 1993.
7. R. J. W. Henry, *Rep. Prog. Phys.* **56,** 327 (1993).
8. P. G. Burke, *Adv. At. Mol. Opt. Phys.* **32,** 39 (1994).
9. M. Allan, *J. Electron Spectrosc. Relat. Phen.* **48,** 219 (1989).
10. W. E. Kauppila and T. S. Stein, *Adv. At. Mol. Opt. Phys.* **26,** 1 (1989).
11. F. B. Dunning, *J. Phys. B* **28,** 1645 (1995).
12. H. Ehrhardt and L. A. Morgan, Eds., *Electron Collisions with Molecules, Clusters, and Surfaces,* Plenum, New York, 1994.
13. D. M. Chase, *Phys. Rev.* **104,** 838 (1956).
14. A. Herzenberg, in [3], p. 191.
15. D. W. Norcross and N. T. Padial, *Phys. Rev. A* **25,** 226 (1982).
16. L. I. Schiff, *Quantum Mechanics,* 3rd ed., McGraw-Hill, New York, 1968, Chapters 5 and 9.
17. R. G. Newton, *Scattering Theory of Waves and Particles,* 2nd ed., Springer, New York, 1982.

18. J. R. Taylor, *Scattering Theory,* Wiley, New York, 1972.
19. N. F. Mott and H. S. W. Massey, *The Theory of Atomic Collisions,* 3rd ed., Oxford University Press, Oxford, 1965.
20. L. I. Schiff, [16], p. 115.
21. P. A. M. Dirac, *The Principles of Quantum Mechanics,* 4th ed. (revised), Oxford University Press, Oxford, 1967, pp. 34 ff.
22. B. A. Lippmann and J. Schwinger, *Phys. Rev.* **79,** 469 (1950).
23. J. Schwinger, *Phys. Rev.* **72,** 742 (1947).
24. W. Kohn, *Phys. Rev.* **74,** 1763 (1948).
25. E. P. Wigner, *Phys. Rev.* **70,** 15 (1946).
26. C. Bloch, *Nucl. Phys.* **4,** 503 (1957).
27. M. H. F. Bettega, L. G. Ferreira, and M. A. P. Lima, *Phys. Rev. A* **47,** 1111 (1993).
28. R. G. Newton, [17], pp. 187 f.
29. C. Schwartz, *Ann. Phys. (N.Y.)* **16,** 36 (1961).
30. R. K. Nesbet, *Variational Methods in Electron-Atom Scattering Theory,* Plenum, New York, 1980.
31. R. R. Lucchese, *Phys. Rev. A* **40,** 6879 (1989).
32. W. H. Miller and B. M. D. D. Jansen op de Haar, *J. Chem. Phys.* **86,** 6213 (1987).
33. M. Kamimura, *Prog. Theor. Phys. (Kyoto) Suppl.* **62,** 236 (1977).
34. T. N. Rescigno, B. H. Lengsfield, and C. W. McCurdy, in [5], Part I, p. 501.
35. C. Winstead and V. McKoy, *Phys. Rev. A* **41,** 49 (1990).
36. K. Takatsuka, R. R. Lucchese, and V. McKoy, *Phys. Rev. A* **24,** 1812 (1981).
37. N. S. Ostlund, *Chem. Phys. Lett.* **34,** 419 (1975); D. K. Watson and V. McKoy, *Phys. Rev. A* **20,** 1474 (1979).
38. K. Takatsuka and V. McKoy, *Phys. Rev. A* **24,** 2473 (1981); **30,** 1734 (1984).
39. C. Winstead and V. McKoy, *Phys. Rev. A* **47,** 1514 (1993).
40. C. Winstead and V. McKoy, in [5], Part II, p. 1375.
41. W. J. Hunt and W. A. Goddard, *Chem. Phys. Lett.* **3,** 414 (1969).
42. A. Szabo and N. S. Ostlund, *Modern Quantum Chemistry,* Macmillan, New York, 1982, Chapter 2.
43. M. A. P. Lima, L. M. Brescansin, A. J. R. da Silva, C. Winstead, and V. McKoy, *Phys. Rev. A* **41,** 327 (1989).
44. V. I. Lebedev, *Zh. Vychisl. Mat. Mat. Fiz.* **15** (1), 48 (1975) [*USSR Comp. Math. Math. Phys.* **15** (1), 44 (1975)]; V. I. Lebedev, *Zh. Vychisl. Mat. Mat. Fiz.* **16** (2), 293 (1976) [*USSR Comp. Math. Math. Phys.* **16** (2), 10 (1976)]; V. I. Lebedev, *Sib. Mat. Zh.* **18** (1), 132 (1977) [*Sib. Math. J.* **18,** 99 (1976)]; V. I. Lebedev and A. L. Skorokhodov, *Ross. Akad. Nauk Dokl.* **324** (3), 519 (1992) [*Russ. Acad. Sci. Dokl. Math.* **45,** 587 (1992)]; V. I. Lebedev, *Dokl. Akad. Nauk* **338** (4), 454 (1994) [*Russ. Acad. Sci. Dokl. Math.* **50,** 283 (1995)]; see also O. Treutler and R. Ahlrichs, *J. Chem. Phys.* **102,** 346 (1995).
45. Y. Itikawa, *J. Phys. Soc. Jpn.* **27,** 444 (1969); O. H. Crawford and A. Dalgarno, *J. Phys. B* **4,** 494 (1971).
46. S. Altshuler, *Phys. Rev.* **107,** 114 (1957); O. H. Crawford, A. Dalgarno, and P. B. Hays, *Mol. Phys.* **13,** 181 (1967); Y. Itikawa and K. Takayanagi, *J. Phys. Soc. Jpn.* **26,** 1254 (1969).

47. T. N. Rescigno and B. I. Schneider, *Phys. Rev. A* **45,** 2894 (1992).
48. T. Gibson, M. A. P. Lima, V. McKoy, and W. M. Huo, *Phys. Rev. A* **35,** 2473 (1987).
49. M. J. Seaton, *Proc. Phys. Soc.* **79,** 1105 (1962).
50. K. J. Miller and M. Krauss, *J. Chem. Phys.* **47,** 3754 (1967).
51. H. P. Pritchard, Ph.D. Thesis, California Institute of Technology, Pasadena, Calif., 1995.
52. P. G. Hipes, C. Winstead, M. A. P. Lima, and V. McKoy, in *Proceedings of the Fifth Distributed Memory Computing Conference, Vol. I: Applications,* D. W. Walker and Q. F. Stout, Eds., IEEE Computer Society, Los Alamitos, CA, 1990, p. 498.
53. G. Fox, M. A. Johnson, G. A. Lyzenga, S. W. Otto, J. K. Salmon, and D. W. Walker, *Solving Problems on Concurrent Processors, Vol. I,* Prentice Hall, Englewood Cliffs, NJ, 1988, p. 57 ff.
54. R. B. Brode, *Phys. Rev.* **25,** 636 (1925).
55. E. Brüche, *Ann. Phys. (Leipzig)* **IV 83,** 70 (1927).
56. E. Brüche, *Ann. Phys. (Leipzig)* **4,** 387 (1930).
57. C. Ramsauer and R. Kollath, *Ann. Phys. (Leipzig)* **4,** 91 (1930).
58. R. Müller, K. Jung, K.-H. Kochem, W. Sohn, and H. Ehrhardt, *J. Phys. B* **18,** 3971 (1985).
59. I. Shimamura, *Phys. Rev. A* **28,** 1357 (1983).
60. A. Jain and D. G. Thompson, *J. Phys. B* **16,** 3077 (1983).
61. N. Abusalbi, R. A. Eades, T. Nam, D. Thirumalai, D. A. Dixon, D. G. Truhlar, and M. Dupuis, *J. Chem. Phys.* **78,** 1213 (1983).
62. F. A. Gianturco and S. Scialla, *J. Phys. B* **20,** 3171 (1987).
63. F. A. Gianturco, A. Jain, and L. C. Pantano, *J. Phys. B* **20,** 571 (1987).
64. P. McNaughten, D. G. Thompson, and A. Jain, *J. Phys. B* **23,** 2405 (1990).
65. L. M. Brescansin, M. A. P. Lima, and V. McKoy, *Phys. Rev. A* **40,** 5577 (1989).
66. A. Jain, *Z. Phys. D* **21,** 153 (1991).
67. T. W. Shyn, *J. Phys. B* **24,** 5169 (1991).
68. B. Mapstone and W. R. Newell, *J. Phys. B* **27,** 5761 (1994).
69. S. L. Lunt, J. Randell, J. P. Ziesel, G. Mrotzek, and D. Field, *J. Phys. B* **27,** 1407 (1994).
70. B. Schmidt, *J. Phys. B* **24,** 4809 (1991).
71. E. Barbarito, M. Basta, M. Calicchio and G. Tessari, *J. Chem. Phys.* **71,** 54 (1979).
72. R. K. Jones, *J. Chem. Phys.* **82,** 5424 (1985).
73. J. Ferch, B. Granitza, and W. Raith, *J. Phys. B* **18,** L445 (1985).
74. B. Lohmann and S. J. Buckman, *J. Phys. B* **19,** 2565 (1986).
75. K. Floeder, D. Fromme, W. Raith, A. Schwab, and G. Sinapius, *J. Phys. B* **18,** 3347 (1985), and private communication.
76. O. Sueoka and S. Mori, *J. Phys. B* **19,** 4035 (1986).
77. M. S. Dababneh, Y.-F. Hsieh, W. E. Kauppila, C. K. Kwan, S. J. Smith, T. S. Stein, and M. N. Uddin, *Phys. Rev. A* **38,** 1207 (1988).
78. H. Nishimura and T. Sakae, *Jpn. J. Appl. Phys.* **29,** 1372 (1990).
79. A. Zecca, G. Karwasz, R. S. Brusa, and C. Szmytkowski, *J. Phys. B* **24,** 2747 (1991).
80. I. Kanik, S. Trajmar, and J. C. Nickel, *Chem. Phys. Lett.* **193,** 281 (1992).

81. I. Kanik, S. Trajmar, and J. C. Nickel, *J. Geophys. Res.* **98,** 7447 (1993).
82. A. Jain and K. L. Baluja, *Phys. Rev. A* **45,** 202 (1992).
83. J. H. Wang, A. N. Tripathi, and V. H. Smith, *J. Chem. Phys.* **101,** 4842 (1994).
84. D. Raj, *Ind. J. Pure Appl. Phys.* **32,** 867 (1994).
85. Y. H. Jiang, J. F. Sun, and L. Wan, *Phys. Rev. A* **52,** 398 (1995).
86. H. Tanaka, T. Okada, L. Boesten, T. Suzuki, T. Yamamoto, and M. Kubo, *J. Phys. B* **15,** 3305 (1982).
87. P. J. Curry, W. R. Newell, and A. C. H. Smith, *J. Phys. B* **18,** 2303 (1985).
88. W. Sohn, K.-H. Kochem, K.-M. Scheuerlein, K. Jung, and H. Ehrhardt, *J. Phys. B* **19,** 3625 (1986).
89. T. W. Shyn and T. E. Cravens, *J. Phys. B* **23,** 293 (1990).
90. B. Mapstone and W. R. Newell, *J. Phys. B* **25,** 491 (1992).
91. P. McNaughten and D. G. Thompson, *J. Phys. B* **21,** L703 (1988).
92. F. A. Gianturco, J. A. Rodriguez-Ruiz, and N. Sanna, *J. Phys. B* **28,** 1287 (1995).
93. F. A. Gianturco, J. A. Rodriguez-Ruiz, and N. Sanna, *Phys. Rev. A* **52,** 1257 (1995).
94. T. Nishimura and Y. Itikawa, *J. Phys. B* **27,** 2309 (1994).
95. M. A. P. Lima, T. L. Gibson, W. M. Huo, and V. McKoy, *Phys. Rev. A* **32,** 2696 (1985).
96. M. A. P. Lima, K. Watari, and V. McKoy, *Phys. Rev. A* **39,** 4312 (1989).
97. C. W. McCurdy and T. N. Rescigno, *Phys. Rev. A* **39,** 4487 (1989).
98. B. H. Lengsfield, T. N. Rescigno, and C. W. McCurdy, *Phys. Rev. A* **44,** 4296 (1991).
99. B. M. Nestmann, K. Pfingst, and S. D. Peyerimhoff, *J. Phys. B* **27,** 2297 (1994).
100. R. G. Newton, [17], p. 353.
101. A. Jain and D. G. Thompson, *J. Phys. B* **24,** 1087 (1991).
102. H. Tanaka, L. Boesten, H. Sato, M. Kimura, M. A. Dillon, and D. Spence, *J. Phys. B* **23,** 577 (1990).
103. M. Tronc, A. Hitchcock, and F. Edard, *J. Phys. B* **22,** L207 (1989).
104. R. Nagpal and A. Garscadden, *J. Appl. Phys.* **75,** 703 (1994), and references therein.
105. H.-X. Wan, J. H. Moore, and J. A. Tossell, *J. Chem. Phys.* **91,** 7340 (1989).
106. O. Sueoka, S. Mori, and A. Hamada, *J. Phys. B* **27,** 1453 (1994).
107. A. Zecca, G. P. Karwasz, and R. S. Brusa, *Phys. Rev. A* **45,** 2777 (1992).
108. A. K. Jain, A. N. Tripathi, and V. H. Smith, *J. Phys. B* **23,** 2869 (1990).
109. J. A. Tossell and J. W. Davenport, *J. Chem. Phys.* **80,** 813 (1984).
110. A. Jain, *J. Chem. Phys.* **86,** 1289 (1987).
111. A. K. Jain, A. N. Tripathi, and A. Jain, *J. Phys. B* **20,** L389 (1987).
112. J. Yuan, *J. Phys. B* **21,** 2737 (1988).
113. J. Yuan, *J. Phys. B* **22,** 2589 (1989).
114. A. Jain and D. G. Thompson, *J. Phys. B* **20,** 2861 (1987).
115. F. A. Gianturco, L. C. Pantano, and S. Scialla, *Phys. Rev. A* **36,** 557 (1987).
116. F. A. Gianturco, V. Di Martino, and A. Jain, *Nuovo Cimento D* **14,** 411 (1992).
117. A. Jain, *Phys. Rev. A* **44,** 772 (1991).
118. C. Winstead and V. McKoy, *Phys. Rev. A* **42,** 5357 (1990).
119. C. Winstead, H. P. Pritchard, and V. McKoy, *J. Chem. Phys.* **101,** 338 (1994).

120. W. Sun, C. W. McCurdy, and B. H. Lengsfield, *Phys. Rev. A* **45,** 6323 (1992).
121. H. Soejima and Y. Nakamura, *J. Vac. Sci. Tech. A* **11,** 1161 (1993).
122. M. A. Dillon, L. Boesten, H. Tanaka, M. Kimura, and H. Sato, *J. Phys. B* **26,** 3147 (1993).
123. G. P. Karwasz, *J. Phys. B* **28,** 1301 (1995).
124. A. Jain, K. L. Baluja, V. Di Martino, and F. A. Gianturco, *Chem. Phys. Lett.* **183,** 34 (1991).
125. K. L. Baluja, A. Jain, V. Di Martino, and F. A. Gianturco, *Europhys. Lett.* **17,** 139 (1992).
126. P. Kumar, A. K. Jain, and A. N. Tripathi, *J. Phys. B* **28,** L387 (1995).
127. C. Winstead, P. G. Hipes, M. A. P. Lima, and V. McKoy, *J. Chem. Phys.* **94,** 5455 (1991).
128. O. Sueoka, S. Mori, and Y. Katayama, *J. Phys. B* **20,** 3237 (1987).
129. C. Szmytkowski, K. Maciąg, G. Karwasz, and D. Filipović, *J. Phys. B* **22,** 525 (1989).
130. X. Ling, M. T. Frey, K. A. Smith, and F. B. Dunning, *Phys. Rev. A* **48,** 1252 (1993).
131. M. Ben Arfa and M. Tronc, *J. Phys. B* **18,** L269 (1985).
132. M. Ben Arfa and M. Tronc, *J. Chim. Phys.* **85,** 889 (1988).
133. M. Furlan, M.-J. Hubin-Franskin, J. Delwiche, and J. E. Collin, *J. Chem. Phys.* **92,** 213 (1990).
134. R. J. Gulley, M. J. Brunger, and S. J. Buckman, *J. Phys. B* **25,** 2433 (1992).
135. D. T. Alle, R. J. Gulley, S. J. Buckman, and M. J. Brunger, *J. Phys. B* **25,** 1533 (1992).
136. T. N. Rescigno, B. H. Lengsfield, C. W. McCurdy, and S. D. Parker, *Phys. Rev. A* **45,** 7800 (1992).
137. A. Jain, *J. Phys. B* **21,** 905 (1988).
138. A. Jain and D. G. Thompson, *J. Phys. B* **16,** 2593 (1983).
139. F. A. Gianturco, *J. Phys. B* **24,** 4627 (1991).
140. J. Yuan and Z. Zhang, *Phys. Rev. A* **45,** 4565 (1992).
141. H. P. Pritchard, M. A. P. Lima, and V. McKoy, *Phys. Rev. A* **39,** 2392 (1989).
142. J. A. Tossell, J. H. Moore, and J. C. Giordan, *Inorg. Chem.* **24,** 1100 (1985).
143. M. Ben Arfa and M. Tronc, *Chem. Phys.* **155,** 143 (1991).
144. J. Yuan and Z. Zhang, *Z. Phys. D* **28,** 207 (1993).
145. C. Winstead, Q. Sun, V. McKoy, J. L. da Silva Lino, and M. A. P. Lima, *Z. Phys. D* **24,** 141 (1992).
146. K. Jung, Th. Antoni, R. Müller, K.-H. Kochem, and H. Ehrhardt, *J. Phys. B* **15,** 3535 (1982).
147. F. A. Gianturco, *J. Phys. B* **24,** 3837 (1991).
148. R. Greer and D. Thompson, *J. Phys. B* **27,** 3533 (1994).
149. G. Seng and F. Linder, *J. Phys. B* **9,** 2539 (1976).
150. T. W. Shyn, S. Y. Cho, and T. E. Cravens, *Phys. Rev. A* **38,** 678 (1988).
151. M. Furlan, M.-J. Hubin-Franskin, J. Delwiche, and J. E. Collin, *J. Chem. Phys.* **95,** 1671 (1991).
152. D. Cvejanović, L. Andrić, and R. I. Hall, *J. Phys. B* **26,** 2899 (1993).
153. A. Jain and D. G. Thompson, *J. Phys. B* **16,** L347 (1983).

154. T. Nishimura and Y. Itikawa, *J. Phys. B* **28,** 1995 (1995).
155. C. Szmytkowski, *Chem. Phys. Lett.* **136,** 363 (1987).
156. H. Nishimura and K. Yano, *J. Phys. Soc. Jpn.* **57,** 1951 (1988).
157. Z. Sağlam and N. Aktekin, *J. Phys. B* **23,** 1529 (1990); **24,** 3491 (1991).
158. A. Zecca, G. Karwasz, S. Oss, R. Grisenti, and R. S. Brusa, *J. Phys. B* **20,** L133 (1987).
159. E. Brüche, *Ann. Phys. (Leipzig)* **1,** 93 (1929).
160. A. Danjo and H. Nishimura, *J. Phys. Soc. Jpn.* **54,** 1224 (1985).
161. T. W. Shyn and S. Y. Cho, *Phys. Rev. A* **36,** 5138 (1987).
162. W. M. Johnstone and W. R. Newell, *J. Phys. B* **24,** 3633 (1991).
163. H. Sato, M. Kimura, and K. Fujima, *Chem. Phys. Lett.* **145,** 21 (1988).
164. Y. Okamoto, K. Onda, and Y. Itikawa, *J. Phys. B* **26,** 745 (1993).
165. L. M. Brescansin, M. A. P. Lima, T. L. Gibson, V. McKoy, and W. M. Huo, *J. Chem. Phys.* **85,** 1854 (1986).
166. L. E. Machado, M.-T. Lee, L. M. Brescansin, M. A. P. Lima, and V. McKoy, *J. Phys. B* **28,** 467 (1995).
167. R. R. Lucchese, G. Raşeev, and V. McKoy, *Phys. Rev. A* **25,** 2572 (1982).
168. T. N. Rescigno and B. H. Lengsfield, *Z. Phys. D* **24,** 117 (1992).
169. V. F. Sokolov and Y. A. Sokolova, *Pisma Zh. Tech. Fiz.* **7,** 627 (1981) [*Sov. Tech. Phys. Lett.* **7,** 268 (1981).]
170. C. Szmytkowski and K. Maciąg, *Chem. Phys. Lett.* **129,** 321 (1986).
171. K. Rohr, *J. Phys. B* **11,** 4109 (1978).
172. R. J. Gulley, M. J. Brunger, and S. J. Buckman, *J. Phys. B* **26,** 2913 (1993).
173. F. A. Gianturco and D. G. Thompson, *J. Phys. B* **13,** 613 (1980).
174. A. Jain and D. G. Thompson, *J. Phys. B* **17,** 443 (1983).
175. B. H. Lengsfield, T. N. Rescigno, C. W. McCurdy, and S. Parker, unpublished, 1992; see [172].
176. L. E. Machado, E. P. Leal, M.-T. Lee, and L. M. Brescansin, *J. Mol. Struct. (Theochem)* **335,** 37 (1995).
177. M. J. W. Boness, I. W. Larkin, J. B. Hasted, and L. Moore, *Chem. Phys. Lett.* **1,** 292 (1967).
178. L. Sanche and G. Shulz, *J. Chem. Phys.* **58,** 479 (1973).
179. P. D. Burrow and K. D. Jordan, *Chem. Phys. Lett.* **36,** 594 (1975).
180. C. W. Duncan and I. C. Walker, *J. Chem. Soc. Faraday Trans. II* **68,** 1800 (1972).
181. I. C. Walker, A. Stamatovic, and S. F. Wong, *J. Chem. Phys.* **69,** 5532 (1978).
182. E. Brüche, *Ann. Phys. (Leipzig)* **2,** 909 (1929).
183. O. Sueoka and S. Mori, *J. Phys. B,* **19,** 4035 (1986).
184. H. Nishimura and H. Tawara, *J. Phys. B* **244,** L363 (1991), and private communication.
185. Y. H. Jiang, J. F. Sun, and L. Wan, *Z. Phys. D* **34,** 29 (1995).
186. B. I. Schneider, T. N. Rescigno, B. H. Lengsfield, and C. W. McCurdy, *Phys. Rev. Lett.* **66,** 2728 (1991).
187. K. D. Jordan and P. D. Burrow, *J. Am. Chem. Soc.* **102,** 6882 (1980).
188. C. Winstead, Q. Sun, and V. McKoy, *J. Chem. Phys.* **96,** 4246 (1992).

189. A. E. Howard and S. W. Staley, in D. G. Truhlar, Ed., *Resonances in Electron-Molecule Scattering, van der Waals Complexes, and Reactive Chemical Dynamics,* ACS Symposium Series **263,** American Chemical Society, Washington, 1984, p. 183.
190. M. Allan, *J. Am. Chem. Soc.* **115,** 6418 (1993).
191. M. Allan, in [12], p. 105.
192. L. Boesten, H. Tanaka, M. Kubo, H. Sato, M. Kimura, M. A. Dillon, and D. Spence, *J. Phys. B* **23,** 1905 (1990).
193. D. L. McCorkle, L. G. Christophorou, D. V. Maxey, and J. G. Carter, *J. Phys. B* **11,** 3067 (1978).
194. H. Tanaka, L. Boesten, D. Matsunaga, and T. Kudo, *J. Phys. B* **21,** 1255 (1988).
195. W. Sun, C. W. McCurdy, and B. H. Lengsfield, *J. Chem. Phys.* **97,** 5480 (1992).
196. M. A. Dillon, L. Boesten, H. Tanaka, M. Kimura, and H. Sato, *J. Phys. B* **27,** 1209 (1994).
197. L. Boesten, M. A. Dillon, H. Tanaka, M. Kimura, and H. Sato, *J. Phys. B* **27,** 1845 (1994).
198. R. A. Bonham, *Jpn. J. Appl. Phys.* **33,** 4157 (1994).
199. L. Boesten, H. Tanaka, A. K. Kobayashi, M. A. Dillon, and M. Kimura, *J. Phys. B* **25,** 1607 (1992).
200. A. Mann and F. Linder, *J. Phys. B* **25,** 533 (1992).
201. A. Mann and F. Linder, *J. Phys. B* **25,** 545 (1992).
202. R. K. Jones, *J. Chem. Phys.* **84,** 813 (1986).
203. H. Nishimura, unpublished, quoted in [199].
204. W. M. Huo, *Phys. Rev. A* **38,** 3303 (1988).
205. C. Winstead, Q. Sun, and V. McKoy, *J. Chem. Phys.* **98,** 1105 (1993).
206. A. P. P. Natalense, M. H. F. Bettega, L. G. Ferreira, and M. A. P. Lima, *Phys. Rev. A* **52,** R1 (1995).
207. F. A. Gianturco, R. R. Lucchese, and N. Sanna, *J. Chem. Phys.* **100,** 6464 (1994).
208. H. F. Winters, J. W. Coburn, and E. Kay, *J. Appl. Phys.* **48,** 4973 (1977).
209. H. F. Winters and M. Inokuti, *Phys. Rev. A* **25,** 1420 (1982).
210. J. A. Sheehy and W. M. Huo, unpublished; see [207].
211. F. A. Gianturco, R. R. Lucchese, and N. Sanna, *J. Chem. Phys.* **102,** 5743 (1995).
212. R. E. Kennerly, R. A. Bonham, and M. McMillan, *J. Chem. Phys.* **70,** 2039 (1979).
213. T. N. Rescigno, *Phys. Rev. A* **52,** 329 (1995).
214. C. Winstead and V. McKoy, unpublished.
215. H. P. Pritchard, V. McKoy, and M. A. P. Lima, *Phys. Rev. A* **41,** 546 (1990).
216. M.-T. Lee, S. E. Michelin, L. E. Machado, and L. M. Brescansin, *J. Phys. B* **26,** L203 (1993).
217. M.-T. Lee, S. E. Michelin, T. Kroin, L. E. Machado, and L. M. Brescansin, *J. Phys. B* **28,** 1859 (1995).
218. T. J. Gil, T. N. Rescigno, C. W. McCurdy, and B. H. Lengsfield, *Phys. Rev. A* **49,** 2642 (1994).
219. K. Becker, B. Stumpf, and G. Schulz, *Chem. Phys. Lett.* **73,** 102 (1980).
220. T. N. Rescigno, C. W. McCurdy, and B. I. Schneider, *Phys. Rev. Lett.* **63,** 248 (1989).

221. T. N. Rescigno, B. H. Lengsfield, and C. W. McCurdy, *Phys. Rev. A* **41,** 2462 (1990).
222. Q. Sun, C. Winstead, V. McKoy, J. S. E. Germano, and M. A. P. Lima, *Phys. Rev. A* **46,** 2462 (1992).
223. E. H. van Veen, W. L. van Dijk, and H. H. Brongersma, *Chem. Phys.* **16,** 337 (1976).
224. G. J. Verhaart and H. H. Brongersma, *Chem. Phys. Lett.* **71,** 345 (1980).
225. D. C. Cartwright, S. Trajmar, W. Williams, and D. L. Huestis, *Phys. Rev. Lett.* **27,** 704 (1971).
226. A. J. Merer and R. S. Mulliken, *Chem. Rev.* **69,** 639 (1969).
227. R. S. Mulliken, *J. Chem. Phys.* **66,** 2448 (1977).
228. Q. Sun, C. Winstead, V. McKoy, and M. A. P. Lima, *J. Chem. Phys.* **96,** 3531 (1992).
229. M. Allan, *Chem. Phys. Lett.* **225,** 156 (1994).
230. D. E. Love and K. D. Jordan, *Chem. Phys. Lett.* **235,** 479 (1995).
231. J. E. Jackson and L. C. Allen, *J. Am. Chem. Soc.* **106,** 591 (1984).
232. D. Feller and E. R. Davidson, *J. Am. Chem. Soc.* **109,** 4133 (1987).
233. O. Schafer, M. Allan, G. Szeimies, and M. Sanktjohanser, *J. Am. Chem. Soc.* **114,** 8180 (1992).
234. M. Allan, *J. Chem. Phys.* **101,** 844 (1994).
235. C. Winstead, Q. Sun, and V. McKoy, *J. Chem. Phys.* **97,** 9483 (1992).
236. C. Winstead, Q. Sun, V. McKoy, J. L. S. Lino, and M. A. P. Lima, *J. Chem. Phys.* **98,** 2132 (1993).
237. T. J. Gil, B. H. Lengsfield, C. W. McCurdy, and T. N. Rescigno, *Phys. Rev. A* **49,** 2251 (1994).
238. M. A. Dillon, R.-G. Wang, and D. Spence, *J. Chem. Phys.* **80,** 5581 (1984).
239. G. R. J. Williams and D. Poppinger, *Mol. Phys.* **30,** 1005 (1975).
240. M. S. Gordon, *Chem. Phys. Lett.* **44,** 507 (1976).
241. M. S. Gordon, *Chem. Phys. Lett.* **52,** 161 (1977).
242. M. S. Gordon and J. W. Caldwell, *J. Chem. Phys.* **70,** 5503 (1979).
243. C. Larrieu, D. Liotard, M. Chaillet, and A. Dargelos, *J. Chem. Phys.* **88,** 3848 (1988).
244. L. Chantranupong, G. Hirsch, R. J. Buenker, and M. A. Dillon, *Chem. Phys.* **170,** 167 (1993).
245. W. Sun, M. A. Morrison, W. A. Isaacs, W. K. Trail, D. T. Alle, R. J. Gulley, M. J. Brennan, and S. J. Buckman, *Phys. Rev. A* **52,** 1229 (1995).
246. Q. Sun, C. Winstead, and V. McKoy, *Phys. Rev. A* **46,** 6987 (1992).
247. J. Zobel, U. Mayer, K. Jung, H. Ehrhardt, H. Pritchard, C. Winstead, and V. McKoy, *J. Phys. B* **29,** 839 (1996).
248. J. Perrin, J. P. M. Schmitt, G. de Rosny, B. Drevillon, J. Huc, and A. Lloret, *Chem. Phys.* **73,** 383 (1982).

TIME-DEPENDENT SEMICLASSICAL MECHANICS

MIGUEL ANGEL SEPÚLVEDA*

Institute for Fundamental Chemistry, 34-4 Takano-Nishihiraki-Cho, Sakyo-ka Kyoto 606, Japan

FRANK GROSSMANN

Universität Freiburg, Fakultät für Physik, D-79104 Freiburg, Germany

CONTENTS

*Present address: Department of Chemistry, Princeton University, Princeton, NJ 08544.

Advances in Chemical Physics, Volume XCVI, Edited by I. Prigogine and Stuart A. Rice.
ISBN 0-471-15652-3

I. INTRODUCTION

During the last 30 years significant efforts have been made to understand and apply classical mechanics in order to gain physical insight into atomic and molecular systems [1–6]. As a result, many methods for the calculation of spectra, eigenfunctions, scattering cross sections, and so on are available. For all these problems it is possible to solve Hamilton's equations of motion for the system under study either analytically or numerically and use the trajectories as a skeleton to construct wave functions and eigenvalues [7]. Generally, one distinguishes two different approaches in this respect: *energy* and *time* domain semiclassics.

Since the early days of the old quantum theory, energy domain WKB (Wentzel-Keller-Brillouin) theory has been a valuable tool for understanding a variety of physical phenomena and bridging the gap between the classical and quantum worlds. First devised for one-degree-of-freedom systems it was later corrected and completed in the Einstein–Brillouin–Keller theory for the quantization of quasi-integrable systems [8], and by Maslov [9] during the 1960s with his seminal work on multidimensional asymptotic approximations.

In addition, in the late 1960s Gutzwiller gave access, through his trace formula [10], to a completely new kind of systems: those with fully chaotic behavior. The lack of invariant tori in phase space makes EBK (Einstein Brillouin Keller) theory inapplicable for classically chaotic systems. Gutzwiller, however, realized that even for these systems, certain invariant structures capable of supporting quantization do remain: unstable periodic orbits. The trace formula is a sum over unstable periodic orbits describing the energy Green function for the system, and exhibits singularities on the energy axis at the eigenvalues of the system.

For a long time the energy-domain approach overshadowed time-dependent approaches, which lack any kind of simple quantization condition. Even the use of the full time-dependent Schrödinger equation, before the introduction of FFT (fast Fourier-transform), DVR (discrete variable representation) and Chebyshev techniques (see e.g. [11]), was more of an academic issue than a practical one. Variational methods reigned as tools for calculating spectra and other stationary properties. Based on the increasing interest in Feynman's concept of path integrals [12], however, a new impetus was given to the formulation of quantum problems in the time domain and to the understanding of the classical–quantum connection. In fact time-dependent semiclassics is based on the $\hbar \rightarrow 0$ limit of the quantum propagator, $\langle \vec{q} | \exp(-i\mathcal{H}t/\hbar) | \vec{q}\,' \rangle$, which can also be written as a path integral [13].

It is historically interesting that the semiclassical propagator had already been introduced by Van Vleck in the late 1920s long before the advent of path integrals [14]. For quite some time it has served merely as an intermediary result in the derivation of the semiclassical energy Green function and consequently also the trace formula. It took about 60 years before the first successful direct applications of this most basic semiclassical object to nontrivial (i.e., chaotic or mixed dynamical systems) appeared in the literature [15–19].

It is the purpose of this work to review some of the progress that has been made in the semiclassical time-domain formulation of nonrelativistic quantum mechanics. We will cover both the physical understanding that can be gained by doing semiclassics and the technical background that is necessary for its application.

The motivation for using classical mechanics to accomplish what exact quantum methods can provide is, in most cases both fundamental and practical. First, when classical trajectories are used to extract the wave function one gains physical insight into aspects of the dynamics that are often hidden by the uncertainty principle. Classical trajectories not only carry the probability amplitude for transitions between different configurations but also the phase information that allows the creation of interference and unique quantum effects. Large densities of states, moreover, often cause great trouble for numerical quantum methods; even with recent advances in computer hardware there is still a limitation on the size of matrices that can be diagonalized. From a more practical point of view, it seems almost absurd to attempt to tackle a system that is known to behave asymptotically using a purely quantum mechanical method. Perhaps approaching the problem from the other "end of the rope," mainly from the perspective of classical mechanics, and adding the phase information required for quantum interference is more efficient.

Going into the energy domain by doing a Laplace transformation on the semiclassical propagator was a necessity due in part to the apparent numerical difficulty of setting up the Van Vleck propagator directly. Perhaps it was also due to the extreme focus given to the derivation of eigenvalues. There are nevertheless many instances, especially in spectroscopy, where the semiclassical eigenvalues are not as important as the construction of wave functions or correlation functions, as we will see. Furthermore, the periodic orbit sum only allows the treatment of integrable systems (see Chapter 17.3 of [1] and references therein) or purely chaotic systems (see Chapter 17.4 of [1] and references therein), and it only provides us with certain eigenvalues of the system. For the calculation of the density of states as a sum over periodic orbits the reader is also referred to [20]. Most molecules, however, do not belong to either class of systems; it is very rare to find a molecule whose classical behavior is completely chaotic or integrable. The most usual case is the so-called mixed dynamical system, in which part of the phase space has a quasi-integrable structure (meaning that its classical orbits can be represented in a Fourier series with a finite number of fundamental frequencies) and other part is chaotic or stochastic. In the latter case, trajectories are not restricted by any fundamental frequencies to explore a finite volume of all of phase space available in the energy shell; instead they appear to explore almost randomly and uniformly all the phase space available. Neither EBK nor periodic orbit sums are developed to treat this kind of system. The Van Vleck propagator, however, has the great advantage of being "dynamics independent." Since it is an asymptotic approximation to the quantum time-evolution operator, it does not rely on any assumption about the nature of the classical system. There are no quantization conditions underlying the Van Vleck propagator, and this makes it perhaps the most widely applicable tool of semiclassical theory.

This review describes several implementations of the semiclassical propagator developed over the last years. First there is "Gaussian wavepacket dynamics" (GWD). This procedure is based on the expansion of the semiclassical propagator using a *single* classical orbit we call the "guiding orbit." It starts at the average position and momentum of the initial state $\phi_0(\mathbf{q})$ and follows the classical dynamics on a time-dependent local quadratic approximation to the true potential. The first account of that method was given in [21]. Since then it has seen many modifications, which have had different levels of success [22, 23]. The most intricate of the alterations of GWD was the extension of the method into complex phase space. This is a complication we want to avoid in this review. The reader is therefore referred to the original literature on that subject [24]. In essence it turns out that simple GWD is very efficient for the propagation of Gaussian wavepackets over relatively short times or in closely harmonic potentials. It has been set up to calculate photoabsorption spectra, photodissociation rates (see e.g., Section 4.3 in [25], and [26]), Raman spectra [27], surface scattering [28], inelastic scattering [21, 29], and so on. In order to overcome the limitations of GWD we will describe a more recent procedure that is based not on a single guiding orbit expansion but on a *multiple* orbit expansion, where each orbit runs now on the real molecular or atomic potential. This latest method, called "cellular dynamics" (CD) [17, 30], perhaps represents the most accurate implementation of the Van Vleck formula, and its applications have been shown to yield precise results up to extremely long times of propagation, making it possible to semiclassically calculate high-resolution spectra of molecules like benzophenone [18], and the spectrum of the Barbanis potential [17]. It seems worthwhile to mention also that, for dissociative systems like the photodissociation dynamics of CO_2 progress has been made using cellular dynamics [31]. Another improvement on GWD, which has proven useful in the study of the stadium billard will not be dealt with explicitly herein and the reader is referred to the literature. This method comes close to doing a full root search and for practical reasons seems to be limited to lower dimensional systems or systems with energy scaling properties [15].

After this introduction a short review of time-dependent concepts in quantum mechanics is given in Section II, which shows how experimentally relevant information can be extracted from the Fourier transform of an auto- or cross-correlation function and how even quantum mechanical eigenfunctions can be uncovered from the dynamics. Section III gives an introduction to the semiclassical propagator through the study of Van Vleck determinants. Sections IV and V are then devoted to review the basic principles of the numerical implementation of semiclassical propagation used in the main part of this paper. The cellular dynamics method, described in Section V, turns out to be the favorable approach because, while it avoids the root search that has hampered the direct application of the semiclassical propagator for so many years, nevertheless includes all the dynamically important parts of classical phase space in the calculation. In Section V.E on benzophenone the excellent agreement between semiclassical and quantum results demonstrates the validity of the cellular dynamics method for the treatment of molecular Hamiltonians whose classical limit exhibit mixed dynamics (i.e., chaos and quasi-integrable motion).

The applications considered so far were characterized by a single potential surface. The time evolution of the initial wave function takes place on a single Born–Oppenheimer surface. In Section VI we will describe an extension of the standard semiclassical tools that makes it possible to overcome this limitation. A great deal of modern nonlinear spectroscopy can then be calculated and interpreted semiclassically through the methodology presented. At the core of this new treatment stands a generalization of the well-known Van Vleck propagator that will allow us to describe the time evolution of a system by consideration of the dynamics in simultaneous electronic states.

Section VII describes how another limitation of the semiclassical methodology: the restriction to conservative systems. We are interested in a (nonharmonic) test system coupled to a bath of harmonic oscillators. For reasons of simplicity the bath is assumed to have a linear spectral density and to be at high temperatures. After averaging over the bath coordinates in the influence functional formalism of dissipative quantum dynamics a semiclassical propagator in the spirit of Van Vleck has been derived for the dynamics of the reduced density matrix of the test system. Using this procedure the vibrational relaxation of a Morse oscillator can be treated using real classical trajectories (instead of the more oftenly used complex paths). The influence of the harmonic bath as characterized by its temperature and the damping strength can be studied.

Another important test of the validity of the Van Vleck formula is summarized in Section VIII, where detailed rendering of eigenstates and eigenvalues in a very complicated system, the stadium billard, will be reviewed. In this system the dynamics is completely chaotic, and phase space does not show the intricate mixture of quasiperiodic and stochastic motion (as in even so benzophenone); the direct application of the Van Vleck propagator provides very accurate results. The issue of when, as a function of time and $\hbar$, the Van Vleck propagator breaks down is nevertheless still unsolved.

Finally, in Section IX two rather new developments will be briefly sketched. First, recent attempts to carry out the Monte Carlo integration over initial phase space of the semiclassical propagator will be reviewed. This is followed by a report on recent progress on the long-standing issue of tunneling and its semiclassical description. In the conclusion we summarize the results presented and give some outlook on possible future research.

II. TIME-DOMAIN FORMULATION OF QUANTUM PROPERTIES

Before dealing with the semiclassical aspects of time-dependent theory in greater depth, let us first review some of the basic relationships between time-dependent and time-independent objects; in particular, we will focus on the extraction of eigenvalues, eigenfunctions, and scattering cross sections from the dynamics. Several applications of time-dependent methods and numerical studies can be found in a recent thematic issue of *Computer Physics Communications* [11].

One of the most prominent uses of the time-dependent approach in molecular applications is the calculation of absorption spectra from the Fourier transform of an overlap, a method that will frequently be used throughout this review. One starts from the autocorrelation function

$$c_{\alpha\alpha}(t) = \langle \Phi_\alpha | \Phi_\alpha(t) \rangle \tag{2.1}$$

of the propagated state $|\Phi_\alpha(t)\rangle$ with itself at $t = 0$, where $|\Phi_\alpha(0)\rangle = \mu|\Omega_g^L\rangle$ is composed of the dipole matrix element times the ground state wave function on a lower potential surface. The initial state is then propagated under the Hamiltonian of the upper potential surface. The absorption spectrum is now given by [5]

$$\Sigma(\omega) = A \int dt \exp(i\omega t) c_{\alpha\alpha}(t) \tag{2.2}$$

with a constant depending on the field strength. For short-time dynamical processes, where $c_{\alpha\alpha}(t)$ vanishes rapidly, this formalism is the method of choice for the calculation of spectra because the time-consuming step is the propagation of the wavepacket. The absorption spectrum shows sharp peaks at the eigenvalues of the system covered by $|\Phi_\alpha(0)\rangle$. The same methodology has recently been used to calculate resonance properties (i.e., local spectra and resonance widths) of reactive scattering systems [32, 33].

Furthermore, eigenfunctions can be also extracted from wavepacket dynamics by Fourier transforming the propagated packet at one of the eigenvalues already resolved

$$\Psi_\nu(x) \sim \int \langle x | \Phi_\alpha(t) \rangle \exp(iE_\nu t/\hbar)\, dt \tag{2.3}$$

For a more detailed exposition on this subject, see for example Section 2.5 of [7].

The most recent progress in the field of time-dependent methodology has been made with the calculation of elements of the S-matrix from wave packet propagation [34]. Let us give a brief description of what has been achieved. For a collinear reactive scattering situation, for example, one propagates a wavepacket $|\Phi_{\alpha\nu}\rangle = |\phi_\alpha\rangle|\Psi_\nu\rangle$, which is given as a product of a Gaussian, $|\phi_\alpha\rangle$ in the asymptotically free direction times the ν-th eigenfunction in the bound motion (in Jacobi coordinates). From its time-dependent overlap with a final state $|\Phi_{\beta\nu'}\rangle$, (cross-correlation function)

$$c_{\beta\nu',\alpha\nu}(t) = \langle \Phi_{\beta\nu'} | \Phi_{\alpha\nu}(t) \rangle \tag{2.4}$$

it is possible to extract a scattering matrix element by Fourier transform and a proper normalization

$$S_{\beta\nu',\alpha\nu}(E) = \frac{(2\pi\hbar)^{-1}}{\eta_\beta^*(E)\eta_\alpha(E)} \int dt \exp(iEt/\hbar) c_{\beta\nu',\alpha\nu}(t) \tag{2.5a}$$

$$\eta_\alpha(E) = \sqrt{\frac{m}{2\pi p}} \int dq \exp\left(-\frac{i}{\hbar}pq\right) \phi_\alpha(q) \tag{2.5b}$$

$$p = \sqrt{2m(E - E_\nu)} \tag{2.5c}$$

For symmetry and computational reasons Tannor and Weeks [34] have formulated their approach by using the Moller states of scattering theory. We were able to take this simplified view by noting that the Moller operator becomes the unit operator when the initial free part of the wave function is located far enough in the asymptotic region. An alternative methodology to determine the scattering matrix uses the final propagated state and has been given by Heller [35]. Let us note that in the Tannor approach one typically calculates a single element of the S-matrix at many different energies by Fourier transform, while in the Heller approach a whole column of the S-matrix can be calculated at specific energies.

We have not given a complete list of quantities that can be calculated in a time-dependent fashion. For example, throughout this review, no use will be made of the correlation function formulation of reaction rates [36]. We want to mention, however, that especially for semiclassical applications, i.e., where the time propagation has been done by using the Van Vleck–Gutzwiller propagator, we have some choice in how to perform the final Fourier transform that extracts the observables. By calculating it by stationary phase we would arrive at the energy-dependent point of view presented most clearly by Miller [6]. Doing the Fourier transform exactly, however, it turns out that the results differ from the stationary phase ones. This can be seen most dramatically in recent studies of the tunneling effect [37, 38], which will be discussed in Section IX. In general it can be said that by refraining from an additional approximation new insights might be gained as compared to the well-known energy domain results.

III. VAN VLECK–GUTZWILLER PROPAGATOR

In all the applications to be described here, the basic step is the solution of an initial condition problem. We will start with the initial state of the system, represented by a wave function ϕ_0 ($\mathbf{q}$), and apply the semiclassical propagator G_{SC} ($\mathbf{q}$, $\mathbf{q}'$; t) to determine the time evolution of the system

$$\phi_t(\mathbf{q}) = \int d\mathbf{q}' G_{SC}(\mathbf{q}, \mathbf{q}'; t)\phi_0(\mathbf{q}') \tag{3.1}$$

It was derived by Van Vleck [14] that, in the asymtotic limit, the modulus square of a transition amplitude like $G_{SC}(\mathbf{q}, \mathbf{q}'; t)$ must correspond to the classical probability density for the system to go from configuration $\mathbf{q}'$ to $\mathbf{q}$ after time t. The classical transition probability densities are given by the Van Vleck determinants, Δ

$$\Delta = \left| \det \frac{\partial^2 S}{\partial \mathbf{q} \partial \mathbf{q}'} \right| \tag{3.2}$$

where $S(\mathbf{q}, \mathbf{q}'; t)$ is the action along the classical trajectory departing from $\mathbf{q}'$ and arriving at $\mathbf{q}$ in time t. As in the path-integral formalism the action S is calculated as the line integral of the Langrangian along the path, in our case a classical orbit

$$S(\mathbf{q}, \mathbf{q}'; t) = \int_0^t \mathcal{L}(\dot{\mathbf{q}}_0, \mathbf{q}_0, t')\, dt' \tag{3.3}$$

The modulus of the semiclassical transition element $G_{SC}(\mathbf{q}, \mathbf{q}'; t)$ is then given as the square root of a classical probability density. Van Vleck also demonstrated that this type of matrix element can carry quantum information in the form of a phase, so that the complex transition element is given by

$$G_{SC}(\mathbf{q}, \mathbf{q}'; t) \sim C^{1/2}\Delta^{1/2}e^{iS/\hbar} \tag{3.4}$$

At first glance the calculation of this semiclassical approximation seems to be very difficult. Little difficulty is presented by C, which is just a normalization constant that will be determined later, but the major obstacle lies in the double-valued boundary condition defining the transition probability. The classical orbit whose action needs to be calculated is restricted by an initial and final condition. Moreover if G_{SC} is going to be used to propagate a wave function, as in Eq. (3.1), the orbit search has to be performed for every pair $(\mathbf{q}, \mathbf{q}')$ and for every time. For this reason, direct applications of the Van Vleck propagator was very rare until we had the computational means to calculate classical orbits and the methodological tools to avoid this "root search" problem.

Before passing on to describe the single-orbit expansion of the Van Vleck propagator (Section IV), we will describe several topics in classical mechanics that form an important part of the semiclassical machinery to be reviewed later.

A. Van Vleck Determinants

We will devote some time to explain Van Vleck's contribution to semiclassics in modern language. Van Vleck determinants establish the link between the classical and quantum probability densities of finding the system in $\mathbf{q}$ after being initially prepared in a certain state α. They are present in the asymptotic limit of the quantum propagator and in the evaluation of averages.

Since the pioneering work of Wentzel, Brillouin, and Kramers [39–41], who formulated the WKB theory for one-dimensional systems, we know that the correspondence principle imposes a connection between the Hamilton–Jacobi equation of classical mechanics and Schrödinger's wave equation. As $\hbar \to 0$ the WKB solution to the wave equation behaves as $\phi(\mathbf{q}) \sim \exp[(S_0 + \hbar S_1 + \ldots)/i\hbar]$, where S_0 is the solution to a Hamilton–Jacobi equation and $\exp(S_1/i)$ is the solution to a classical flow equation. It was not until the 1928 paper of J. H. Van Vleck [14] however, that a more solid foundation for this limit was established. His work was a first step in the calculation of the semiclassical time-evolution operator. The paper was concerned with understanding how the Dirac–Jordan formulas [42, 43] for calculating expectation values of observables and probabilities merge into the analogous classical expressions for large quantum numbers.[1]

Let us take f to be a function of the canonical variables $\mathbf{p}, \mathbf{q}$. According to the transformation theory of Dirac, $\langle \mathbf{q}|\boldsymbol{\alpha}\rangle$ is the probability amplitude associated with the transformation from the $\mathbf{p}, \mathbf{q}$ set of coordinates to the new $\boldsymbol{\alpha}, \boldsymbol{\beta}$. Quantum mechanics postulates that the average value of f when the value of the $\boldsymbol{\alpha}$'s is specified is

$$\langle \boldsymbol{\alpha}|f(\mathbf{p}, \mathbf{q})|\boldsymbol{\alpha}\rangle = \int \cdots \int \langle \boldsymbol{\alpha}|\mathbf{q}\rangle f\left(i\hbar \frac{\partial}{\partial \mathbf{q}}, \mathbf{q}\right) \langle \mathbf{q}|\boldsymbol{\alpha}\rangle \, d\mathbf{q} \qquad (3.5)$$

The values of $\boldsymbol{\beta}$, conjugate to $\boldsymbol{\alpha}$, are taken as unknown and distributed with equal probability, according to the uncertainty principle.

Van Vleck considers the case where the new set of coordinates $\boldsymbol{\alpha}, \boldsymbol{\beta}$ diagonalizes the Hamiltonian. This means that

$$\left[\mathcal{H}\left(i\hbar \frac{\partial}{\partial \mathbf{q}}, \mathbf{q}\right) + i\hbar \frac{\partial}{\partial t}\right] \langle \mathbf{q}|\boldsymbol{\alpha}\rangle = 0 \qquad (3.6)$$

Let us assume $\boldsymbol{\alpha}, \boldsymbol{\beta}$ to be good action-angle variables. The old quantization condition required the action to be an integer multiple of $\hbar$; therefore the

[1] This limit is equivalent to $1/\hbar \to \infty$ and $\hbar \to 0$.

action variable can, in principle, be used to characterize eigenstates. Van Vleck, however, missed a very fundamental point that Einstein had already pointed out in his 1917 paper [8]. Einstein was perhaps the first physicist to recognize that there are two important kinds of classical solutions in systems with two or more degrees of freedom. They are identified today as integrable motion and chaotic motion. Only for the first kind is it possible to identify good action-angle variables capable of quantization in Bohr's fashion; the second kind escapes this quantization scheme. From his original paper, we can only speculate that Van Vleck was not aware, like many other physicists at the time, of the existence of chaotic solutions to the equations of motion. Therefore, strictly speaking Van Vleck's derivation only applies to integrable systems. It was up to Morette, Gutzwiller, and others to complete the derivation of the semiclassical propagator correctly.

In the limit of large quantum numbers, it was already known that the Schrödinger equation transforms into

$$\mathcal{H}\left(\frac{\partial S}{\partial \mathbf{q}}, \mathbf{q}\right) + \frac{\partial S}{\partial t} = 0 \tag{3.7}$$

which is the Hamilton–Jacobi equation of classical mechanics. Van Vleck, however, questioned the validity of the commonly used limit $\langle \mathbf{q}|\boldsymbol{\alpha}\rangle \sim C \exp[S/i\hbar]$; this classical limit of the eigenstates of the system does not transform the Dirac–Jordan formula for the calculation of expectation values into its classical counterpart. According to the quantum postulates the expectation value of f is given by Eq. (3.5), while classically

$$\langle \boldsymbol{\alpha}|f(p, q)|\boldsymbol{\alpha}\rangle = A \int \ldots \int f\left(\frac{\partial S}{\partial \mathbf{q}}, \mathbf{q}\right) \Delta dq_1 \ldots dq_s \tag{3.8}$$

The classical action depends on the coordinates and the specified actions, $S(\mathbf{q}, \boldsymbol{\alpha}; t)$. $\Delta d\mathbf{q}$ gives the classical probability of finding the system prepared in state α around certain coordinate configuration. As defined in Eq. (3.2), Δ is the determinant of the Hessian of S.

Using a simplified one-dimensional argument, still based on Van Vleck's demonstration, we will show that as $\hbar \rightarrow 0$ the transition amplitude $\langle q|\alpha\rangle$ should go as $C^{1/2}\Delta^{1/2} \exp[S/i\hbar]$. A and C are just normalization constants

$$\frac{1}{A} = \int \Delta dq \tag{3.9}$$

Let us start by considering the following *ansatz* for the stationary state

$$\langle q|\alpha\rangle = Ge^{S/i\hbar} \tag{3.10}$$

Both functions $G(q)$ and $S(q, \alpha; t)$ are independent of $\hbar$. In the case of a stationary state, we expect to see a linear time dependence in the exponent. Then G should be time independent and $S \sim Et$. Substitution of Eq. (3.10) in Eq. (3.6) gives the following relationships to zeroth and first order in $\hbar$

$$\hbar^0: \quad \mathcal{H}\left(\frac{\partial S}{\partial q}, q\right) + \frac{\partial S}{\partial t} = 0 \tag{3.11a}$$

$$\hbar^1: \quad \frac{\partial G}{\partial q}\frac{\partial S}{\partial q} + \frac{1}{2}G\frac{\partial^2 S}{\partial q^2} + \frac{\partial G}{\partial t} = 0 \tag{3.11b}$$

The lowest order term, Eq. (3.11a), shows that S corresponds to the classical action, the solution to Hamilton–Jacobi's equation [Eq. (12)]. The eigenenergies of the system can be obtained by quantization of this classical action. Then we will assume that S is evaluated at a certain eigenvalue.

From the first-order term, the time derivative of G drops out because we are taking G to be time independent. Then, for Hamiltonians with a quadratic momentum dependence, we can rewrite the Hamilton–Jacobi equation as

$$\frac{\partial S}{\partial q}\frac{\partial^2 S}{\partial q\,\partial\alpha} + \frac{\partial^2 S}{\partial t\,\partial q} = 0 \tag{3.12}$$

By combining Eqs. (3.12) and (3.11b) we can arrive in a straightforward fashion at a differential equation that defines the prefactor G

$$\frac{\partial G}{\partial q}\Delta = \frac{1}{2}G\frac{\partial \Delta}{\partial q} \tag{3.13}$$

whose solution gives the desired classical probability density term of Van Vleck's asymptotic formula, $G = A^{1/2}\Delta^{1/2}$.

Using the correct limit for the wave function we can apply the quantum operator f to obtain the appropriate semiclassical expectation value

$$f\left(i\hbar\frac{\partial}{\partial q}, q\right)(\Delta^{1/2}e^{S/i\hbar}) = \Delta^{1/2}e^{S/i\hbar}f\left(\frac{\partial S}{\partial q}, q\right) + O(\hbar) \tag{3.14}$$

In the original paper the asymptotic limit for off-diagonal elements like $\langle\alpha_1|f(p, q)|\alpha_2\rangle$ was also contemplated. These terms are proportional to

$\Delta^{1/2} \exp[(S_1 - S_2)/i\hbar]$, where the action difference in the exponent is equivalent to the classical action to go from state α_2 to α_1. Van Vleck's expression is an early precursor of the semiclassical time-evolution operator because the set $\{\alpha, \beta\}$ can be interpreted as a new system of coordinates, e.g., position and momentum, at a later time.

B. Stability Analysis

Classical trajectories are the solutions to a system of differential equations subject to initial value conditions. In a Hamiltonian formulation, the Hamilton function $\mathcal{H}$, interpreted as the total mechanical energy of the system, is the starting point for the construction of the equations of motion. If $\mathbf{q}$ is the position and $\mathbf{p}$ the momentum vector, then Hamilton equations specify their time derivatives

$$\frac{d\mathbf{p}}{dt} = -\nabla_{\mathbf{q}} \mathcal{H} \tag{3.15a}$$

$$\frac{d\mathbf{q}}{dt} = \nabla_{\mathbf{p}} \mathcal{H} \tag{3.15b}$$

Given a set of positions and momentum $(\mathbf{p}_0, \mathbf{q}_0)$ at time t_0, the existence and uniqueness theorems of Hamilton equations tell us that if the solution exists there is one and only one trajectory that passes through a given point in phase space. At a time t the classical trajectory will end at a point $(\mathbf{p}_t, \mathbf{q}_t)$ in phase space.

In semiclassical mechanics it is usually necessary to calculate the classical probability to go from a coordinate location, say $\mathbf{q}_0$, to another $\mathbf{q}_t$, in time t. Section III.A has already presented the idea that the transition amplitude for such processes varies as the square root of its classical probability density, in Dirac notation $\langle \mathbf{q}_t | \mathbf{q}_0 \rangle \sim$ (classical probability density)$^{1/2}$. How does one calculate this classical probability density? First, only classically allowed orbits that go from $\mathbf{q}_0$ to $\mathbf{q}_t$ contribute to the probability amplitude, so in fact one has to consider a sum over classical orbits satisfying the boundary conditions. Second, the classical probability density can be found by examining the dynamics around each of the classical orbits. $\mathbf{q}_0$ and $\mathbf{q}_t$ are fixed and the initial momenta $\mathbf{p}_0$ are decided by the boundary conditions. Nonetheless, we can imagine launching a dense set of orbits at $\mathbf{q}_t$ with initial momenta infinitesimally close to $\mathbf{p}_0$; then all of them will fan out following the central orbit until reaching the neighborhood of $\mathbf{q}_t$. The classical probability density along the orbit in question is therefore proportional to the ratio $|\partial \mathbf{p}_0 / \partial \mathbf{q}_t|$ (Fig. 1).

Partial derivatives of either final position or momentum along a given

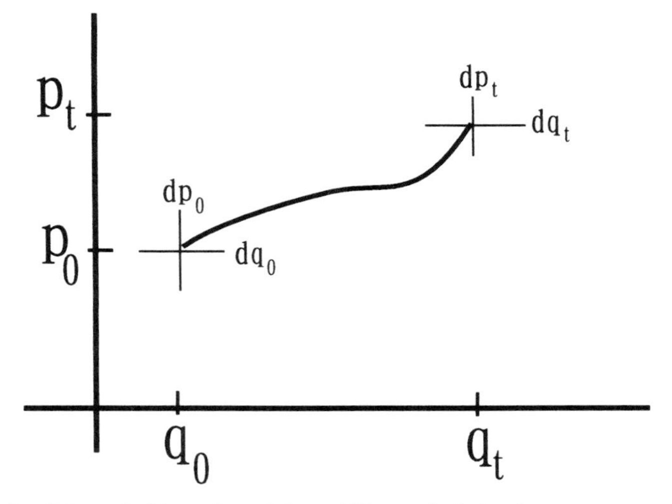

Figure 1. Schematic illustration of the stability analysis in phase space. p_0, q_0 refer to the initial conditions, while p_t, q_t to the final destination of the central guiding orbit. The small variation dp_0 is propagated along the orbit into the diagonal line defined by dp_t, dq_t.

classical orbit appear constantly in semiclassical analysis. These are the objects that allow us to calculate the classical probability densities along particular orbits. The most efficient way to evaluate these derivatives is through a linear stability analysis of Hamilton equations of motion along a given solution or classical orbit.

Consider the orbit connecting ($\mathbf{p}_0$, $\mathbf{q}_0$) and ($\mathbf{p}_t$, $\mathbf{q}_t$) in time t. A small variation about its initial conditions ($\mathbf{p}_0 + \delta\mathbf{p}_0$, $\mathbf{q}_0 + \delta\mathbf{q}_0$) leads to a classical solution whose final phase space location is infinitesimally close to the orbit considered, ($\mathbf{p}_t + \delta\mathbf{p}_t$, $\mathbf{q}_t + \delta\mathbf{q}_t$). Variations like $\delta\mathbf{p}_t$, $\delta\mathbf{q}_t$ can be viewed also as an initial value problem; the "equations of motion" satisfied by these infinitesimal quantities can be derived from Hamilton equations of motion [Eq. (3.15)] by linearization

$$\frac{d}{dt}\begin{pmatrix}\delta\mathbf{p}_t \\ \delta\mathbf{q}_t\end{pmatrix} = \begin{pmatrix} -\partial^2\mathcal{H}/\partial\mathbf{p}^2 & -\partial^2\mathcal{H}/\partial\mathbf{p}\partial\mathbf{q} \\ \partial^2\mathcal{H}/\partial\mathbf{q}\partial\mathbf{p} & \partial^2\mathcal{H}/\partial\mathbf{q}^2 \end{pmatrix}\begin{pmatrix}\delta\mathbf{p}_0 \\ \partial\mathbf{q}_0\end{pmatrix} \tag{3.16}$$

Solution of this linear differential equation leads to a matrix relating the variations of the boundary conditions for the classical orbit

$$\begin{pmatrix}\delta\mathbf{p}_t \\ \delta\mathbf{q}_t\end{pmatrix} = M(t)\begin{pmatrix}\delta\mathbf{p}_0 \\ \delta\mathbf{q}_0\end{pmatrix} \tag{3.17}$$

The matrix $M(t)$, also known as *stability* or *monodromy* matrix, plays a crucial role in semiclassical mechanics. Its elements provide the classical transition probabilities for going from one extreme to the other of the classical orbit, while keeping either the initial momentum or position constant.

The elements of the stability matrix are calculated along the classical orbit $(\mathbf{p}_t, \mathbf{q}_t)$; therefore the evaluation of Eqs. (3.16) requires the simultaneous integration of Hamilton equations of motion, Eqs. (3.15). From the definition of M it is obvious that $M(0) = I$, the identity matrix, and this is the most convenient initial condition to use when integrating the linearized equations of motion numerically.

Every element of the stability matrix will become important when we study the semiclassical Green function, or time-evolution operator, in different representations Section III.E. So far we have only referred to the coordinate representation form of the propagator, $\langle \mathbf{q}_t | \mathbf{q}_0 \rangle$. Our previous analysis and the stability matrix enables us to recast the Van Vleck propagator as $\langle \mathbf{q}_t | \mathbf{q}_0 \rangle \sim |M_{21}(t)|^{-1/2} \exp(iS/\hbar)$.

Let us assume now that a classical orbit $(\mathbf{p}_0, \mathbf{q}_0) \rightarrow (\mathbf{p}_t, \mathbf{q}_t)$ and its stability matrix at time t are given. Then the position-momentum at time t of any other orbit, say $(\mathbf{p}, \mathbf{q})$, initially launched at $(\mathbf{p}', \mathbf{q}')$ (in the vicinity of $(\mathbf{p}_0, \mathbf{q}_0)$) can be approximated in terms of the following linear expansion

$$\mathbf{p} \approx M_{11}(t)\,(\mathbf{p}' - \mathbf{p}_0) + M_{12}(t)\,(\mathbf{q}' - \mathbf{q}_0) \tag{3.18a}$$

$$\mathbf{q} \approx M_{21}(t)\,(\mathbf{p}' - \mathbf{p}_0) + M_{22}(t)\,(\mathbf{q}' - \mathbf{q}_0) \tag{3.18b}$$

The validity of this linear approximation depends on the proximity of $(\mathbf{p}', \mathbf{q}')$ to $(\mathbf{p}_0, \mathbf{q}_0)$, the magnitude of the stability matrix elements, and on the kind of dynamics under study, i.e., whether the classical orbits are solutions to an integrable, quasi-integrable or nonintegrable system.

Another interesting property of the matrix $M(t)$ is its group property. Continuing with the classical path $(\mathbf{p}_0, \mathbf{q}_0) \rightarrow (\mathbf{p}_t, \mathbf{q}_t)$, let us select a point along the orbit at half time during the propagation, $(\mathbf{p}_{t/2}, \mathbf{q}_{t/2})$. Using the same definition Eq. (3.17) we can introduce now two stability matrices, $M_1(t/2)$ and $M_2(t/2)$ such that

$$\begin{pmatrix} d\mathbf{p}_{t/2} \\ d\mathbf{q}_{t/2} \end{pmatrix} = M_1(t/2) \begin{pmatrix} d\mathbf{p}_0 \\ d\mathbf{q}_0 \end{pmatrix} \tag{3.19a}$$

$$\begin{pmatrix} d\mathbf{p}_t \\ d\mathbf{q}_t \end{pmatrix} = M_2(t/2) \begin{pmatrix} d\mathbf{p}_{t/2} \\ d\mathbf{p}_{t/2} \end{pmatrix} \tag{3.19b}$$

where $M_i(0)$ $i = 1, 2$ are unity.

The total stability matrix satisfies then the following relationship in terms of the M_1, M_2

$$M(t) = M_2(t/2) \cdot M_1(t/2) \tag{3.20}$$

A similar relation holds for propagations backwards in time. The stability matrix has an inverse, $M^{-1}(t) = M(-t)$, with $M(-t) \cdot M(t) = I$. The inverse stability matrix can be calculated either by integrating the equations of motion backwards in time, along with the linearized equations Eq. (3.16), or by obtaining the inverse of the stability matrix evolved forwards in time.

The linearized analysis just sketched comprises a huge area in the theory of differential equations, and there are several good reviews specifically on classical mechanics and dynamical systems [44]. We will not treat the subject of stability analysis in greater depth, and refer the reader to several of the excellent standard books available on the subject. The definition of the stability matrix and its ability to expand any orbit about a central guiding path will be sufficient to understand the work presented in this review.

C. Action Expansions

Together with the stability matrix, $M(t)$, the next important classical object necessary in the construction of semiclassical objects is the action, S; defined as the line integral of the classical Lagrangian of the system along the classical orbit

$$S(\mathbf{q}_2, \mathbf{q}_1, t) = \int_{\mathbf{q}_1}^{\mathbf{q}_2} \mathcal{L}(\mathbf{q}, \mathbf{q}; \tau)\, d\tau \tag{3.21}$$

where the classical Langrangian in most common situations is the difference between the kinetic energy, $T = \mathbf{p}^2/2m$, and the potential energy $V(\mathbf{q})$. This function is an extremum along classical orbits. The Lagrangian depends on positions, velocities, and time; while the Hamiltonian function is formulated in terms of momenta, positions, and time. Both functions can be related through the following well-known relationships

$$p_i = (\partial\mathcal{L}/\partial\dot{q}_i) \tag{3.22a}$$

$$\mathcal{H}(\mathbf{p}, \mathbf{q}, t) = \sum \dot{\mathbf{q}}_i(\partial\mathcal{L}/\partial\dot{\mathbf{q}}_i) - \mathcal{L} \tag{3.22b}$$

The expansion of the classical action in terms of initial conditions or the extreme positions of a given classical orbit is a crucial step in the development of numerical and analytical schemes to apply the Van Vleck–Gutzwiller propagator. For almost any atomic or molecular Hamiltonian, the exact dependency of the action in terms of the position and momenta is not known. If it were, this would be equivalent to solving the equations of motion exactly. However, we may have knowledge about one or several classical orbits of the system; then the action can be expanded, at least locally, in terms of the given set of orbits.

Consider a classical orbit in a one-degree-of-freedom system that starts at $q' = q_0$ and after time t ends up at $q = q_t$. We will call this orbit a guiding orbit. Assume that the classical action along this path is S_0. Any small variation of the extreme positions of the guiding orbit leads to an accumulation of action. It is easy to show that this variation is

$$dS = p'\,dq' + p\,dq \tag{3.23}$$

Let us say now that we want to expand the action of any trajectory in the neighborhood of the guiding orbit in terms of S_0 and the stability properties of the guiding orbit. Then

$$\begin{aligned} S \approx S_0 &+ \left(\frac{\partial S}{\partial q'}\right)(q' - q_0) + \left(\frac{\partial S}{\partial q}\right)(q - q_t) + \frac{1}{2}\left(\frac{\partial^2 S}{\partial (q')^2}\right)(q' - q_0)^2 \\ &+ \frac{1}{2}\left(\frac{\partial^2 S}{\partial q^2}\right)(q - q_t)^2 + \left(\frac{\partial^2 S}{\partial q'\,\partial q}\right)(q' - q_0)(q - q_t) \end{aligned} \tag{3.24}$$

The partial derivatives of the action are calculated along the guiding orbit. They are easily recast in terms of the elements of the stability matrix M for this system

$$\begin{pmatrix} dp \\ dq \end{pmatrix} = M \begin{pmatrix} dp' \\ dq' \end{pmatrix} \tag{3.25}$$

It is implicitly assumed that the variables held constant for the evaluation of the partial derivatives in Eq. (3.24) are the initial and final positions, therefore

$$\left(\frac{\partial S}{\partial q'}\right) = -p_0 \qquad \left(\frac{\partial S}{\partial q}\right) = p_t \tag{3.26a}$$

$$\left(\frac{\partial^2 S}{\partial (q')^2}\right) = \frac{M_{22}}{M_{21}} \qquad \left(\frac{\partial^2 S}{\partial q^2}\right) = \frac{M_{11}}{M_{21}} \tag{3.26b}$$

$$\left(\frac{\partial^2 S}{\partial q\,\partial q'}\right) = -\frac{1}{M_{21}} \tag{3.26c}$$

These relationships provide a convenient way to evaluate the coefficients of the expansion in terms of the stability matrix along the guiding orbit. As we discussed earlier, the numerical integration of the stability matrix must be carried out simultaneously with the integration of Hamilton equations of motion.

There are certain situations in which an expansion of the action in terms of the initial and final positions of a guiding orbit is not desirable. Instead, we look to express the action as a function of only initial values

$$S \approx S_0 + \left(\frac{\partial S}{\partial q'}\right)(q' - q_0) + \left(\frac{\partial S}{\partial p'}\right)(p' - p_0) + \frac{1}{2}\left(\frac{\partial^2 S}{\partial (q')^2}\right)(q' - q_0)^2 + \frac{1}{2}\left(\frac{\partial^2 S}{\partial (p')^2}\right)(p' - p_0)^2 + \left(\frac{\partial^2 S}{\partial q'\, \partial p'}\right)(q' - q_0')(p' - p_0) \quad (3.27)$$

where, p_0 and q_0 are the initial momentum and position of the guiding orbit.

In Eq. (3.24) the action S was represented using $\{\mathbf{q}', \mathbf{q}\}$ as independent variables. Now s is considered a function of $\{\mathbf{q}', \mathbf{p}'\}$, therefore the meaning of partial derivation changes accordingly from Eq. (3.24) to Eq. (3.27). As with Eq. (3.26), it is useful to cast the partial derivatives in terms of the elements of the stability matrix. Substitution of dq in Eq. (3.23) gives

$$dS = (pM_{22} - p')\, dq' + pM_{21}\, dp' \quad (3.28)$$

Consequently the parameters of the quadratic expansion in Eq. (3.27) take now the following form:

$$\left(\frac{\partial S}{\partial q'}\right) = p_t M_{22} - p_0 \qquad \left(\frac{\partial S}{\partial p'}\right) = p_t M_{21} \quad (3.29a)$$

$$\left(\frac{\partial^2 S}{\partial (q')^2}\right) = M_{22} M_{12} \qquad \left(\frac{\partial^2 S}{\partial (p')^2}\right) = M_{11} M_{21} \quad (3.29b)$$

$$\left(\frac{\partial^2 S}{\partial q\, \partial p'}\right) = M_{12} M_{21} \quad (3.29c)$$

Notice that the partial derivatives are still evaluated along the guiding classical orbit.

Both expansions allow representation of the action function along a "tube" in phase space localized along the chosen quiding orbit. In general, this tube of convergence shrinks as a function of time; the rate depends on the nature of the guiding orbit and of the system. It depends on whether the orbit is stable (integrable phase space) or unstable (nonintegrable, dissociative, or chaotic phase space). Therefore caution will be required in using either of the action expansions for integrating the semiclassical propagator. For a matrix version of the exposition given above we refer the reader to the appendix of [45].

D. Van Vleck–Gutzwiller Propagator

The semiclassical limit of the time-evolution operator, named after Van Vleck, was rederived and corrected by Gutzwiller in 1967 [46] during his studies of the semiclassical hydrogen atom [1, 47]. His derivation was based on the asymptotic evaluation of Feynman's path integral representation of the propagator. Apparently, the first of such derivations was due to Morette [13]; unfortunately she failed to resolve the problem of extending the validity of the semiclassical formula beyond the first singularity (caustic). It was Gutzwiller's use of Morse's theorem [48] that allowed him to arrive at the correct expression for the semiclassical propagator. A very rigorous account of the mathematical foundation of the semiclassical propagator can be found in [49].

To illustrate the connection between path integrals and semiclassical approximations, we will give an outline of a derivation of the Van Vleck–Gutzwiller formula (see also [1]).

Let us consider the representation of the quantum propagator $\langle q|\exp[-i\mathcal{H}t/\hbar]|q'\rangle$ in terms of a Feynman path integral

$$\lim_{N\to\infty}\left\{\prod_{k=1}^{N}\left[\frac{m}{2\pi i\hbar(t_k - t_{k-1})}\right]\right\}^{1/2}\int dq_1\, dp_2 \ldots dp_{N-1}\exp\left(\frac{i}{\hbar}S_N\right) \tag{3.30}$$

where

$$S_N = \sum_{k=1}^{N}(t_k - t_{k-1})\,\mathcal{L}\left(\frac{q_k - q_{k-1}}{t_k - t_{k-1}}, q_{k-1}, t_{k-1}\right) \tag{3.31}$$

In the limit $N \to \infty$ (or $t_k - t_{k-1} \to 0$), the function S_N transforms into the integral of the classical Lagrangian $\mathcal{L}$ along the piecewise linear path $q', q_1, \ldots, q$. This path is not a classical one, however. The integrals run over *all* possible paths that connect positions q' and q. However, as $\hbar \to 0$, the contributions of the nonclassical paths start to cancel out because of their rapid variation in phase. This can be easily seen by evaluating the $(N - 1)$-fold integral by stationary phase approximation.

The stationary phase approximation is concerned with the integration of oscillatory functions like

$$I = \int \exp[i\lambda\phi(\mathbf{x})]\, d\mathbf{x} \tag{3.32}$$

When λ is a large number, the integrand in Eq. (3.32) becomes highly oscillatory. Then it is a good approximation to consider only the regions of $\mathbf{x}$ where the exponent is stationary, $\phi'(\mathbf{x}_0) = 0$. Under these conditions it can be easily proven that

$$I \approx \left(\frac{2\pi}{\lambda}\right)^{n/2} |\det \phi''(\mathbf{x}_0)|^{1/2} \exp\left(i\lambda\phi(\mathbf{x}_0) + \frac{i\pi}{4}\, \mathrm{Sgn}\phi''(\mathbf{x}_0)\right) \tag{3.33}$$

where n is the dimensionality of the integral. The Hessian of the exponent at the stationary point, $\phi''(\mathbf{x}_0)$, has in principle ν_+ positive and ν_- negative eigenvalues. The signature of this matrix, Sgn, is the difference $\nu_+ - \nu_-$. Alternatively one can write

$$I \approx \left(\frac{2\pi i}{\lambda}\right)^{n/2} |\det \phi''(\mathbf{x}_0)|^{1/2} \exp\left(i\lambda\phi(\mathbf{x}_0) - \frac{i\pi}{2}\nu_-\right) \tag{3.34}$$

Applying then this formula to the $(N - 1)$-fold integral in Eq. (3.30) (now the large parameter is $1/\hbar$) one gets

$$\lim_{N\to\infty} \left\{ \prod_{k=1}^{N} \left[\frac{m}{2\pi i\hbar(t_k - t_{k-1})}\right]\right\}^{1/2} \left(\frac{2\pi\hbar i}{|\delta^2 S|}\right)^{(N-1)/2} \exp\left(\frac{i}{\hbar} S - \frac{i\pi}{2}\nu_-\right) \tag{3.35}$$

The "stationary point" for the path integral is a classical orbit. S is the action along that classical path that now connects q' and q in time t.

At this stage we arrive at the two most sticky points of the derivation. First, it is necessary to show that the prefactor is equal to the Van Vleck determinant times a constant

$$\lim_{N\to\infty} \left\{ \prod_{k=1}^{N} \left[\frac{m}{2\pi i\hbar(t_k - t_{k-1})}\right]\right\}^{1/2} \left(\frac{2\pi\hbar i}{|\delta^2 S_N|}\right)^{(N-1)/2}$$
$$= (2\pi i\hbar)^{-1/2} \left|\frac{\partial^2 S}{\partial q\, \partial q'}\right|^{1/2} \tag{3.36}$$

The derivation of Eq. (3.36) was accomplished even before publication of Gutzwiller's 1967 paper [13]. In the case of a simple quantum map it can be shown very simply [50]. The determination of the number of negative eigenvalues was not resolved, however, until Gutzwiller applied Morse's theorem, which says that the matrix of $\delta^2 S$ reduces its rank by one each time the classical orbit goes through a focal point. A focal point, by analogy with geometric optics, is where the stationary phase approximation diverges due to the singularity in $|\partial^2 S/\partial q\, \partial q'|$. There, the classical probability goes to infinity because myriads of orbits focus on a very small region of configuration space. Before Gutzwiller's work it was not known how to carry the semiclassical approximation beyond those singularities. Now we know that, although the stationary phase approximation breaks down at these singularities (also known as caustics), nevertheless it does fine again later on. The orbit passing through the caustic just picks up an additional $\exp(-i\pi/2)$ phase. Then ν_-, or using the most common notation μ (Morse index), is an integer that counts the number of caustics along the classical path. For the most common type of Hamiltonians, $\mathcal{H} = \mathbf{p}^2/2m + V(\mathbf{q})$, μ is the number of times $(\partial^2 S/\partial q \partial q')$ changes sign.

Finally putting all the pieces together we arrive at the celebrated cornerstone of semiclassical mechanics, the Van Vleck–Gutzwiller (VVG) propagator:

$$\langle q|\exp(-i\mathcal{H}t/\hbar)|q'\rangle \approx (2\pi i\hbar)^{-1/2} \sum \left|\frac{\partial^2 S}{\partial q\, \partial q'}\right|^{-1/2} \exp\left(\frac{i}{\hbar} S(q, q') - \frac{i\pi}{2}\mu\right) \quad (3.37)$$

E. Change of Representations: Morse Indices for Mixed Representations

There is no fundamental reason for doing semiclassical approximations in the configuration representation. Moreover, semiclassical calculations can be strongly representation dependent (see Section VIII). The asymptotic propagator shown so far is written in the q, q' representation; however, it is a very simple exercise to derive the corresponding asymptotic expression in a different representation. Of particular interest to us will be the propagator in the mixed representation $\langle p|\exp(-i\mathcal{H}t/\hbar)|q'\rangle$. This propagator plays a principal role in the development of the "cellular dynamics" (CD) method, Section V.

Changes of representation are done by Fourier transformation of the propagator, or Green function[2]

$$\langle p|\exp(-i\mathcal{H}t/\hbar)|q'\rangle = \int dp\, \langle p|q\rangle\, \langle q|\exp(-i\mathcal{H}t/\hbar)|q'\rangle \quad (3.38)$$

If the Green function in the configuration representation is substituted by its semiclassical limit and the integral is performed by stationary phase evaluation, the final result is an asymptotic form to the same order in $\hbar$

$$\langle p|\exp(-i\mathcal{H}t/\hbar)|q'\rangle \approx (2\pi i\hbar)^{-1/2} \left|\frac{\partial^2 S}{\partial p\, \partial q'}\right|^{1/2} \exp\left(i(S - pq_t)/\hbar - i\pi\bar{\mu}/2\right) \quad (3.39)$$

where

$$\bar{\mu} = \mu - \tfrac{1}{2}\,\mathrm{Sgn}(M_{11}M_{21}) \quad (3.40)$$

This new representation for the propagator depends on classical orbits subject to double-valued boundary conditions: p, q'. The position at time t, q_t, is also a function of initial position and final momentum.

IV. SINGLE-ORBIT EXPANSION: GAUSSIAN WAVEPACKET DYNAMICS METHOD

In this section we will consider the propagation of a Gaussian wavepacket given by

$$\Psi(q) = \left(\frac{\alpha}{\pi}\right)^{1/4} \exp\left[-\frac{\alpha}{2}(q - q_0)^2 + \frac{i}{\hbar}p_0(q - q_0)\right] \quad (4.1)$$

Using the VVG propagator whose derivation was outlined in Section III, we can write the wavepacket at time t as

$$\Psi(q, t) = \left(\frac{\alpha}{\pi}\right)^{1/4} (2\pi i\hbar)^{-1/2} \int_{-\infty}^{+\infty} dq' \left|\frac{\partial^2 S}{\partial q\, \partial q'}\right|^{1/2} \cdot \exp\left[\frac{iS}{\hbar} - \frac{i\pi}{2}\mu - \frac{\alpha}{2}(q' - q_0)^2 + \frac{i}{\hbar}p_0(q' - q_0)\right] \quad (4.2)$$

In the previous equation, $S(q, q', t)$ is the action accumulated by the classical orbit connecting the initial q' with the final position q. Often it is not possible to solve this integral exactly due to the complex dependence of the action function on the boundary conditions; however, if the initial state is sufficiently narrow (large α) and the dynamics is not considered for too long a time period, it is legitimate to expand the action about the classical orbit departing from the center of the initial state, (q_0, p_0). We have already seen this action expansion in Eq. (3.24). Upon substitution of the quadratic expansion of S in initial and final positions it is easy to evaluate Eq. (4.2) by a Gaussian quadrature

$$\Psi(q; t) = \left(\frac{\alpha}{\pi}\right)^{1/4} [i\alpha\hbar|M_{21}| + M_{22}\, \mathrm{Sgn}(M_{21})]^{1/2} \cdot \exp\left[\frac{iS_0}{\hbar} + \frac{ip_t}{\hbar}(q - q_t) - \frac{\alpha_t}{2}(q - q_t)^2\right] \tag{4.3}$$

where

$$\alpha_t = \frac{i}{\hbar}\frac{M_{11}\alpha\hbar - iM_{12}}{M_{21}\alpha\hbar - iM_{22}} \tag{4.4}$$

Note that the signature "Sgn" of a number is the same to 1 times its sign. This formula was first derived by Heller [21] in the context of semiclassical scattering theory. The derivation described above, however, differs from the one used originally, which was not based on the application of the Van Vleck propagator. For harmonic potentials and, to a certain degree of approximation, for potentials with small anharmonicity, the time-evolving state preserves its Gaussian shape. This is easy to prove directly from the time-dependent Schrödinger equation. Even when the anharmonic terms in the potential are large, the approximation Eq. (4.3) is still valid for short times.

A. Harmonic Potential

As a simple example let us apply the semiclassical approximation Eq. (4.3) to a coherent state [Eq. (4.1) with $\alpha = m\omega$] moving on a harmonic potential. Say the initial state has average position and momentum (q_0, p_0). The equations of motion for the guiding orbit are

$$p_t = p_0 \cos \omega t - m\omega q_0 \sin \omega t \tag{4.5}$$

$$q_t = q_0 \cos \omega t + \frac{p_0}{m\omega} \sin \omega t \tag{4.6}$$

Obtaining the stability matrix from the previous equations is then a simple task. Upon substitution of the M_{ij} in Eq. (4.4), one gets that $\alpha_t = m\omega$ (m is the mass of the particle). In other words, on a harmonic potential, the spread of the evolving Gaussian we chose is time independent. This is a well-known property of coherent states. Moreover, the average position and momentum of the evolving semiclassical state follow the classical equations of motion.

As it turns out, harmonic potentials represent one of the few cases where semiclassics is exact. In general, potentials with a maximally quadratic dependence on coordinates and momenta are semiclassically exact. Let us remember that the GWD formula is not strictly speaking the exact semiclassical solution. We have invoked an additional approximation, the quadratic expansion of the action function $S(q, q')$ about a *single* guiding orbit. Failure of Eq. (4.3) is more often caused by a breakdown of the linearization of the dynamics (which is exact in the harmonic case) about the guiding orbit than by the failure of semiclassics.

B. Successes and Limitations

Besides a more rigorous theoretical foundation than that given here [3], the GWD method also has a broad spectrum of applications. They range from scattering cross section calculations [29, 35] to photodissociation [51, 26], time-dependent Raman scattering [27], atom–surface scattering [28], reactive scattering calculations in the interaction picture [52], incoherent neutron scattering [52], and a very recent calculation on control by tailored light fields [54].

In many cases the simple one-orbit method can provide physical insight and give good qualitative results. Photodissociation and some types of scattering, in particular, have in common a very short interaction time. Therefore it is likely that the nonlinear terms in the potential do not have enough time to manifest themselves. Consequently the GWD is a reasonable approximation.

When the interaction time scale is longer, as in high-resolution absorption spectra or in low-energy atom–surface scattering or reactive scattering, then many orbits with completely different topology contribute to the final cross section. It is unrealistic to assume that a single-orbit expansion like Eq. (4.3) can represent all this complexity. In other words, the function $S(q, q')$ in the semiclassical propagator becomes strongly dependent on small variations of initial and final positions.

There have been many attempts to improve on the original GWD. They range from the so-called frozen Gaussian approximation [5] (see also [23]), where the initial state is expanded into smaller Gaussians whose width is held constant throughout the evolution, all the way to a generalization of GWD into complex phase space [24], which is computationally cumbersome. It has become clear through all the efforts that have been made that the nonlinear structure of phase space has to be accounted for in a generic way. The road towards that goal was paved through an article by Heller [30] who introduced the cellular dynamics method that will be the subject of the next section.

V. MULTIPLE ORBIT EXPANSION: "CELLULAR DYNAMICS" METHOD

The main obstacle encountered by the semiclassical method described in Section IV is the inability of a single orbit to account for the full dynamics of the system. As described earlier, the classical orbit launched at the center of the initial wavepacket was taken as a guiding orbit. The action function $S(q, q')$ was then expanded about this guiding orbit. This is equivalent to assuming that any orbit started in the phase space region covered by the initial state follows a path topologically similar to the guiding orbit. Of course this is not the common situation, particularly for strongly nonlinear systems (e.g., strongly anharmonically coupled oscillators) and for dynamics over long time periods.

Figure 2 shows a diagram representing the breakdown of the linearization about a single orbit. The grey circle shows the phase space region covered by the initial state $|\Psi(0)\rangle$, which is assumed to be located on a periodic orbit. The approximate semiclassical dynamics moves along the central guiding orbit (solid line) for almost one full period. The expansion of the dynamics about a single guiding orbit leads to a moving distribution of orbits covering the phase space region enclosed by the dashed ellipse; however, the real dynamics may exhibit a completely different distribution due to the

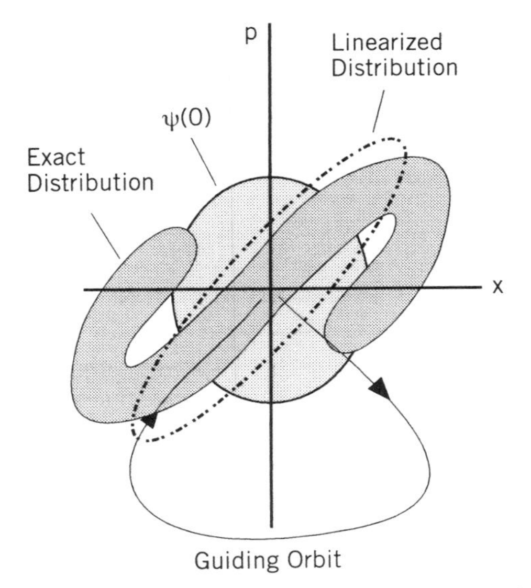

Figure 2. Phase space portrait of an initial state $\Psi(0)$, and two classically propagated distributions: (dark gray) exact solution, (dashed line) linear approximation about the center guiding orbit.

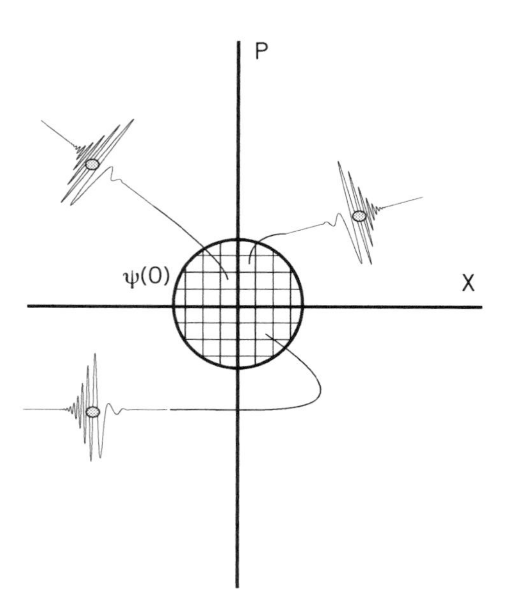

Figure 3. Schematic illustration of the cellular dynamics (CD) method. The initial state is divided into a grid of small subdomians from which small pieces of the wave function evolve following the local linearized dynamics.

special sensitivity of the classical dynamics with respect to initial conditions (s-shaped dark distribution). When the goal of our calculation is the determination of a correlation function, $\langle \Psi(0) | \Psi(t) \rangle$, a linearized approximation like that in Fig. 2 does not predict the quantum interferences that exist due to the three intersections of the initial and evolving distributions of classical orbits.

The obvious solution for this problem is sketched in Fig. 3. In the ''cellular dynamics'' method an ensemble of orbits launched in the region of the initial state is used to propagate the dynamics. Each orbit carries a small piece of the wave function and the final superposition of all these fragments creates the semiclassically propagated state.

The ''splitting'' of the dynamics is done by first representing the semiclassical propagator as an integral that unravels the ''root searching'' problem, and then discretizing the resulting expression in terms of a set of initial conditions in phase space.

A. Initial Value Representation

Let us interchange the roles of the Green functions in Eq. (3.38) by performing the appropriate inverse Fourier transform

$$\langle q|\exp(-i\mathcal{H}t/\hbar)|q'\rangle = \int dp\, \langle q|p\rangle\langle p|\, \exp(-i\mathcal{H}t/\hbar)|q'\rangle \qquad (5.1)$$

It is legitimate to pass to the $\hbar \to 0$ limit by inserting the mixed representation Green function of Eq. (3.39). Substituting the variable of integration p by $p_t\,(p', q')$, which is the solution to Hamilton equations of motion and then changing integration variable to p', being the initial momentum, one finds

$$\langle q|\, \exp(-i\mathcal{H}t/\hbar)|q'\rangle = (4\pi^2 i\hbar^2)^{-1/2} \int dp' \left|\frac{\partial p_t}{\partial p'}\right|^{1/2} \cdot \exp\left\{\frac{i}{\hbar}\,[S + p_t(q - q_t)] - i\pi\bar{\mu}/2\right\} \qquad (5.2)$$

The classical action $S(q_t, q')$ depends exclusively on the initial conditions (p', q'), the position and momentum of the classical orbit at time $t = 0$. Final positions and momenta, q_t p_t are also dependent on the variable of integration. The position q where the Green function is to be calculated enters Eq. (5.2) only as a plane-wave dependence. The integration over all initial momentum launches a series of classical orbits, each of which creates a rapidly oscillating interference pattern which is stationary only if the final position q_t is near q. Here we see how the double-ended boundary conditions enter again "through the back door." If Eq. (5.2) was evaluated by stationary phase approximation it would lead us immediately to the normal Van Vleck–Gutzwiller propagator. Instead, we allow a manifold of orbits with different initial momenta to run forward in time and let the plane wave overlap "select" which contribution is relevant at time t.

Using a representation like Eq. (5.2) for the propagator not only avoids the problem of searching for "good orbits" to perform the propagation. A more careful comparison of Eq. (5.2) and the standard Van Vleck–Gutzwiller formula reveals different behavior at the caustics. The standard semiclassical propagator becomes singular each time a classical orbit crosses a focal point. This makes the numerical implementation of the usual semiclassical propagator inconvenient. On the other hand the integral representation derived here lacks the usual singularities! The integrand is simply an oscillatory function times a prefactor that may become zero but never singular. Does this mean that we have somehow unfolded the caustics and thus eliminated the singularities? Unfortunately the answer is negative. Caustics are still a source of error in the semiclassical expansion of the propagator. It is not the caustics in the coordinate representation that cause problems, however, but the ones in the mixed representation $\langle p|\, \exp(-i\mathcal{H}t/\hbar)|q'\rangle$ because this is our starting point for the asymptotic expansion. If the Green function in the integrand of Eq. (5.1) is substituted by its $\hbar \to 0$ limit, errors due to the breakdown of the asymptotic expansion Eq. (3.39) immediately appear. Because we are now working in a different Fourier space, the singularities of Eq. (3.39) appear in the coordinate representation as rapidly oscillatory components. Later we will show several examples of these oscillatory errors in our semiclassical results.

The integral expression studied here is an example of an "initial value representation" for the semiclassical propagator [6]. It will also be the starting point of the "cellular dynamics method," which is a numerical scheme for propagating localized initial states based on a discretization of Eq. (5.2).

B. "Cellularization"

As in the cases studied in Sections IV and VIII, we will consider the semiclassical propagation of an initially localized state $|\Psi(0)\rangle$ that either has a Gaussian form *a priori* or can be written as a sum of narrow Gaussians

$$\langle q|\Psi(0)\rangle = \left(\frac{\alpha}{\pi}\right)^{1/4} \exp\left[-\frac{\alpha}{2}(q-q_0)^2 + \frac{i}{\hbar}p_0(q-q_0)\right] \tag{5.3}$$

After time t, the semiclassical propagator Eq. (5.2) applied on the Gaussian state yields

$$\langle q|\Psi(t)\rangle \approx \eta \int\int dq'\, dp' \left|\frac{\partial p_t}{\partial p'}\right|^{1/2} \exp\left(\Xi(q', p')\right) \tag{5.4a}$$

$$\eta = \left(\frac{\alpha}{\pi}\right)^{1/4} (4\pi^2 i\hbar^2)^{-1/2} \tag{5.4b}$$

$$\begin{aligned}\Xi = {} & \frac{i}{\hbar} S(q_t, q'; t) - i\bar{\mu}\pi/2 + \frac{i}{\hbar} p_t(q-q_t) \\ & - \frac{\alpha}{2}(q'-q_0)^2 + \frac{i}{\hbar} p_0(q'-q_0)\end{aligned} \tag{5.4c}$$

Equally important in a typical absorption-emission spectrum is the linear response of the system to the electromagnetic field. In the Condon approximation this response is given by a correlation function, $\langle\Psi(0)|\Psi(t)\rangle$. Its Fourier transform gives the absorption or emission spectra of the system (compare Section II). Within our semiclassical approximation

$$\langle\Psi(0)|\Psi(t)\rangle \approx (2\pi^2 i\hbar^2)^{-1/2} \int\int dq'\, dp' \left|\frac{dp_t}{dp'}\right|^{1/2} \exp(\Xi'(q', p')) \tag{5.5a}$$

$$\Xi'(q', p') = \frac{i}{\hbar} S(q_t, q'; t) - i\bar{\mu}\pi/2 - \frac{i}{\hbar} p_t(q_t - q_0) + \frac{i}{\hbar} p_0(q' - q_0)$$

$$- \frac{\alpha}{2} (q' - q_0)^2 - \frac{\alpha^{-1}}{2\hbar^2} (p_t - p_0)^2 \qquad (5.5b)$$

The exponential $\exp(\Xi')$ is the product of real Gaussians, strongly centered at $q' \approx q_0$ and $p_t \approx p_0$, times an oscillatory function. When the parameter α is large, only a small region of initial positions around the center of the initial state contribute to the integration. Then, to the lowest order approximation, one can expand the action and orbits q_t p_t about $q' \approx q_0$ and $p' \approx p_0$ (the average initial momentum of the system). As a result we would obtain a different derivation of the GWD method of Section IV. Beyond this approximation we can discretize the integral over initial momentum (and initial position if necessary) and perform local expansions of S, q_t, and p_t. For this purpose let us consider the following sum over N Gaussians with equidistant centers along the momentum axis

$$E_{g1}(p', p_1) + E_{gN}(p', p_N) + 2\sqrt{\frac{1}{\pi}} \sum_{k=2}^{N-1} \exp[-\beta(p' - p_k)^2] \approx 1 \qquad (5.6)$$

$$E_{g1}(p', p_1) = \begin{cases} 1 & \text{if } p' \leq p_1 \\ 2\sqrt{1/\pi} \exp[-\beta(p' - p_1)^2] & \text{otherwise} \end{cases} \qquad (5.7a)$$

$$E_{gN}(p', p_N) = \begin{cases} 1 & \text{if } p' \geq p_N \\ 2\sqrt{1/\pi} \exp[-\beta(p' - p_N)^2] & \text{otherwise} \end{cases} \qquad (5.7b)$$

The grid $p_1, \ldots, p_N$ is set up in the region covered by the initial state, and the two special ''half-Gaussians'', E_{g1}, E_{gN}, extrapolate the behavior of the classical manifolds beyond the grid. These two ending-cells are not necessary when the high-momentum cells lie in a dissociative part of phase space because then these fragments of the manifold stay in the initial state region for a relatively short time and thus have a negligible contribution to the correlation function.

There is freedom with respect to the selection of the parameter β and the separation between cells $\Delta p = p_k - p_{k-1}$. The total range in momentum to be covered is fixed by the initial state, and the total number of cells N allocated is a function of the stability of the system and the total propagation time. Then β controls the approximation of unity along the p-axis. In our calculations we take $\beta = 4/\Delta p^2$.

Inserting Eq. (5.6) into Eq. (5.5a) will immediately lead, through a local expansion of the action (as in Eq. (3.27)) and of the classical orbits (as in Eq. (3.18)), to a sum of Gaussian quadratures. The number of "guiding orbits" for the expansion is equal to the number of Gaussian cells inserted. The dynamics is expanded up to quadratic order around the center of the cell (q_0, p_k) for each term in the sum.

The correlation function $\langle \Psi(0)|\Psi(t)\rangle$ can be then written as a sum over cell contributions

$$\langle \Psi(0)|\Psi(t)\rangle \approx C_1 + C_N + \sum_{k=2}^{N-1} C_k \tag{5.8}$$

where

$$C_k = \left(\frac{2}{\pi i \hbar^2}\frac{|M_{11}|}{\det \mathcal{C}}\right)^{1/2} \exp\left(Z_0 + \frac{1}{4}\mathbf{B}\mathcal{C}^{-1}\mathbf{B}\right) \tag{5.9}$$

Both $\mathcal{C}$ (2×2 matrix), Z_0 and $\mathbf{B}$ are evaluated along the classical orbits $(q_0, p_k) \rightarrow (q_{kt}, p_{kt})$

$$Z_0 = \frac{i}{\hbar} S_{kt} - \frac{i\pi}{2}\bar{\mu} - \frac{i}{\hbar} p_{kt}(q_{kt} - q_0) - \frac{1}{2\alpha\hbar^2}(p_{kt} - p_0)^2 \tag{5.10a}$$

$$B_1 = \frac{i}{\hbar} M_{11}(q_{kt} - q_0) - \frac{1}{\alpha\hbar^2}(p_{kt} - p_0)M_{11} \tag{5.10b}$$

$$B_2 = \frac{i}{\hbar} M_{12}(q_{kt} - q_0) - \frac{1}{\alpha\hbar^2}(p_{kt} - p_0)M_{12} + \frac{i}{\hbar}(p_0 - p_k) \tag{5.10c}$$

$$\mathcal{C}_{11} = \frac{M_{11}^2}{2\alpha\hbar^2} + \beta + \frac{i}{2\hbar} M_{11}M_{21} \tag{5.10d}$$

$$\mathcal{C}_{22} = \frac{M_{12}^2}{2\alpha\hbar^2} + \frac{\alpha}{2} + \frac{i}{2\hbar} M_{12}M_{22} \tag{5.10e}$$

$$\mathcal{C}_{12} = \frac{M_{11}M_{12}}{2\alpha\hbar^2} + \frac{i}{2\hbar} M_{11}M_{22} \tag{5.10f}$$

Also the action S_{kt} as well as the stability matrix are calculated along the classical orbit for the kth cell.

Similar expressions can be derived for the C_1 and C_N contributions [17]. They involve the evaluation of error functions instead of Gaussians, but the mechanics of the numerical method for these other cells is completely analogous.

C. Morse One-Dimensional Example

As a first simple example, consider launching a narrow initial Gaussian wavepacket on a Morse potential surface with only a dozen or so bound states. This case is very far from the harmonic limit. The potential surface and the initial state are plotted in Fig. 4. We will focus on the calculation of the autocorrelation function $\langle \Psi(0) | \Psi(t) \rangle$. As presented in Section II, this problem is equivalent to the calculation of the photoabsorption spectrum of a molecule whose ground state is harmonic and whose first excited state is a Morse potential

$$V(q) = D(1 - e^{-\lambda q})^2 \tag{5.11}$$

The initial state is given by Eq. (5.3) with $\alpha = 20.0$, $q_0 = 0.0$, $p_0 = 0$ and $\hbar = 1$. Its propagation on the upper Morse surface is far from harmonic as showed by the comparison between the quantum autocorrelation function versus the Gaussian wavepacket (semiclassical single-orbit approximation) Fig. 5. The early decay is well described by the single-orbit approximation, but before one period of the motion of the central orbit is complete this semiclassical approximation already fails.

In Fig. 6 we show the corresponding result for the cellular dynamics implementation of the propagation. One hundred classical orbits were

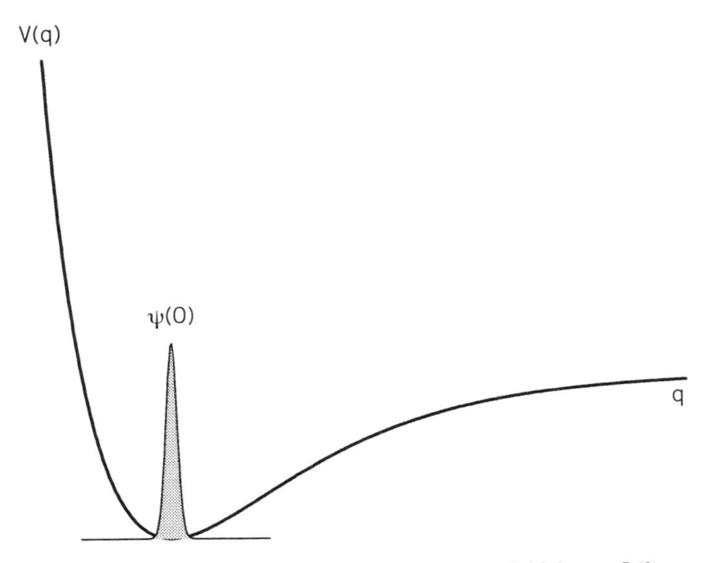

Figure 4. Morse potential with $D = 5.0$ and $\lambda = 0.2$. The initial state $\Psi(0)$ corresponds to the ground state of a harmonic potential ($\alpha = 20.0$).

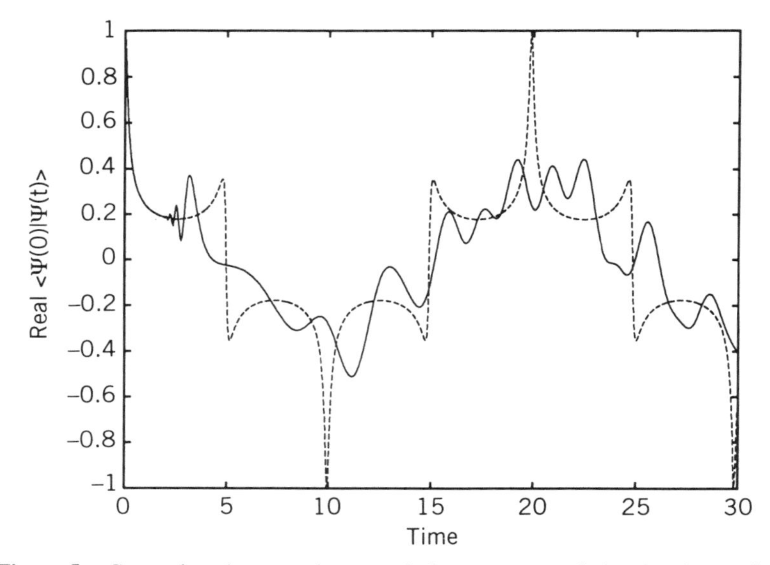

Figure 5. Comparison between the numerical quantum correlation function (solid line) and the Gaussian wavepacket method (dashed line).

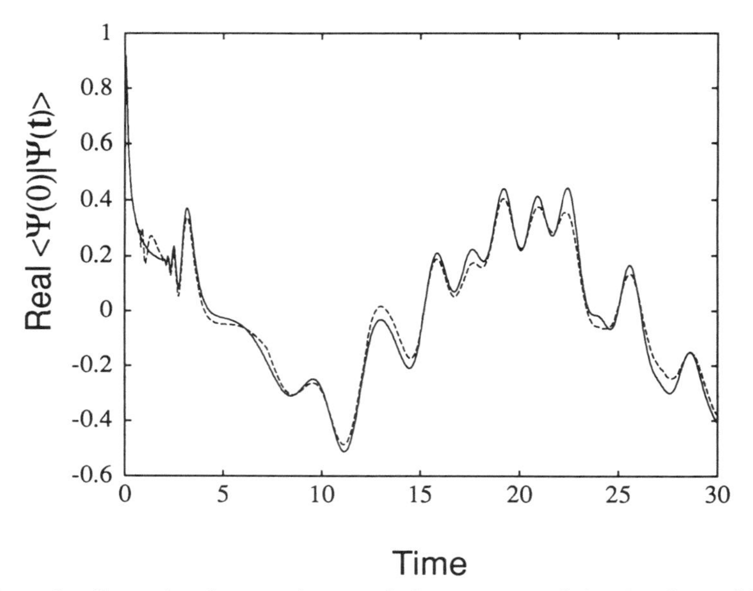

Figure 6. Comparison between the numerical quantum correlation function (solid line) and the cellular dynamics method (dashed line).

launched at $q = q_0$, with initial momentum spread over a grid covering the initial state $|\Psi(0)\rangle$. As seen from the comparison between the multiorbit semiclassical and the FFT-implemented quantum correlation function, the agreement is excellent. Similar accuracy can be reached over 50 periods of the central orbit (or average period of the distribution).

As mentioned earlier, the use of a mixed representation for the propagator eliminates the occasional singularities from the semiclassical wave function and correlation functions. The fast oscillatory pattern observed after the initial decay ($t \sim 0$ to 3) corresponds to a typical error caused by a caustic in the mixed representation. Because $\hbar$ is not too large compared with the action accumulated per period there are small disagreements with the numerical quantum result. Overall, the performance of the cellular dynamics is vastly superior to the single-orbit expansion. The reason for the difference between these two methods can be further understood by looking at the structure of the classical manifold at time $t = 30$ (Fig. 7). All the orbits in the manifold shown were launched at q_0; a large number of the cells dissociated into the continuum. From the set that remained bounded we can immediately see that there is a finite number of orbits (paths in phase space) that come back to the initial state at $t = 30.0$; these particular paths, each carrying an independent phase, build the correlation function at this instant.

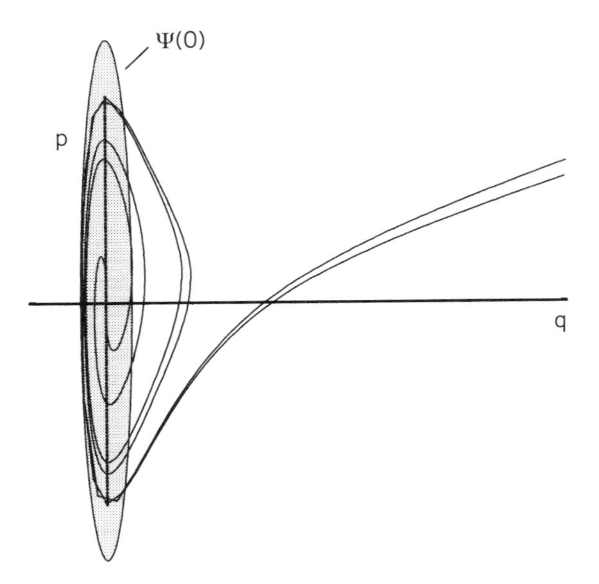

Figure 7. Phase space portrait of the initial state $\Psi(0)$ and the classical manifold at $t = 30.0$.

As a result, no single-orbit expansion is capable of describing multiple classical paths with distinct topology.

D. Barbanis Potential: Mixed Dynamical Systems

Any example we may study with one degree of freedom suffers from the lack of normal dynamical features that occur in real molecular systems. Only Hamiltonian models with two or more degrees of freedom exhibit resonances, which are common in molecular vibrations. In general, the classical phase space of these systems is composed of two distinct regions, the quasiperiodic and the irregular or chaotic regions. In the first region, the classical dynamics can be easily understood in terms of a finite set of fundamental frequencies and any dynamical variable (i.e., position, action) can be expanded as a Fourier series; thus the name quasiperiodic. In the irregular region, the validity of the Fourier series expansion breaks down and no finite set of frequencies allows the representation of a typical orbit.

The semiclassical calculation of this kind of system escapes the traditional EBK methods, and thus it was important for any new semiclassical approximation to demonstrate its validity for such systems. Our favorite paradigm is the Barbanis potential, a model of two nonlinearly coupled harmonic oscillators [17]

$$V(q_1, q_2) = \frac{1}{2}\,\omega_1 q_1^2 + \frac{1}{2}\,\omega_2 q_2^2 + \gamma q_1^2 q_2 \tag{5.12}$$

For $\omega_1 = 1.1$, $\omega_2 = 1.0$, and $\gamma = -0.11$ the two vibrational modes are in almost 1 : 1 resonance. It is obvious that this potential has a periodic orbit along the $q_1 = 0$ axis; this orbit becomes unstable due to the classical resonance above energy $\sim$ 7.9. This resonance creates two new stable tori that can be associated with local modes of the system. The potential also has a dissociative threshold around $E \sim 15.125$ (see Fig. 8).

The easiest way to visualize the dynamical nature of the system is through the aid of surfaces of section. Also known as Poincaré sections, we will define them for this problem as the mapping onto the (p_1, q_1) plane defined by $q_2 = 4.0$, $p_2 \geq 0$ at a given energy. In the case of the Barbanis model, Figs. 9–11 show the Poincaré sections at energies $E = 10, 14$, and 16. The first figure shows a typical image of a bifurcated orbit; the periodic orbit is given by the point at the origin of the plot. Any small distance away from this point the classical orbit falls into either of the two stable islands (local modes) to the left and right sides in the picture. Both local modes are stable at this energy, but at higher energy they become unstable and part of them "flood" the dissociative region of phase space. In the second picture we also see that what before was a large region of stable quasiperiodic motion

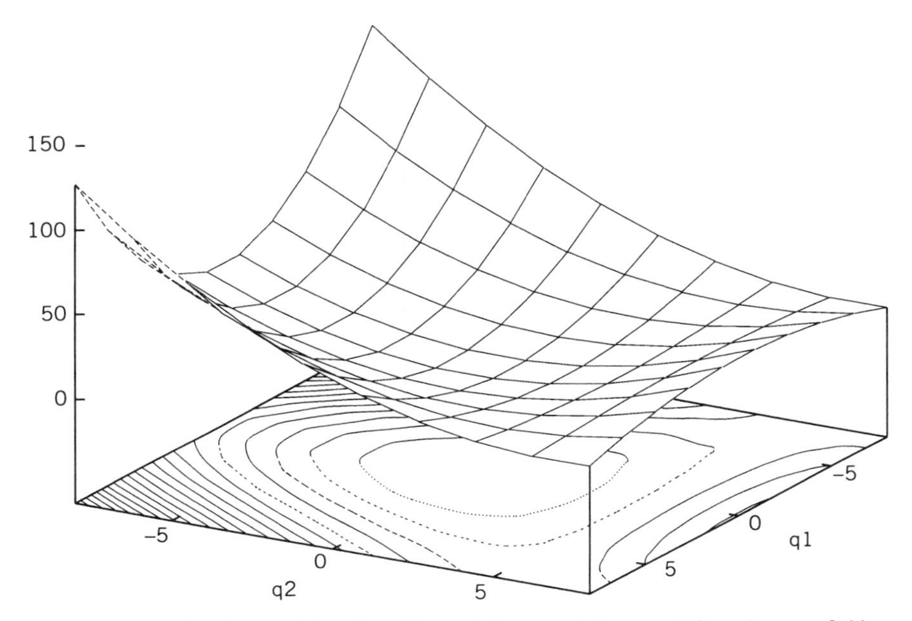

Figure 8. Barbanis potential surface $V(q_1, q_2)$ for $\omega_1 = 1.1$, $\omega_2 = 1.0$, and $\gamma = -0.11$.

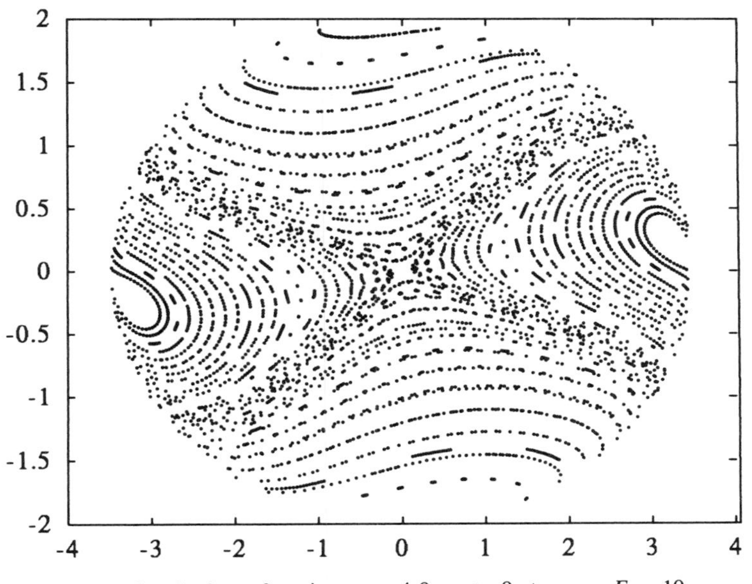

Figure 9. Surface of section $q_2 = 4.0$, $p_2 \geq 0$ at energy $E = 10$.

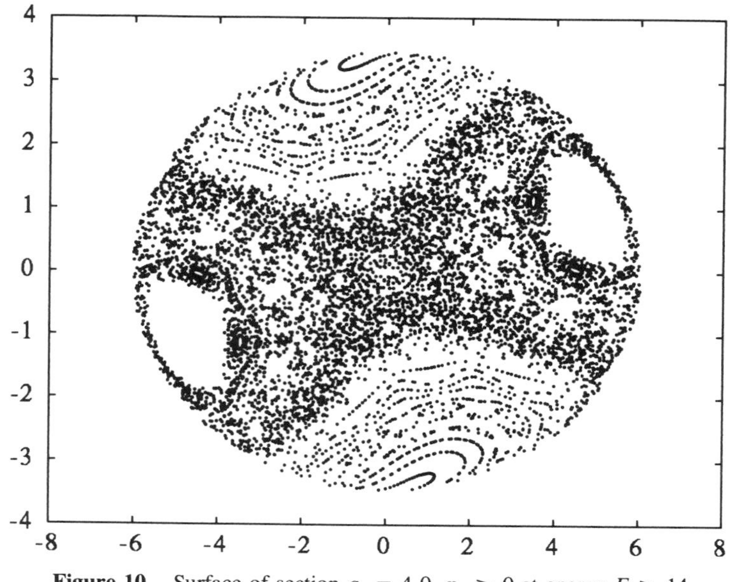

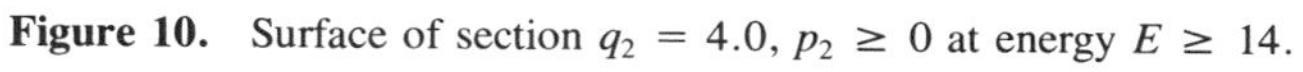

Figure 10. Surface of section $q_2 = 4.0$, $p_2 \geq 0$ at energy $E \geq 14$.

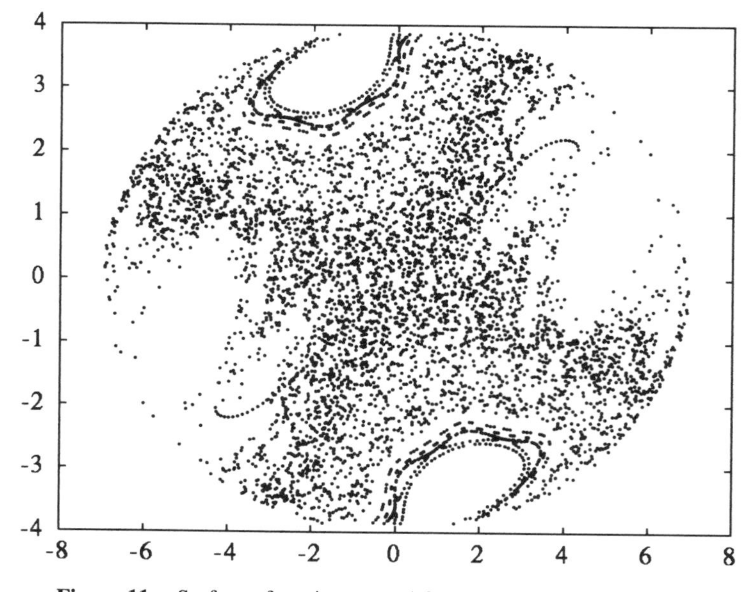

Figure 11. Surface of section $q_2 = 4.0$, $p_2 \geq 0$ at energy $E = 16$.

is now substituted by a uniform tube of stochastic-like motion. The classical orbits trapped in this region exhibit chaotic motion and they are not able to escape because this phase space domain is separated from the remaining stable regions; however, quantum mechanically it is possible to penetrate into the disconnected part of phase space by tunneling. This means that if a local excitation is carefully prepared in the chaotic region it will eventually transfer part of its energy to the global vibrational mode.

Taking into account the existence of a chaotic part in phase space is important when calculating and interpreting semiclassical spectra. The example we are going to discuss corresponds to an initial Gaussian state of the form Eq. (5.3), with $\alpha_1 = 6.6$, $\alpha_2 = 6.0$, and centered at ($q_1 = 0.1$, $q_2 = 5.0$). The energy spectrum covered by this state ranges from $E = 8$ to 24, which gives us a complete scan of the resonance region and of the quantum resonances, above dissociation. Both quantum and semiclassical spectra were calculated for this model, and they are compared in Fig. 12; two versions of semiclassical spectra are shown in order to compare the old single orbit expansion (*a*) and the new multiorbit expansion (*b*). In the first version only one classical orbit was needed; in the second we use a grid in $p_1 - p_2$ located at the center of the initial Gaussian state of 61×61 trajectories. The quantum spectrum was calculated by using an FFT propagation of the time-dependent Schrödinger equation. The agreement between quantum and CD ("Cellular Dynamics") results is remarkable; the underlying classical dynamics coverted by the spectum shown is a mixture of integrable and chaotic motion. Some peaks in the spectrum are associated with stable regions of phase space, and some (especially close to dissociation) with the purely chaotic regions. Notice that more than a third of the structures in this spectrum exist above dissociation. There is good agreement in both frequencies and relative intensities of the peaks in the spectra, something that semiclassical methods based on periodic orbit expansions of the energy Green function cannot provide, due to the lack of information on the eigenfunctions.

Now we have established a numerical connection between an ensemble of classical trajectories and a quantum spectrum. This correspondence should go beyond the prediction of eigenvalues and intensities, it should take us to the core of the underlying physical interpretation of the spectrum.

The spectrum shown have a simple structure at the lowest resolution. It consists of a series of peaks separated by approximately $2\pi/\omega_1$, which is the average period of the periodic orbit along the q_2-axis. Shortly after the launching of the initial state, $|\Psi(t)\rangle$ follows closely the dynamics of the periodic orbit and returns to its initial position after one period, thus creating a first recurrence in $\langle\Psi(0)|\Psi(t)\rangle$. Consequently, the resolved spectra up to this moment must show a series of equidistant peaks. The GWD spectra

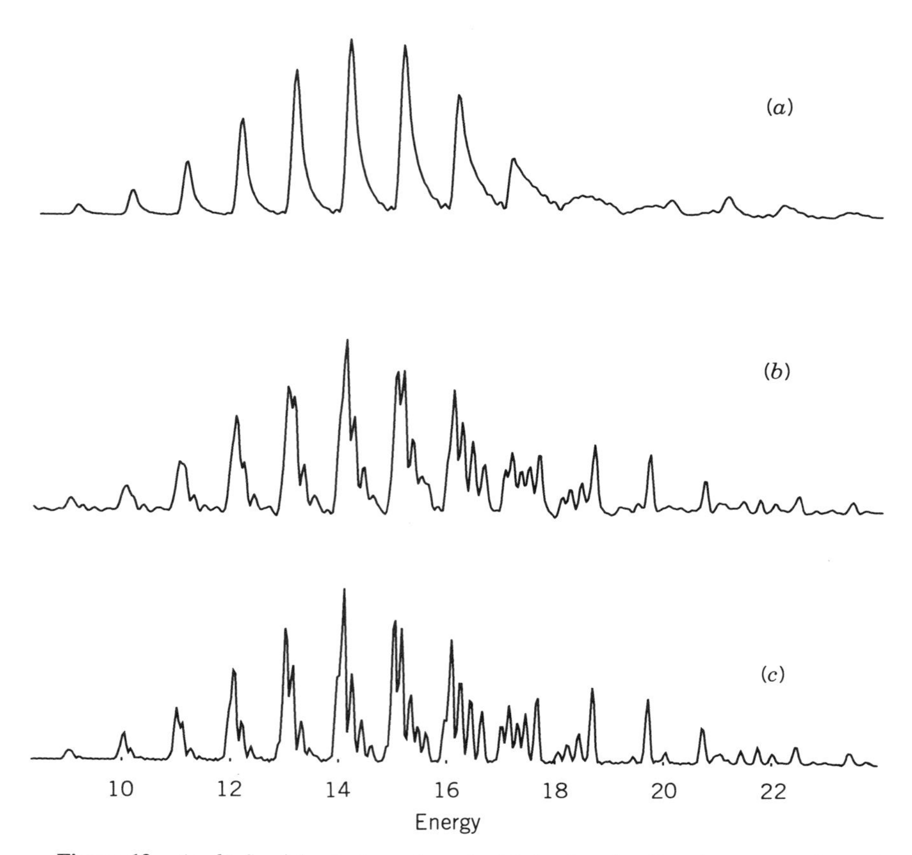

Figure 12. (*a*, *b*) Semiclassical spectra calculated using the Gaussian wavepacket and cellular dynamics methods respectively. (*c*) Quantum spectra obtained by FFT propagation.

shows exactly that, because the only classical orbit entering the calculation is a central unstable periodic orbit.

After running the calculation for many periods the approximate semiclassical mechanics predicts a spectrum of individual transitions with a line width that roughly covers the true eigenstates underlying each clump. This is predicted thanks to the information contained in the stability matrix, as analyzed in [55, 56]. However, the true dynamics of the wave function so far is not only a spreading Gaussian moving along the q_2-axis. Part of the amplitude that leaves the central guiding orbit eventually returns to create additional recurrences in the correlation function $\langle \Psi(0) | \Psi(t) \rangle$. The nonlinear coupling in the potential is responsible for bringing back wavepocket amplitude along the q_2-axis. There are precise time scales for the return of this probability amplitude. These time scales can be identified classically and are responsible for the additional splitting of the spectral features resolved by the central periodic orbit. By using a variety of classical orbits, with a wide range of energies and initial conditions, it is possible to have a better prediction of each of the resolved features and their approximate line widths. The stability of the classical orbits in the energy shell of the spectral feature determines the spectral width. The cellular dynamics method can now discriminate between the stability of the periodic orbit at the low end of the spectrum, where it is just becoming unstable (and therefore makes the moving wave function take a long time before it spreads into other regions of phase space), or the high end of the spectrum, in which the same periodic orbit is extremely unstable and the moving wave function requires only a short travel before it escapes.

The grid of classical orbits used in our semiclassical calculation allow the resolution of a number of dynamical processes which are highly initial condition dependent; most of these processes have a direct connection to quantum signatures in the spectrum: dissociative orbits, trapped trajectories in local modes, and quasi-integrable and stochastic orbits. It also makes it possible to link small stable tori at energies above dissociation with very long-lived quantum resonances in the spectrum. Incidentally the long lifetimes of the resonances have a classical origin: The stable tori existing at this high energy, $E = 20$ to 22, are analogous to the libratory motion of the pendulum; however, the dissociation channels are always transverse to the direction of motion in phase space of the orbits trapped in this torus. This means there is never enough energy in the direction of dissociation for the classical orbit to escape and consequently for the nuclei to dissociate. Of course, as we mentioned earlier, the true quantum resonances have a finite lifetime, while the semiclassical ones are infinite. The total time of propagation shown in Fig. 12 is not long enough to measure the true line width of these features. It is possible that a semiclassical approximation based on complex trajectories (or going to higher order in $\hbar$) would predict the line width of this kind of resonance, but we have not tried that yet.

By the time the semiclassical propagation has resolved the features under the equidistant clusters in the spectrum, the wave function $|\Psi(t)\rangle$ looks like anything but a Gaussian wavepacket. Most orbits have spread all over configuration space, and some eventually return to the region covered by $|\Psi(0)\rangle$ to create additional recurrences. These recurrences define time scales that can be attributed to intramolecular energy transfers between the vibrational modes [18].

In the energy domain one can detect these important time scales by gradually changing the resolution of the spectra. As the resolution is increased one typically observes the splitting of broad features into smaller and smaller peaks. The organization of splitting as a function of resolution can be thought of as a hierarchical structure [18, 57, 58]. Each of these bifurcations represents a new dynamical process in the time domain. With the help of classical orbits and the surfaces of section it is possible to organize this sequence of events and simultaneously explain the physical reasons behind the hierarchical spectral structure. For such a structure to exist, a more or less clear differentiation of time scales should be present. We will see that this is the case for the UV spectrum of the benzophenone molecule. Unfortunately, this is not so for the present Barbanis model; its highly chaotic contribution makes many new recurrences occur very close in time, most of them related to the so-called *homoclinic* orbits [44], which are responsible for the great number of features shown in Fig. 12. Therefore it is very difficult and even meaningless to specify any particular time scale in the dynamics.

E. Benzophenone UV Spectra

The lower vibrational band of the $S_1 \leftarrow S_0(n, \pi^*)$ transition [59, 60] shows a close resemblance with the spectra of the Barbanis potential for parameters, such that the two vibrational modes are in 1 : 1 Fermi resonance [56]. Although benzophenone has many degrees of freedom, only two of them have been observed to be relevant in the frequency range from 25,000 to 27,000 cm^{-1}. This vibrational band is part of an excitation induced by the promotion of an electron from the oxygen nonbonding orbital to a π^* orbital of the carbonyl group. The C—O bond gets excited and gives rise to a progression of the C—O stretching from 26,000 to 32,000 cm^{-1}. We are interested in the first bond of the progression that shows two main series of lines separated by approximately 60 cm^{-1}. These modes were assigned [59] to two torsional modes of the benzene rings (Fig. 13). The band origin of the torsional modes is 26,181 cm^{-1}, and the upper electronic surface was fitted to a Barbanis-like model using mass-weighted normal coordinates

$$\mathcal{H} = \frac{1}{2} \sum_{i=1}^{2} \omega_i (p_i^2 + q_i^2) + c_1 q_1^2 q_2 + c_2 q_2^3 \tag{5.13}$$

The parameters that best fit this model of the S_1 state to the experimental results are given in Table I. To a good approximation, the lower surface S_0 can be assumed to be harmonic and uncoupled. The minimum of the ground state is displaced relative to the excited state by Δq_1, Δq_2; therefore the initial state $\Psi(0)$ for the calculation of the absorption spectrum is a two-dimensional Gaussian on the S_1 surface, the lower vibrational state of the S_0 surface.

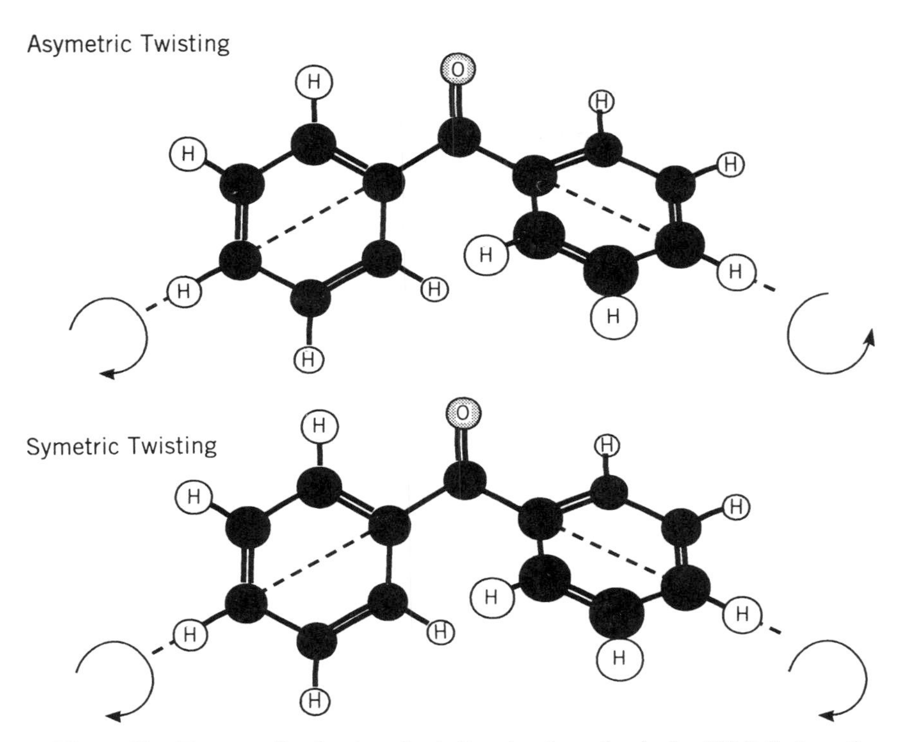

Figure 13. The two vibrational modes believed to be active in the UV 0–0 absorption band of C=O stretching mode of benzophenone.

In Figs. 14 and 15 the quantum and semiclassical calculations of both spectra and correlation functions are compared. The quantum results are obtained, as in the previous section, by integrating the time-dependent Schrödinger equation with an FFT method. The benzophenone potential model has a dissociative part; however, the overlap of the initial state $\Psi(0)$

TABLE I.
Constants for the Benzophenone Potentials and Initial State

Parameters for S_1 Surface	Parameters for S_1 Surface	Initial State
$w_1 = 69.7\ \text{cm}^{-1}$	$w_1^{(0)} = 75.0\ \text{cm}^{-1}$	$\Delta q_1 = 0.0$
$w_2 = 62.0\ \text{cm}^{-1}$	$w_2^{(0)} = 57.0\ \text{cm}^{-1}$	$\Delta q_2 = 4.25$
$c_1 = -5.4$		
$c_2 = -0.85$		

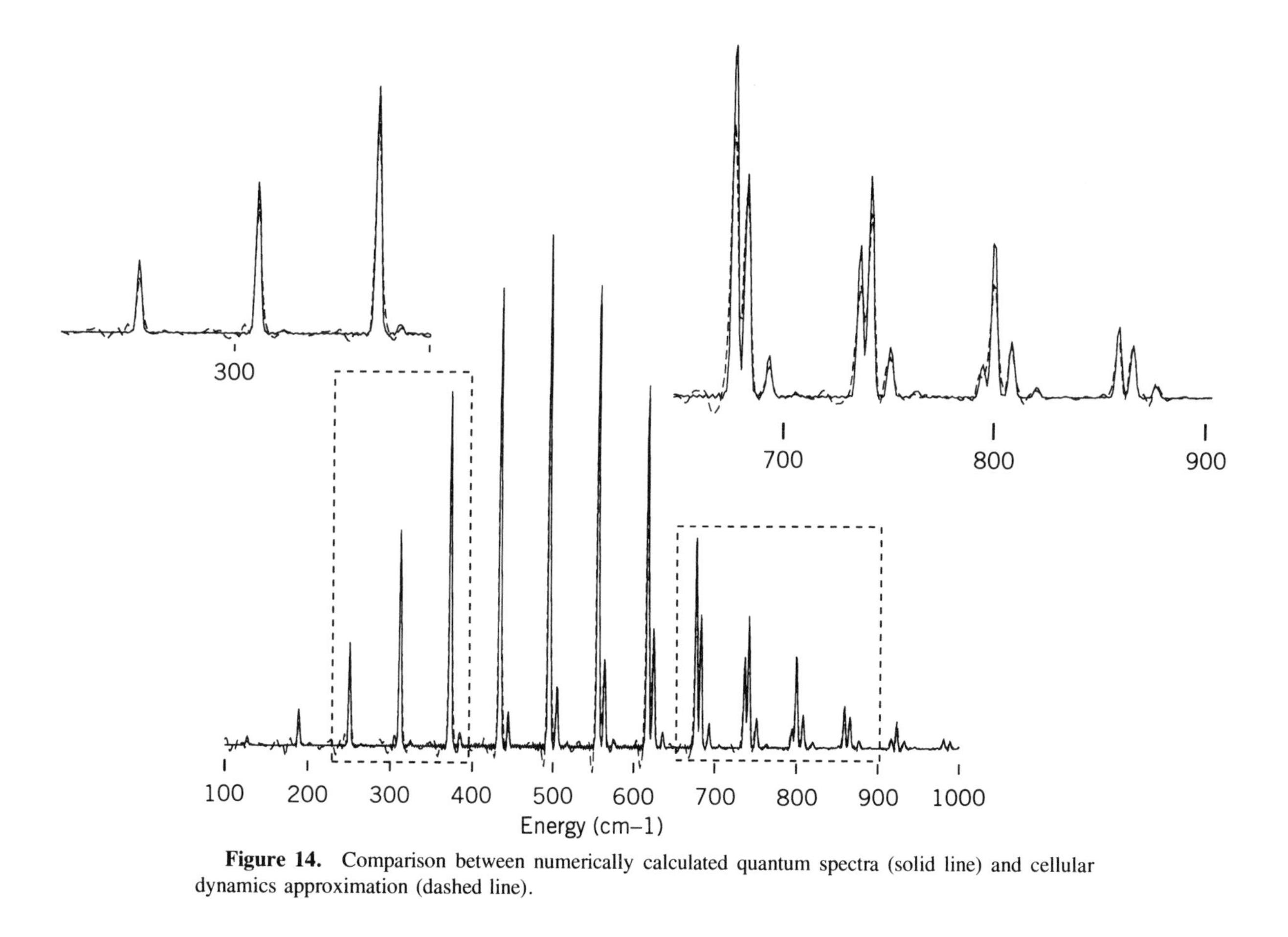

Figure 14. Comparison between numerically calculated quantum spectra (solid line) and cellular dynamics approximation (dashed line).

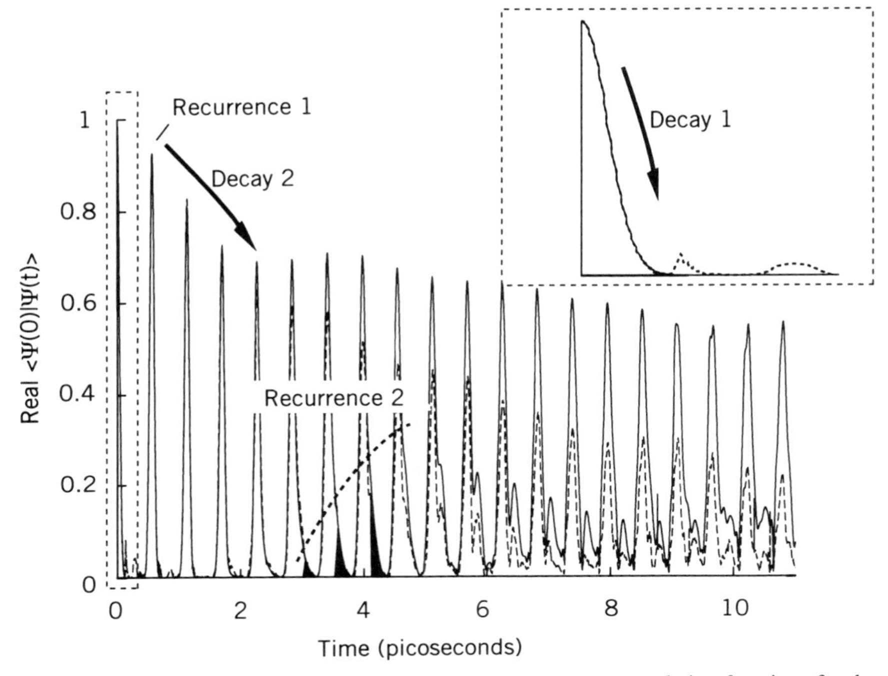

Figure 15. Quantum (solid) and cellular dynamics (dashes) correlation functions for the benzophenone model. Only the real part of the autocorrelation function is shown.

with the states of the continuum is almost negligible. Therefore, this spectra could have been evaluated equally well by a variational method. The semiclassical results were produced by using the cellular dynamics method with a grid of 11^4 classical orbits; we have expanded both coordinates and momenta in terms of cells in order to reach the maximum convergence possible with the quantum calculation. The agreement between both results is satisfactory; therefore we can conclude that all the dynamical features that go into the splitting of peaks in the spectrum can be assigned to classical features.

After the electronic excitation and the rearrangement of electrons has taken place a new nuclear equilibrium position is defined. The initial state $|\Psi(0)\rangle$ finds in a new nonequilibrium configuration of the upper S_1 state, displaced in q_2 degree of freedom (asymmetric twisting mode).

Both Figs. 14 and 15 contain information about the first 11 ps of the dynamics. This comprises roughly 20 periods of the fundamental asymetric twisting mode. In the uncoupled model ($\gamma = 0$) the excitation energy remains trapped in the asymmetric twisting mode. For the actual coupled model, however, we observe a slow energy transfer to the symmetric mode. For instance, the real part of $\langle \Psi(0)|\Psi(t)\rangle$ shows a decay of the total amplitude by nearly 30% in three periods, a clear indication of the unstability of the dynamics and of the transfer rate to other regions of phase space.

In the uncoupled model, two quantum numbers (n_1, n_2), associated with each of the vibrational modes, characterize the eigenstates of the system. The fundamental frequencies of each mode are almost in a 1 : 1 resonance. There is a near degeneracy between the (n_1, n_2) and $(n_1 - 2, n_2 + 2)$ levels excited by the initial state $|\Psi(0)\rangle$. When the coupling is turned on, a level-repulsion is expected between these levels with respect to the uncoupled system and a borrowing of intensity between both transitions. It is instructive to look at the behavior of these pairs of levels as a function of energy across the spectrum. Energy behaves in this case as an external parameter that drives the levels (n_1, n_2) $(n_1 - 2, n_2 + 2)$ into resonance at $E \approx 500$ cm^{-1}. Around that region, a classical resonance between both modes exists, and the asymmetric mode becomes unstable. Quantum mechanically one observes a preresonant attraction of levels, this is the pair of energy levels are put closer together by the coupling. Above resonance the normal repulsion between resonant states sets in.

Both the location of the resonance condition and the behavior of the energy levels nearby the resonance energy can be described using classical mechanics, as demonstrated by the agreement between semiclassical and quantum calculations. This conclusion was first suggested by Uzer, Marcus, and collaborators [61]. They used a very rough approximation: EBK quantization of the normal form approximation to the classical Hamiltonian. Because the Hamiltonian used in their quantization was not the exact one, their conclusions seemed at that time less general. However, in our semiclassical study we have used the exact equations of motion and reached similar results; both for a quasi-integrable system, such as benzophenone, and for a strongly mixed Hamiltonian (previous section). There is a clear connection between classical and quantum resonances; the energy splitting between levels before and after the energy resonance condition can be described semiclassically. This suggests that the time scales for energy transfer between these modes, due to the nonlinear coupling, are also classical. We cannot predict from a purely classical perspective what happens at the resonance energy. In fact we know that the semiclassical approximation that both we and Uzer and Marcus used predicts a crossing of energy levels, while in reality they never cross. The splitting between both levels at resonance is due to tunneling, and as such it cannot be described by the simple, real-orbit, semiclassical approximation at hand.

Classical mechanics can also let us understand the dynamics underlying the correlation function. During the earlier stages of the time evolution, the localized Gaussian wavepacket moves along the periodic orbit $q_1 = 0$. Because this orbit is particularly unstable at high energy, part of the high initial momentum components escape from the periodic orbit. During the first 2 ps of dynamics the amplitude that escapes to make a long excursion in phase space has not had enough time to return to the initial launching site. As a result, the correlation function up to this moment can be described by linearization of the dynamics about the central guiding orbit (GWD). Initially energy was exclusively in the asymmetric mode. Some of it excites the symmetric mode in the first 2 ps; later, the nonlinear coupling of the potential is responsible for the energy transfer back to the asymmetric mode.

Several time scales can be identified with the help of the surface of section $q_2 = 3.0\, p_2 \geq 0$ for energy 800 cm^{-1} [18]:

1. *Initial Decay* (*decay 1*, time scale 5.5×10^{-2} ps). The initial state departs from the launching site and follows the guiding periodic orbit. As a result, the correlation function $\langle \Psi(0) | \Psi(t) \rangle$ decays rapidly to zero. This time scale contains information about the slope of the potential at the launching site. The center of the moving wavepacket initially satisfies the following relations

$$q_2(t) \approx q_2(0) + \frac{1}{2} \left(\frac{\partial V}{\partial q_2} \right) t^2 \tag{5.14}$$

$$p_2(t) \approx -\frac{\partial V}{\partial q_2} t \tag{5.15}$$

The distance travelled varies as t^2, while the momentum gain is linear with respect to time. We can consider the wavepacket as stationary at the launching site and gaining momentum upon departure, thus the rapid decay in the correlation function. This implies

$$\langle \Psi(0) | \Psi(t) \rangle \sim \exp \left[- \left(\frac{\partial V}{\partial q_2} \right)^2 \frac{t^2}{2\hbar^2 \alpha_2} \right] \tag{5.16}$$

where α_1, α_2 are the parameters defining the initial state, Eq. (5.3) [55]. This time scale determines the total width of the spectrum $\Delta\omega \approx 600\ \text{cm}^{-1}$.

2. *First Linear Recurrence* (*recurrence 1*, time scale 5.5×10^{-1} ps). The wavepacket follows the central guiding orbits, and after one full period most of the original packet returns to the launching site. At this moment we observe the first recurrence in the correlation function (see

Fig. 15). The unstability of the periodic orbit in the upper part of the spectrum is responsible for the loss of some original amplitude. From the first mapping onto the surface of section (Fig. 16) it is evident why the overlap between the propagated distribution (dark area) and the original one (white cricle) is smaller. The distribution has stretched along the direction of unstability of the central periodic orbit. The period of the central guiding orbit, t_1, is not very sensitive to energy across the spectrum. At multiple times this period of successive recurrences of the same kind appears, resulting in the resolution of many features in the spectrum separated by $2\pi/t_1 \approx 61\ \text{cm}^{-1}$, which is approximately the fundamental frequency of the q_2 mode.

3. *Decay of the Linear Recurrences* (*decay 2*, time scale 1.6 ps). Each of the linear recurrences carries less amplitude, due to the unstability of the dynamics along the q_2-axis. The time scale for this decay shows the rate of transfer of amplitude, and therefore energy, from the q_2 mode into the symmetric twisting mode, q_1 (Fig. 15). This transfer rate can be measured by the Lyapunov exponent of the unstable periodic orbit, which controls the exponential stretching of the distribution along the unstable manifold of the orbit (Fig. 16). In contrast to the period of the orbit, the stability matrix and the Lyapunov exponent are more sensitive to energy. Therefore a purely linearized analysis based on a single central orbit predicts the wrong width of the spectral features resolved by the linear recurrences. Within the

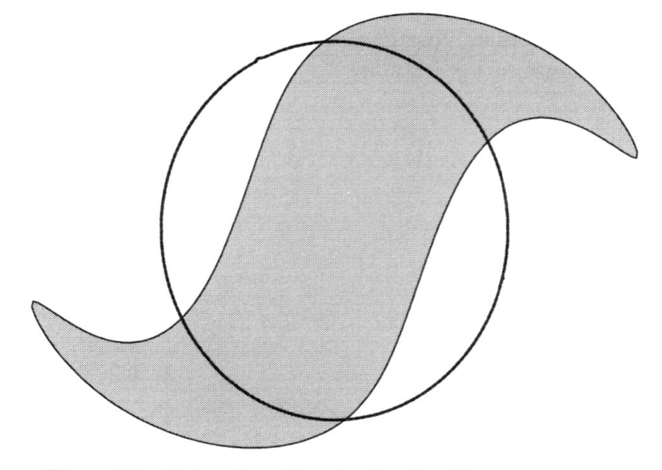

Figure 16. Poincaré section showing first linear recurrence.

linear theory, the decay of the recurrences is given by [18]

$$C_n = \exp(-\lambda n t_1/2) \tag{5.17}$$

where C_n is the maximum amplitude during the nth return along the periodic orbit. Since the Lyapunov exponent λ depends on energy, the Lorentzian width of the features varies across the spectrum. In the present case, the Lyapunov exponent is increasing with energy. The trjectories used as input into the cellular method successfully reproduce the feature widths and their change with energy. At higher energy, the decay is slightly faster and produces wider clumps of unresolved peaks in the spectrum (figure not shown).

4. *First Nonlinear Recurrence* (*recurrence 2*, time scale 2.8 ps). The trajectories that have left the vicinity of the periodic orbit contribute negligibly to the correlation function thereafter. However, due to the nonlinear dynamics, they will eventually return, resulting in a departure from the exponential *decay 2*. Although the dynamics is chaotic near the periodic orbit at energies above 545 cm^{-1}, for short to moderate times, the dynamics perpendicular to the periodic orbit is scarcely different from integrable motion near a barrier top in a double well. The orbits belong to one of the classes in this analogous integrable system—those inside the separatrix (below the barrier), and those outside it (above the barrier). Motion inside is local mode motion, whereas outside we have motion that strongly couples the local modes, with energy flowing from one to the other and back. For this initial nonlinear recurrence, most orbits return at approximately the same time, whether they are inside or outside the separatrix. Figure 17 shows this event happening after approximately five periods in the asymmetric twisting mode (*recurrence 2*). Energy has had time to flow from the asymmetric twisting mode, enter the local mode (twisting of a single benzene ring), and return to the asymetric motion. Thus the first nonlinear recurrence is a classical guiding path for an intramolecular vibrational energy transfer.
5. *Second Nonlinear Recurrence* (*recurrence 3*, time scale 5–7 ps) By this time a second nonlinear recurrence occurs (Fig. 18). The third recurrence type is associated with a new kind of intravibrational energy transfer. Figure 19 depicts this and the previous important time scales in terms of the normal and local modes. Energy from the symmetric mode flows to each of the local modes via this third recurrence.

To summarize: The CD method deals automatically with superposition of classical orbits; it is dynamics independent, fully vectorizable, and fur-

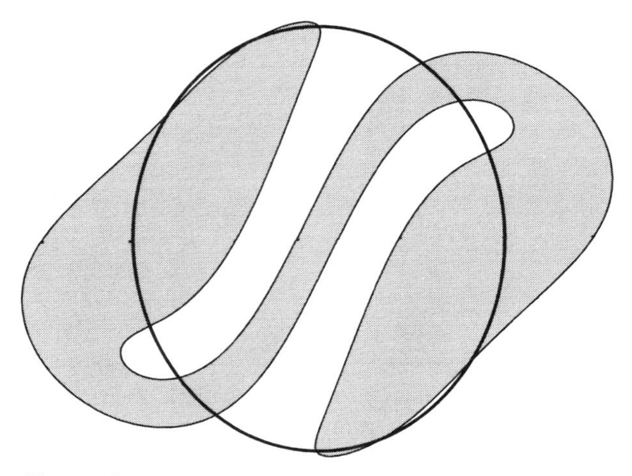

Figure 17. Poincaré section showing first nonlinear recurrence.

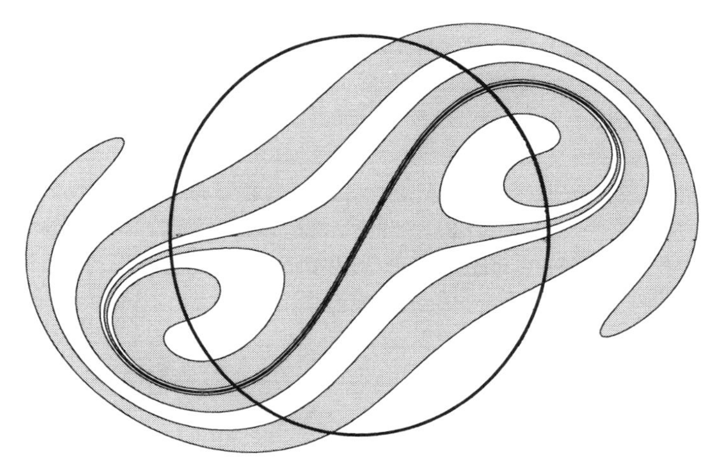

Figure 18. Poincaré section showing a following nonlinear recurrence.

thermore allows a simple identification of dynamical time scales in the system.

VI. ELECTRONIC COHERENCES

All of our examples so far have been limited to the dynamics of a molecular system on a single electronic state. There are, however, important chemical

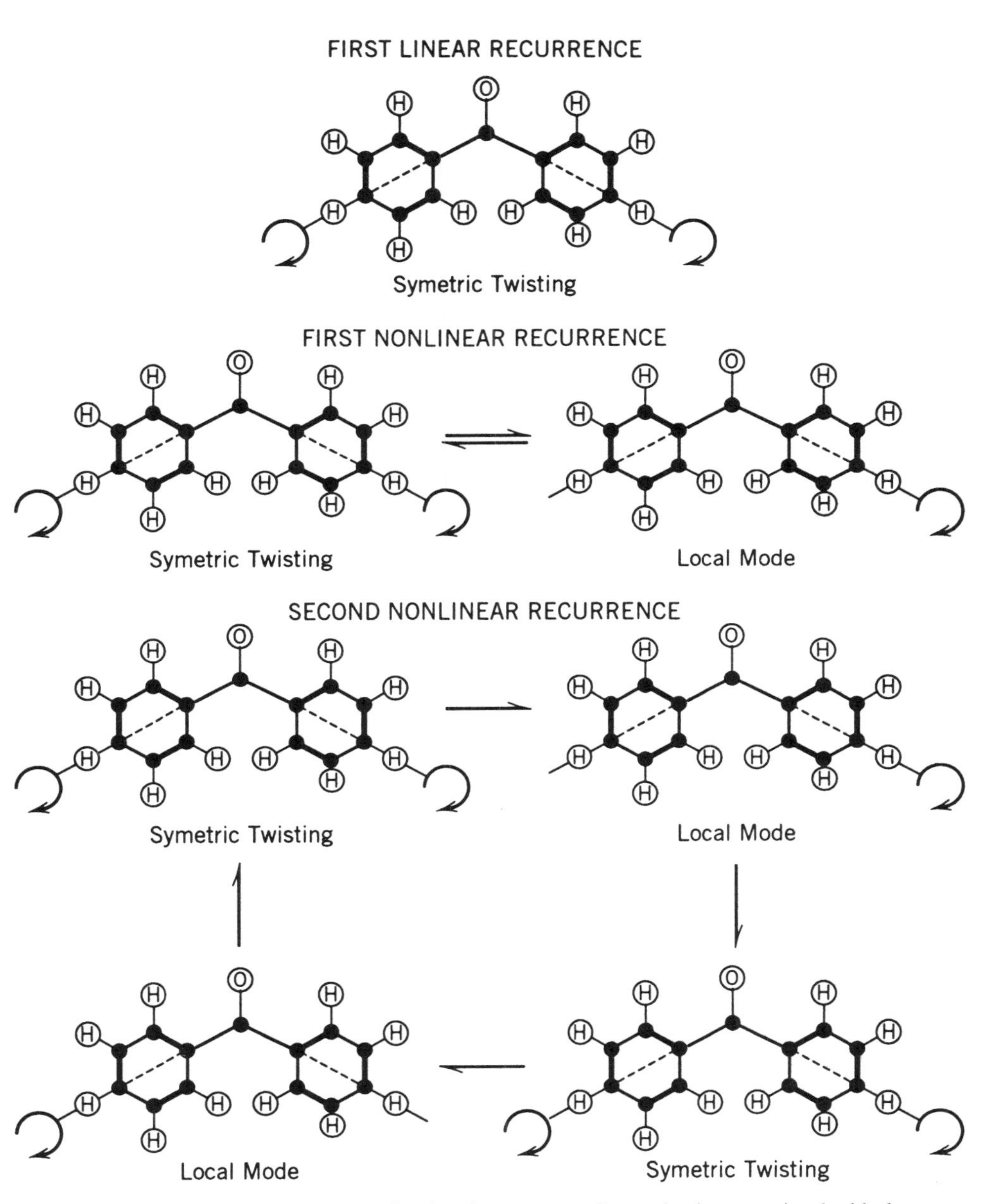

Figure 19. Sketch of the intravibrational energy transfer mechanism associated with the linear and nonlinear recurrences.

and physical processes that involve not only nuclear vibrations but also changes in the electronic configuration of the molecular orbitals. Nonadiabatic transitions [62], reactive scattering [63], pump-probe and photon echo spectroscopies are just few examples of the kind of processes that require the simultaneous integration of two or more coupled Schrödinger equations.

As indicated in the title of this section, in all the examples mentioned above, there exists some kind of electronic coherence induced via multiphoton excitation or dynamically by a breakdown of the Born–Oppenheimer approximation. This vast area of chemical physics has escaped semiclassical treatment for a long time, except for the Landau-Zener and similar models [64]. One of the main difficulties in developing an asymptotic expansion of vector wave functions is that the usual WKB *ansatz* leads to a set of coupled equations for the phases that do not satisfy Hamilton–Jacoby equations; thus any connection with classical mechanics gets lost. Even the WKB attempts to improve the Landau–Zener formula have always been limited to systems only one degree of freedom, and usually have the inconvenience of using complex plane integration to calculate the transition rate between the surfaces. Unfortunately this removes any kind of classical–quantum connection and the whole purpose of such a procedure is questionable.

Coherences of this kind, and tunneling, can be thought of as the most typical quantum effects. Consider for a moment using as a basis set molecular adiabatic states; the diagonal elements of the density matrix then correspond to the dynamics of the system on a given electronic state, while the off-diagonal elements represent the so-called coherences. Only the former have a clear classical meaning; they are the populations. We lack a classical picture for the off-diagonal elements. The motivation to develop a semiclassical method to calculate the nonlinear optical response of molecular systems [65] is that we would like to have a formulation that gives us a classical picture to better understand the semiclassical counterpart of the electronic coherences. It turns out that interpotential processes are less quantum in nature than expected.

The basic argument starts by ordering the dynamics using time-dependent perturbation theory. This is legitimate in the multiphoton-molecule scenario for laser intensities that are not too large, and in the reactive scattering case for small reactive yields. Once the dynamics is sequentially organized in a series of time propagations for several electronic states, then it is easy to find a generalization for the Van Vleck–Gutzwiller propagator that applies to multiple Born surfaces [66]. The method can be developed both in the density matrix or the wave function formalism.

A. Time-Dependent Perturbation Theory

Let us consider the zeroth order Hamiltonian, $\mathcal{H}_0$, of uncoupled electronic states. Then the time evolution of the system is given by the Liouville–Von

Neuman equation

$$\frac{d\rho}{dt} = -\frac{i}{\hbar}[\mathcal{H}_0, \rho] \tag{6.1}$$

whose formal solution is easily found to be

$$\rho(t) = \exp\left[-\frac{i\mathcal{H}_0 t}{\hbar}\right]\rho(0)\exp\left[\frac{i\mathcal{H}_0 t}{\hbar}\right] \tag{6.2}$$

For simplicity in the notation let us write the quantum propagator exp $(-i\mathcal{H}_0 t/\hbar)$ as $U(t)$. The dynamics on the Born surfaces becomes coupled in the presence of an external electromagnetic field. Using the dipole approximation the new Hamiltonian is

$$\mathcal{H} = \mathcal{H}_0 - \lambda\mu E(t) \tag{6.3}$$

where μ is the transition dipole operator, $E(t)$ the external electric field, and $\lambda = 1$ is simply a dummy parameter for the pertubation expansion. If the field–dipole interaction is not too strong the density matrix at time t can be expanded in power series of λ

$$\rho(t) \approx \rho^{(0)} + \rho^{(1)}\lambda + \rho^{(2)}\lambda^2 + \dots \tag{6.4}$$

Substitution in Liouville's equation and comparing terms of equal power in λ one arrives at the following set of coupled equations

$$\dot{\rho}^{(0)} = -\frac{i}{\hbar}(\mathcal{H}_0, \rho^{(0)}) \tag{6.5a}$$

$$\dot{\rho}^{(1)} = -\frac{i}{\hbar}(\mathcal{H}_0, \rho^{(1)}) + \frac{i}{\hbar}(\mu E(t), \rho^{(0)}) \tag{6.5b}$$

$$\vdots \quad \vdots \tag{6.5c}$$

$$\dot{\rho}^{(n)} = -\frac{i}{\hbar}(\mathcal{H}_0, \rho^{(n)}) + \frac{i}{\hbar}(\mu E(t), \rho^{(n-1)}) \tag{6.5d}$$

The couplings act like inhomogeneous terms in the set of differential equations and it is then easy to derive the formal solutions to the full system of equations using the zeroth order Green function,

$$\rho^{(0)}(t) = \rho^{(0)}(-\infty) \tag{6.6a}$$

$$\rho^{(1)}(t) = \left(\frac{i}{\hbar}\right) \int_{-\infty}^{t} dt_1\, L(t, t_1)E(t_1) + \text{h.c.} \tag{6.6b}$$

$$\rho^{(2)}(t) = \left(\frac{i}{\hbar}\right)^2 \int_{-\infty}^{t} dt_2 \int_{-\infty}^{t_2} dt_1\, S(t, t_1, t_2)E(t_1)E(t_2) + \text{h.c.} \tag{6.6c}$$

$$\rho^{(3)}(t) = \left(\frac{i}{\hbar}\right)^3 \int_{-\infty}^{t} dt_3 \int_{-\infty}^{t_3} dt_2 \int_{-\infty}^{t_2} dt_1 \cdot T(t, t_1, t_2, t_3)E(t_1)E(t_2)E(t_3) + \text{h.c.} \tag{6.6d}$$

The L, S, and T terms are operators related to the linear, second-, and third-order response of the molecular system to the external electric field. The term "h.c." refers to the Hermitian conjugate of the previous term. Using the Heisenberg representation they can be written as

$$L(t, t_1) = U(t)\mu(t_1)\rho^{(0)}U^{\dagger}(t) \tag{6.7a}$$

$$S(t, t_1) = U(t)\mu(t_2)\mu(t_1)\rho^{(0)}U^{\dagger}(t) - U(t)\mu(t_1)\rho^{(0)}\mu(t_2)U^{\dagger}(t) \tag{6.7b}$$

$$\begin{aligned} T(t, t_1, t_2, t_3) = {} & U(t)\mu(t_3)\mu(t_2)\mu(t_1)\rho^{(0)}U^{\dagger}(t) \\ & U(t)\mu(t_1)\rho^{(0)}\mu(t_2)\mu(t_3)U^{\dagger}(t) \\ & U(t)\mu(t_2)\rho^{(0)}\mu(t_1)\mu(t_3)U^{\dagger}(t) \\ & U(t)\mu(t_3)\rho^{(0)}\mu(t_1)\mu(t_2)U^{\dagger}(t) \end{aligned} \tag{6.7c}$$

The initial state of the system, $\rho^{(0)}$, will usually correspond to the thermal equilibrium distribution of the nuclear degrees of freedom of the ground electronic state.

B. Nonlinear Optical Response Formalism

Eqs. (6.6) represent the solutions to the time evolution of the system under the external driving field. Any optical measurement on an ensemble of such molecules is going to directly examine the magnitude and time dependence of the polarization induced by the external electromagnetic field. In a uniform sample of chromophores, the total polarization is N/V times the molecular polarization $P(t)$ (where N is the number of chromophores in volume V). Because of the λ expansion of the density matrix, it is also possible to write the molecular polarization in a power series of the external field strength

$$P(t) = P^{(0)} + P^{(1)}\lambda + P^{(2)}\lambda^2 + P^{(3)}\lambda^3 + \ldots \tag{6.8}$$

The zeroth order term corresponds to the average instantaneous polarization, $Tr(\mu\rho^{(0)})$. Higher terms are proportional to the nth power in the electric component of the field. By definition $P^{(n)}(t) = \mathrm{Tr}(\mu\rho^{(n)}(t))$, then

$$P^{(1)}(t) = \int_{-\infty}^{t} dt_1 R^{(1)}(t_1)E(t_1) \tag{6.9a}$$

$$P^{(2)}(t) = \int_{-\infty}^{t}\int_{-\infty}^{t_2} dt_1\, dt_2 R^{(2)}(t_1, t_2)E(t_1)E(t_2) \tag{6.9b}$$

$$P^{(3)}(t) = \int_{-\infty}^{t}\int_{-\infty}^{t_3}\int_{-\infty}^{t_2} dt_1\, dt_2\, dt_3\, R^{(3)}(t_1, t_2, t_3)E(t_1)E(t_2)E(t_3) \tag{6.9c}$$

The form of the above equations is typical for response theory [67]. The observable consists of a convolution of the external driving force (in our case the field $E(t)$) over the response kernels generated by a set of delta function excitations $R^{(n)}$. These latest terms, as expected, are correlation functions of the dipole operator moment along the dynamical path of the system. The explicit formulas for the first to third responses are as follows

$$R^{(1)} = \left(\frac{i}{\hbar}\right)[\langle\mu(t)\mu(t_1)\rho^{(0)}\rangle - \mathrm{c.c}] \tag{6.10a}$$

$$R^{(2)} = \left(\frac{i}{\hbar}\right)^2[\langle\mu(t)\mu(t_2)\mu(t_1)\rho^{(0)}\rangle - \langle\mu(t)\mu(t_1)\rho^{(0)}\mu(t_1)\rangle + \mathrm{c.c}] \tag{6.10b}$$

$$R^{(3)} = \left(\frac{i}{\hbar}\right)^3[\langle\mu(t)\mu(t_3)\mu(t_2)\mu(t_1)\rho^{(0)}\rangle + \langle\mu(t)\mu(t_1)\rho^{(0)}\mu(t_2)\mu(t_3)\rangle$$
$$+ \langle\mu(t)\mu(t_2)\rho^{(0)}\mu(t_1)\mu(t_3)\rangle + \langle\mu(t)\mu(t_3)\rho^{(0)}\mu(t_1)\mu(t_2)\rangle - \mathrm{c.c}] \tag{6.10c}$$

c.c represents the complex conjugate of the previous expression. Alternatively it is also possible to write Eq. (6.9) in terms of time delays rather than absolute times [67]

$$P^{(1)}(t) = \int_{0}^{+\infty} dt_1\, S^{(1)}(t_1)E(t - t_1) \tag{6.11a}$$

$$P^{(2)}(t) = \int_0^{+\infty} dt_2 \int_0^{+\infty} dt_1\, S^{(2)}(t_1, t_2)E(t - t_2)E(t - t_2 - t_1) \tag{6.11b}$$

$$P^{(3)}(t) = \int_0^{+\infty} dt_3 \int_0^{+\infty} dt_2 \int_0^{+\infty} dt_1\, S^{(3)}(t_1, t_2, t_3) \cdot E(t - t_3)E(t - t_3 - t_2)E(t - t_3 - t_2 - t_1) \tag{6.11c}$$

The relation between the $S^{(n)}$ functions and the previous $R^{(n)}$ is simple and can be derived by direct comparison of Eq. (6.9) and Eq. (6.11). The selection of one set over the other is more a matter of personal taste than fundamental importance.

Up to this point we have described the integration of the density matrix, the field expansion of the molecular polarization and its relationship with the so-called nonlinear response functions. How the polarization relates to any particular experimental scheme of detection is an issue beyond the scope of this review (see for example [67–69]).

C. Generalized Van Vleck Propagator

The polarizations are expressed using multidimensional correlation functions as in Eq. (6.10). Their quantum evaluation can be determined using a basis set calculation or directly in the time domain with an FFT–split-operator method or DVR. The question we want to answer, however, is how to find a semiclassical limit for such correlation functions. The problem may appear simple at first glance because the correlation function looks like a classical average of the dipole moment, but this is not so. In general the zeroth order Hamiltonian is a sum over several electronic states

$$\mathcal{H}_0 = \sum_k |k\rangle\, \langle k| h_k \tag{6.12}$$

where h_k is the nuclear Hamiltonian for the kth electronic state.

Calculating the response functions therefore involves the propagation of the initial state under successive potential surfaces. We should distinguish two possible limits, the Condon approximation, in which one considers the nuclear coordinate dependence of the dipole operator to be negligible, and the non-Condon limit. Both cases can be treated semiclassically [70], but in the present review we will deal only with the Condon approximation. Within this limit the nonlinear response functions can be reduced to the propagation of an initial state using a propagator as follows

$$\langle q|U_n(t_n)\ \ldots\ U_2(t_2)U_1(t_1)|q'\rangle \tag{6.13}$$

Again we have used the notation $U_k(t) = \exp(-ih_kt/\hbar)$ for the quantum time-evolution operator. We will call the asymptotic limit $\hbar \to 0$ of Eq. (6.13) the *generalized Gutzwiller propagator* because it involves the time evolution on not one, but multiple potential surfaces. A semiclassical expression for Eq. (6.13) to the same order in $\hbar$ as the Van Vleck formula can be obtained by writing the above expression as a product of normal Van Vleck propagators, and integrating by stationary phase approximation

$$\begin{aligned}\langle q|U_n(t_n)\ \ldots\ U_2(t_2)U_1(t_1)|q'\rangle \\ \approx \int dq_1dq_2\ \ldots\ dq_{n-1}(2\pi i\hbar)^{-n/2}\exp\left(\frac{i}{\hbar}S_T(q,q') - \frac{i\pi}{2}\mu_T\right) \\ \cdot \prod_{j=1}^{n}\left|\frac{\partial^2 S_j}{\partial q_j\,\partial q_{j-1}}\right|^{1/2}\end{aligned} \tag{6.14}$$

S_T is the sum of the actions along the classical paths connecting the pairs q_j, q_{j-1} (clearly $q_n = q$ and $q_0 = q'$). The index μ_T is the total Morse index, i.e., $\mu_1 + \ldots + \mu_n$. The integration runs over all paths from q' to q that make a transition between potential surfaces at $q_1, \ldots q_{n-1}$. In principle, a discontinuity in momentum is allowed at the instant of a potential surface transition, i.e., each path in the integral consists of a set of classical orbits joined only at the extreme coordinate values, Fig. 20.

However, as $\hbar \to 0$, of all the paths contributing to the total integral, only the connected ones are relevant. This means that both coordinate and momentum must be preserved at the crossing between potential surfaces. It is easy to derive this result from Eq. (6.14) by using stationary phase integration

$$\begin{aligned}\langle q|U_n(t_n)\ \ldots\ U_2(t_2)U_1(t_1)|q'\rangle \\ \approx (2\pi i\hbar)^{-1/2}\sum\exp\left[\frac{i}{\hbar}S_T(q,q') - \frac{i\pi}{2}(\mu_T + \delta)\right]\left|\frac{\partial^2 S_T}{\partial q\,\partial q'}\right|^{1/2}\end{aligned} \tag{6.15}$$

The sum goes over all classical paths, taking the system from q' to q in time $t = t_1 + t_2 + \ldots + t_n$ after going through the potential surfaces h_1, $\ldots h_n$. Then S_T is the total action, the sum of the actions accumulated along each of the n classical fragments in the path. As usual the phase shifts are

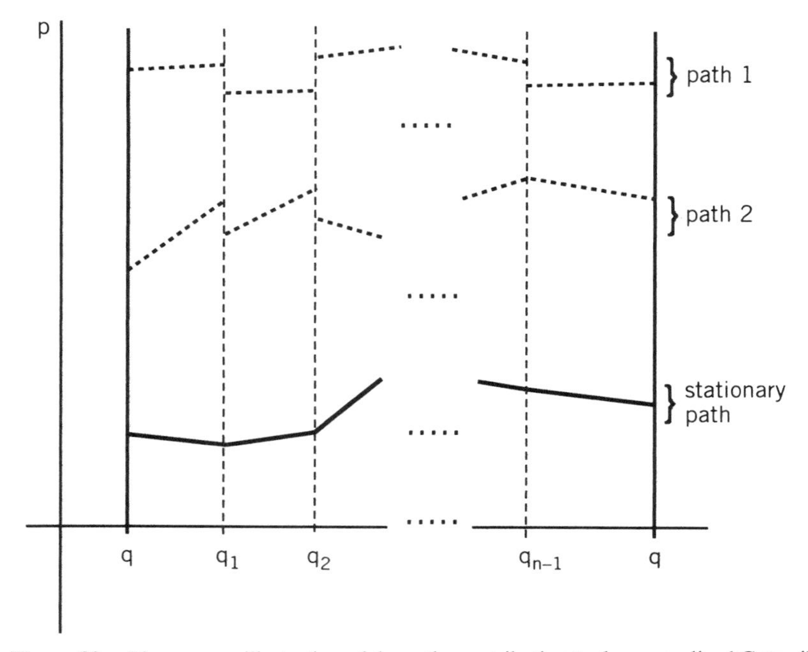

Figure 20. Phase space illustration of the paths contributing to the generalized Gutzwiller propagator. Dark line corresponds to the stationary point on the path integral, a classical path with no jumps in momentum or position.

crucial for the asymptotic formula to yield correct results; μ_T is the sum of the Morse indexes for the n classical orbits and the dephasing δ accounts for the construction of the total amplitude probability from a set of Van Vleck propagators.

Let us now use the stability matrices to define more precisely each element in the total phase shift. First consider the stability matrices $M^{(k)}$ for each of the n classical fragments in the path

$$\begin{bmatrix} dp_k \\ dq_k \end{bmatrix} = M^{(k)} \begin{bmatrix} dp_{k-1} \\ dq_{k-1} \end{bmatrix} \tag{6.16}$$

The matrices $M^{(1)}, M^{(2)}, \ldots, M^{(n)}$ contain all the information needed about the *local* stability of the classical orbits in the path. They are calculated numerically by integrating the linearized equations of motion together with Hamilton's equation for the orbit

$$\dot{p}_k(t) = -(\partial h_k / \partial q_k) \tag{6.17a}$$

$$\dot{q}_k(t) = (\partial h_k/\partial p_k) \tag{6.17b}$$

$$\dot{dp}_k(t) = -(\partial^2 h_k/\partial q_k^2)\, dq_k - (\partial^2 h_k/\partial q_k\, \partial p_k)\, dp_k \tag{6.17c}$$

$$\dot{dq}_k(t) = (\partial^2 h_k/\partial p_k^2)\, dp_k + (\partial^2 h_k/\partial q_k\, \partial p_k)\, dq_k \tag{6.17d}$$

By definition, the 2 × 2 matrices $M^{(k)}$ are unit matrices at the beginning of the classical fragment. As we have mentioned in Section III.B, the stability matrices thus defined are very useful because they provide a simple means of calculating partial derivatives of the actions along the classical orbits; for example $|\partial^2 S_k/\partial q_k\, \partial q_{k-1}|^{1/2}$ can be also written as $|M_{21}^{(k)}|^{-1/2}$. Moreover, if we construct a *total stability matrix* $\mathfrak{M}^{(n)}$ as

$$\begin{bmatrix} dp \\ dq \end{bmatrix} = \mathfrak{M}^{(n)} \begin{bmatrix} dp' \\ dq' \end{bmatrix} \tag{6.18}$$

where $\mathfrak{M}^{(n)} = M^{(n)} \times M^{(n-1)} \times \ldots M^{(1)}$, it becomes possible to write the *generalized Gutzwiller propagator* as

$$\langle q|U_n(t_n) \ldots U_2(t_2)U_1(t_1)|q'\rangle$$
$$\approx (2\pi i\hbar)^{-1/2} \sum |\mathfrak{M}_{21}^{(n)}|^{-1/2} \exp\left[\frac{i}{\hbar} S_T(q, q') - \frac{i\pi}{2}(\mu_T + \delta)\right] \tag{6.19}$$

The following matrix, D, contains the information about the discontinuity of the dynamics when crossing the potential surfaces. By examining the eigenvalues of D we will determine the phase shift δ

$$D = \begin{bmatrix} \frac{M_{11}^{(n-1)}}{M_{21}^{(n-1)}} + \frac{M_{22}^{(n)}}{M_{21}^{(n)}} & -\frac{1}{M_{21}^{(n-1)}} & 0 & 0 & \cdots & 0 \\ -\frac{1}{M_{21}^{(n-1)}} & \frac{M_{11}^{(n-2)}}{M_{21}^{(n-2)}} + \frac{M_{22}^{(n-1)}}{M_{21}^{(n-1)}} & -\frac{1}{M_{21}^{(n-2)}} & 0 & \cdots & 0 \\ 0 & -\frac{1}{M_{21}^{(n-2)}} & \frac{M_{11}^{(n-3)}}{M_{21}^{(n-3)}} + \frac{M_{22}^{(n-2)}}{M_{21}^{(n-2)}} & -\frac{1}{M_{21}^{(n-3)}} & \cdots & 0 \\ \vdots & \vdots & \vdots & \vdots & \ddots & \vdots \\ 0 & 0 & 0 & 0 & -\frac{1}{M_{21}^{(2)}} & \frac{M_{11}^{(1)}}{M_{21}^{(1)}} + \frac{M_{22}^{(2)}}{M_{21}^{(2)}} \end{bmatrix} \tag{6.20}$$

The index δ, as obtained from the stationary phase approximation, is given by the number of negative eigenvalues of D. Furthermore, the structure of the matrix D is so simple that it is possible to write an efficient algorithm for the determination of its eigenvalues. Let us denote the eigenvalues of D by $\lambda_1, \lambda_2, \ldots, \lambda_{n-1}$, then

$$\lambda_1\lambda_2 \ldots \lambda_{k-1} = \frac{\Pi_{i=1}^{k} M_{21}^{(i)}}{\mathfrak{M}_{21}^{(k)}} \tag{6.21}$$

for $k = 2, 3, \ldots, n$.

The determination of the sign of the eigenvalues is easly accomplished by comparing λ_1, $\lambda_1\lambda_2$, etc., along the classical path going from q' to q.

The analogy between the semiclassical expressions Eq. (3.37) and Eq. (6.14) essentially tells us about the basic underlying path-integral structure common to the two types of propagators. In both cases, there are phase shifts each time a classical orbit goes through a focal point. This raises an interesting question: What is the physical origin of the discontinuity phase shift δ? It is easy to show [66] that δ and the normal μ index physically correspond to the same geometrical structures of the Lagrangian manifolds. Both are associated with caustics generated during the folding of the classical manifolds. δ appears in Eq. (6.14) as an index related to the surface crossing because the description given for the calculation of μ_T is *local*! This index is calculated in terms of the separate stability matrices for each of the classical orbits that constitutes the full path in Eq. (6.14). We could use a *global* description, which means evaluating the total stability matrix $\mathfrak{M}^{(k)}$ by integration of the linearized equations of motion across the several Hamiltonian jumps without *resetting the stability matrix* nor relying on any local $M^{(k)}$. From this new perspective it is possible to show [66] that the *generalized Gutzwiller* formula takes now the form

$$\begin{aligned}\langle q|U_n(t_n) \ldots U_2(t_2)U_1(t_1)|q'\rangle \\ \approx (2\pi i\hbar)^{-1/2} \sum |\mathfrak{M}_{21}^{(n)}|^{-1/2} \exp\left(\frac{i}{\hbar} S_T(q, q') - \frac{i\pi}{2}\mu\right)\end{aligned} \tag{6.22}$$

where $\mu = \mu_T + \delta$, the global Morse index, is calculated by finding the number of times $\mathfrak{M}_{21}$ changes sign across the stationary path.

D. Single-Pair Expansion

Now we have the tool for evaluating any kind of multidimensional correlation function for arbitrary systems. After unfolding the Heisenberg $\mu(t_i)$ operators of any of the correlations functions one has

$$C(t_k, t_l, \ldots, t_i) = \mathrm{Tr}\left(\underbrace{U_k(t_k)U_l(t_l)\ldots U_m(t_m)}_{\text{left}}\,\rho^{(0)}(-\infty)\,\underbrace{U_n^\dagger(t_n)U_o^\dagger(t_o)\ldots U_i^\dagger(t_i)}_{\text{right}}\right) \tag{6.23}$$

Notice that we are still using the Condon approximation and that for notational convenience we have set $\mu = 1$.

Next we have to specify the initial state. In a common scenario the chromophore would be at equilibrium with a bath at temperature T, and the harmonic approximation of the ground state is usually not bad for temperatures not too high. Therefore let us consider the following initial state

$$\langle q_l'|\rho^{(0)}(-\infty)|q_r'\rangle = \sqrt{\frac{2(\alpha-\gamma)}{\pi}}\exp\left[-\alpha(q_l')^2 - \alpha(q_r')^2 + 2\gamma q_l' q_r'\right] \tag{6.24}$$

$$\alpha = \frac{m\omega}{2\hbar}\coth(\hbar\beta\omega) \tag{6.25}$$

$$\gamma = \frac{m\omega}{2\hbar\sinh(\hbar\beta\omega)} \tag{6.26}$$

where ω is the fundamental frequency of the ground state and β is the inverse of kT. If the initial state is too far from harmonic because of high temperature or strong nonlinearities, or because the system is far from equilibrium it is always possible to represent the exact initial state as a sum over Gaussians of the form Eq. (6.24).

Using the coordinate representation it is straightforward to see that the trace in Eq. (6.23) is equivalent to the integration of the initial density matrix with two generalized Gutzwiller propagators, one for the left and another for the right

$$C(t_k, t_l, \ldots, t_i) = \int dq \int dq_l'\, dq_r'\, \langle q|U_k(t_k)\ldots U_m(t_m)|q_l'\rangle\, \langle q_l'|\rho^{(0)}|q_r'\rangle \cdot \langle q_r'|U_n^\dagger(t_n)\ldots U_i^\dagger(t_i)|q\rangle \tag{6.27}$$

Direct substitution of the semiclassical propagator leads to [65]

$$C(t_k, t_l, \ldots, t_i) \approx \frac{1}{2\pi\hbar} \int dq \int dq_r' \, dq_l' \, \rho^{(0)}(q_l', q_r') |\mathfrak{M}_{21}^l \mathfrak{M}_{21}^r|^{-1/2} \tag{6.28}$$

$$\cdot \exp\left[\frac{i}{\hbar}(S_l - S_r) - \frac{i\pi}{\hbar}(\mu_l - \mu_r)\right] \tag{6.29}$$

The subindices l and r refer to the "left" and "right" propagators in Eq. (6.23). To each of the semiclassical operators corresponds a "classical" path that goes through the several potentials and take the ket $|q_l'\rangle$ forward in time to $\langle q|$ and the bra $\langle q_r'|$ *forward* to $|q\rangle$. There is a total stability matrix for the left and another for the right propagation. The actions $S_l(q, q_l')$ and $S_r(q, q_r')$ are both dependent on the variables of integration running over the initial density matrix.

Exact integration of Eq. (6.23) is practically impossible except for a few simple examples. Numerically two methods similar to the already described GWD and CD can be applied. Expanding each generalized Gutzwiller propagator by a single path leads to the "single pair" expansion of the dynamics. Both classical paths form a Liouville trajectory that controls the time evolution of the density matrix, Fig. 21.

As with the GWD approximation, let us consider first the case of relatively short propagation time. Expanding the dynamics of the density matrix about its center is then sufficient, and this expansion will preserve the Gaussian structure of the evolved state. The initial average positions and momenta are $q_l' = 0$, $q_r' = 0$, $p_l' = 0$, and $p_r' = 0$. If we denote $(\bar{q}_l, \bar{p}_l, \bar{q}_r, \bar{p}_r)$ as the

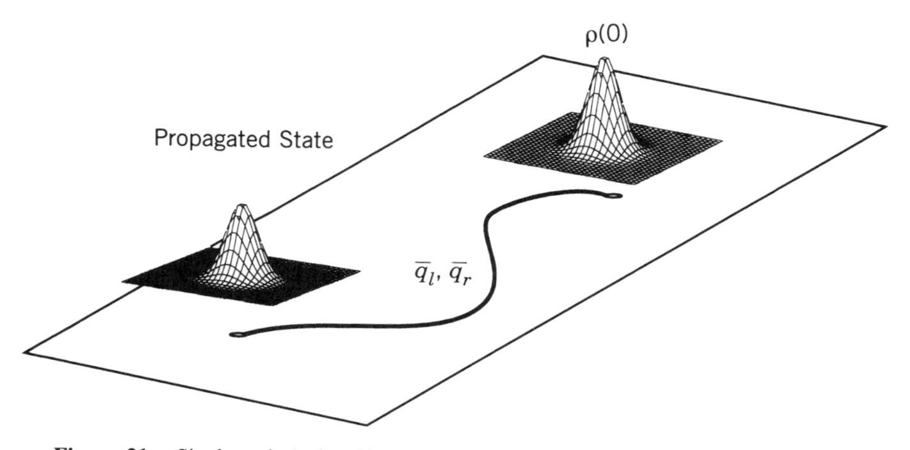

Figure 21. Single-path (pair-orbit) expansion of the Liouville–Von Neuman equation. The initially localized density matrix follows the classical guiding path along configuration space in the same fashion as Gaussians in the GWP method.

two classical paths departing from the center of the initial state then linearization of the dynamics leads to

$$p_l \approx \bar{p}_l + \mathcal{M}^l_{11} p'_l + \mathcal{M}^l_{12} q'_l \tag{6.30a}$$

$$q_l \approx \bar{q}_l + \mathcal{M}^l_{21} p'_l + \mathcal{M}^l_{22} q'_l \tag{6.30b}$$

$$p_r \approx \bar{p}_r + \mathcal{M}^r_{11} p'_r + \mathcal{M}^r_{12} q'_r \tag{6.30c}$$

$$q_r \approx \bar{q}_r + \mathcal{M}^r_{11} p'_r + \mathcal{M}^r_{12} q'_r \tag{6.30d}$$

The classical actions for the left and right propagations can be expanded about the guiding classical path as well

$$S_l(q_l, q'_l) \approx \bar{S}_l + \bar{p}_l(q_l - \bar{q}_l) + \frac{\mathcal{M}^l_{11}}{2\mathcal{M}^l_{21}} (q_l - \bar{q}_l)^2 + \frac{\mathcal{M}^l_{22}}{2\mathcal{M}^l_{21}} (q'_l)^2 - \frac{1}{\mathcal{M}^l_{21}} (q_l - \bar{q}_l) q'_l \tag{6.31a}$$

$$S_r(q_r, q'_r) \approx \bar{S}_r + \bar{p}_r(q_r - \bar{q}_r) + \frac{\mathcal{M}^r_{11}}{2\mathcal{M}^r_{21}} (q_r - \bar{q}_r)^2 + \frac{\mathcal{M}^r_{22}}{2\mathcal{M}^r_{21}} (q'_r)^2 - \frac{1}{\mathcal{M}^r_{21}} (q_r - \bar{q}_r) q'_r \tag{6.31b}$$

Inserting the previous expansions into Eq. (6.28) and performing the integrations, the correlation function problem transforms in a simple Gaussian quadrature

$$C(t_k, t_l, \ldots, t_i) \approx \sqrt{\frac{2(\alpha - \gamma)}{k_2}} \left(\frac{\operatorname{sign}(\mathcal{M}^r_{21} \mathcal{M}^l_{21})}{g} \right)^{1/2} \exp\left(k_0 + \frac{k_1^2}{4k_2} \right) \tag{6.32}$$

with

$$g = (2\hbar \mathcal{M}^l_{21} \alpha - i\mathcal{M}^l_{22})(2\hbar \mathcal{M}^r_{21} \alpha + i\mathcal{M}^r_{22}) - 4\hbar^2 \mathcal{M}^r_{21} \mathcal{M}^l_{21} \gamma^2 \tag{6.33}$$

$$k_2 = \frac{i}{2\hbar} \left(\frac{\mathcal{M}^r_{11}}{\mathcal{M}^r_{21}} - \frac{\mathcal{M}^l_{11}}{\mathcal{M}^l_{21}} \right) - \frac{2\gamma}{g} + \frac{2\alpha\hbar \mathcal{M}^r_{21} + i\mathcal{M}^r_{22}}{2\hbar g \mathcal{M}^l_{21}} + \frac{2\alpha\hbar \mathcal{M}^l_{21} - i\mathcal{M}^l_{22}}{2\hbar g \mathcal{M}^r_{21}} \tag{6.34}$$

$$k_1 = \frac{i}{\hbar}(\bar{p}_l - \bar{p}_r) - \frac{i\mathcal{M}^l_{11}}{\hbar\mathcal{M}^l_{21}}\bar{x}_l + \frac{i\mathcal{M}^r_{11}}{\hbar\mathcal{M}^r_{21}}\bar{x}_r + \frac{(2\alpha\hbar\mathcal{M}^r_{21} + i\mathcal{M}^r_{22})}{g\hbar\mathcal{M}^l_{21}}\bar{x}_l$$
$$+ \frac{(2\alpha\hbar\mathcal{M}^l_{21} - i\mathcal{M}^l_{22})}{g\hbar\mathcal{M}^r_{21}}\bar{x}_r - \frac{2\gamma}{g}(\bar{x}_l + \bar{x}_r) \tag{6.35}$$

$$k_0 = \frac{i}{\hbar}(\bar{S}_l - \bar{S}_r) + \frac{i}{\hbar}(\bar{x}_r\bar{p}_r - \bar{x}_l\bar{p}_l) + \frac{i\mathcal{M}^l_{11}}{2\hbar\mathcal{M}^l_{21}}\bar{x}_l^2 - \frac{i\mathcal{M}^r_{11}}{2\hbar\mathcal{M}^r_{21}}\bar{x}_r^2 + \frac{2\gamma\bar{x}_r\bar{x}_l}{g}$$
$$- \frac{2\alpha\hbar\mathcal{M}^r_{21} + i\mathcal{M}^r_{22}}{2\hbar g\mathcal{M}^l_{21}}\bar{x}_l^2 - \frac{2\alpha\hbar\mathcal{M}^l_{21} - i\mathcal{M}^l_{22}}{2\hbar g\mathcal{M}^r_{21}}\bar{x}_r^2 \tag{6.36}$$

E. Example: Ground plus Excited State

Let us now apply the previous semiclassical approximation to a two-level system. Consider a one-degree-of-freedom-system with a harmonic ground state and a Morse potential as excited state. The full Hamiltonian of the system is given by

$$H = h_1(q)|g\rangle\langle g| + h_2(q)|e\rangle\langle e| - \hat{\mu}_{ge}\cdot E(t)|g\rangle$$
$$\cdot\langle e| - \hat{\mu}_{eg}\cdot E(t)|e\rangle\langle g| \tag{6.37}$$

with

$$h_1(q) = \frac{p^2}{2m} + \frac{1}{2}m\omega^2q^2 \tag{6.38}$$

$$h_2(q) = \frac{p^2}{2m} + D_e\{1 - \exp[-\alpha_e(q - q_e)]\}^2 + E_{ge} \tag{6.39}$$

Figure 22 depicts both potential curves ($m = 1$, $\omega = 6.0$, $D_e = 50.0$, $q_e = -1.0$, $\alpha_e = 0.3086$ a.u.).

The lowest nonzero nonlinear response for the two-level system is the third-order $S^{(3)}(t_1, t_2, t_3)$ (we will use time delays instead of absolute times). From Eq. (6.11c) and Eq. (6.10c) we have

$$S^{(3)}(t_1, t_2, t_3) = \left(\frac{i}{\hbar}\right)^3 \sum_{k=1}^{4}(R_k^{(3)} - \text{c.c}) \tag{6.40}$$

where

$$R_1^{(3)} = \text{Tr}\,[\mu_{12}(t)\mu_{21}(t - t_3 - t_2 - t_1)\rho^{(0)}\mu_{12}(t - t_3 - t_2)\mu_{21}(t - t_3)] \tag{6.41a}$$

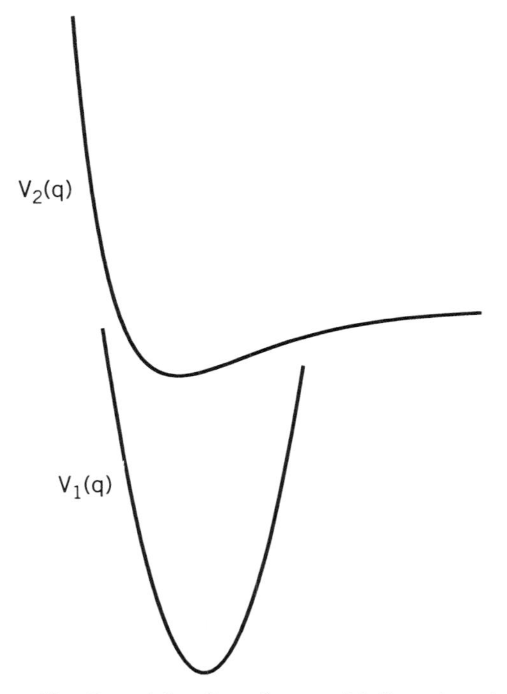

Figure 22. Potential surfaces for a model diatomic molecule.

$$R_2^{(3)} = \mathrm{Tr}\,[\mu_{12}(t - t_3 - t_2)\rho^{(0)}\mu_{12}(t - t_3 - t_2 - t_1)\mu_{21}(t - t_3)] \tag{6.41b}$$

$$R_3^{(3)} = \mathrm{Tr}\,[\mu_{12}(t)\mu_{21}(t - t_3)\rho^{(0)}\mu_{12}(t - t_3 - t_2 - t_1)\mu_{21}(t - t_3 - t_2)] \tag{6.41c}$$

$$R_4^{(3)} = \mathrm{Tr}\,[\mu_{12}(t)\mu_{21}(t - t_3)\mu_{12}(t - t_3 - t_2)\mu_{21}(t - t_3 - t_2 - t_1)\rho^{(0)}] \tag{6.41d}$$

Each of the $R_k^{(3)}$ corresponds to a Liouville pathway for the third-order excitation of the system. Sometimes it becomes hard to remember the expressions of each of the correlation functions and for that reason it is customary [67] to use a graphical representation of the Liouville paths. As we mentioned earlier, each correlation function can be viewed as a time evolution of the density matrix with multiple dipole–field interactions along the time axis. Figure 23 depicts the double-sided Feynman diagram representations of the four terms in the nonlinear response. For each term two vertical bars

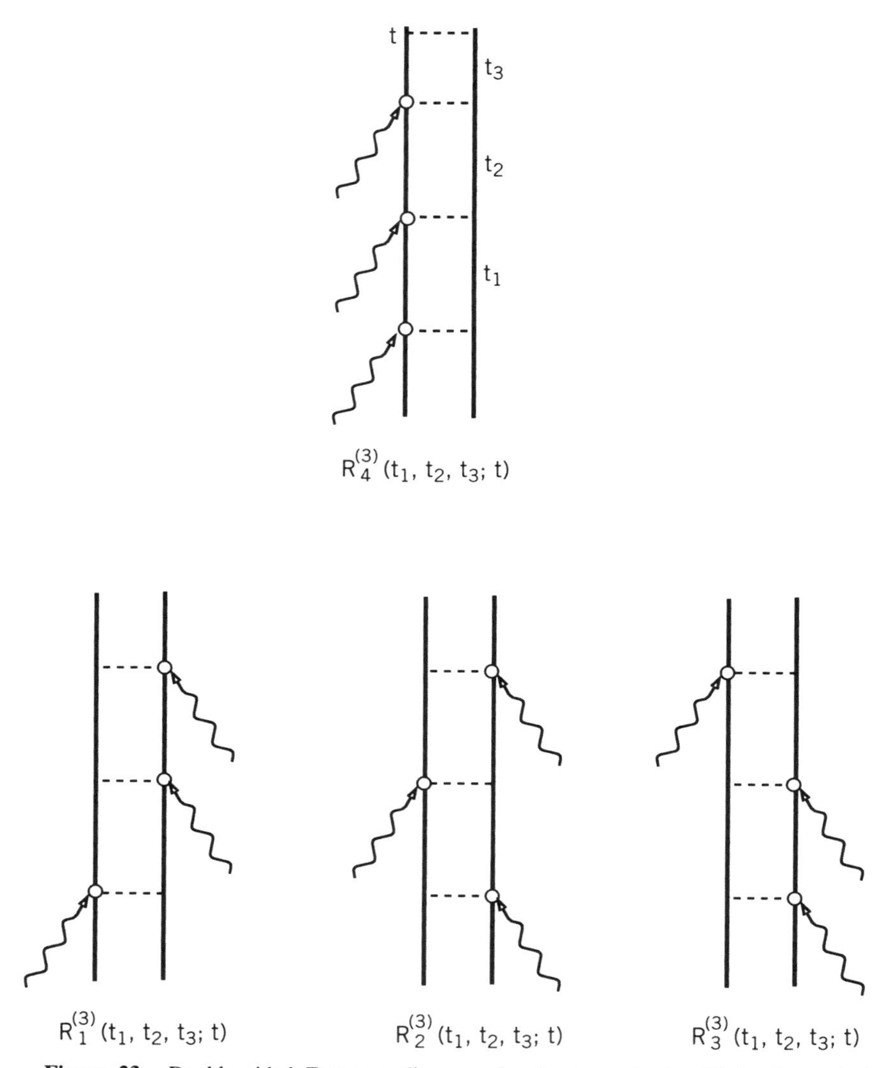

Figure 23. Double sided Feynman diagrams for the terms in the third-order optical response.

are used to denote the time-evolving density matrix: the left bar corresponds to the ket-side and the right bar to the bra-side; the system starts at equilibrium at the bottom of the bars and moves forward in time vertically; along the bars we can draw to the left or right the dipole–field interactions according to the side of the density matrix where they act.

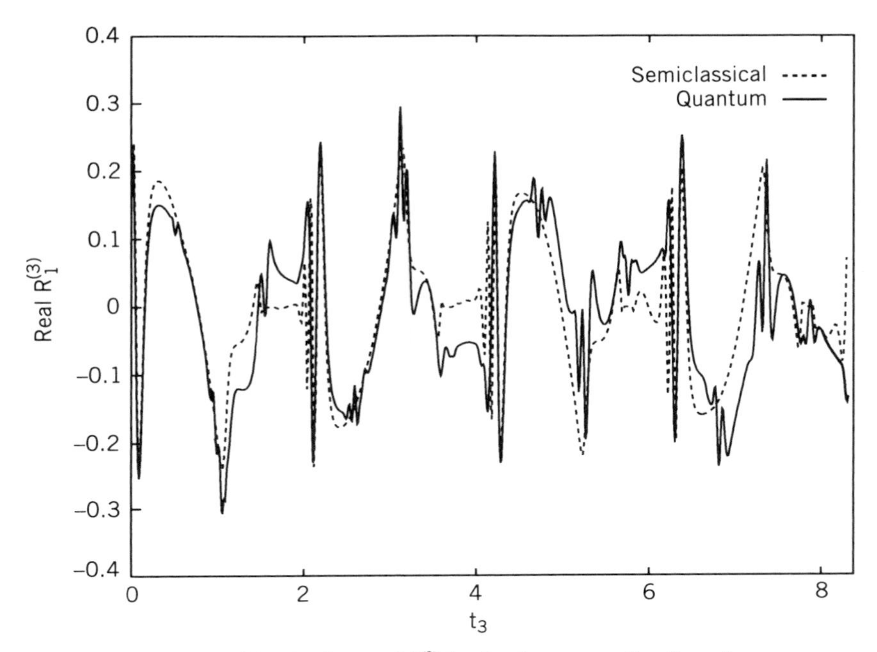

Figure 24. Real part of $R_1^{(3)}$ for fixed t_1, t_2 as a function of t_3.

Figures 24–27 show a comparison between semiclassical and quantum correlation functions. The initial system starts at equilibrium in the harmonic state with $kT = 10\hbar\omega$ (k is Boltzman's constant). The full time-dependence of the $R_k^{(3)}$ is not shown; for this example t_1 and t_2 are fixed to 0.59 and 0.61 (a.u.). The third parameter, t_3, is used as the independent variable. This one describes the evolution of an electronic coherence between the ground and excited states. Performing numerical quantum calculations is not a difficult task. The present example was calculated using a Liouville propagator based on the FFT–Feit and Fleck method. There are other methods available for the integration of the Liouville–Von Neuman equation [71]. As for the semiclassical calculation we run a "single pair" of classical orbits, launched at the center of the initial density matrix and calculated the correlation functions using GWD.

The agreement between this simple semiclassical approximation and the quantum results is very satisfactory, particularly when comparing the CPU times required for each calculation. Performing the FFT propagation took several hours on a workstation, while the simple semiclassical one only took a few seconds. A single-pair approximation is not the "exact" semiclassical solution to the generalized Gutzwiller propagation; by using only two orbits

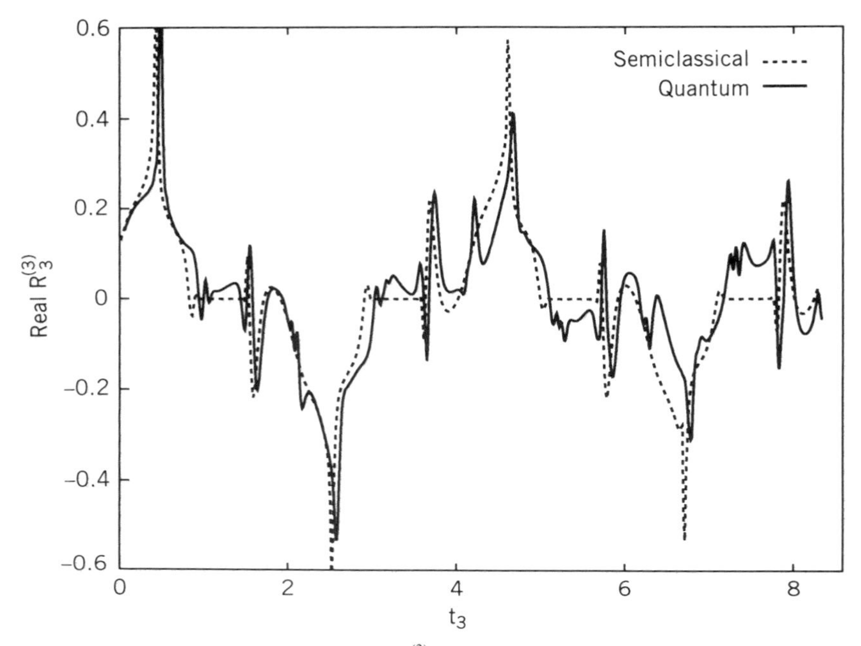

Figure 25. Real part of $R_3^{(3)}$ for fixed t_1, t_2 as a function of t_3.

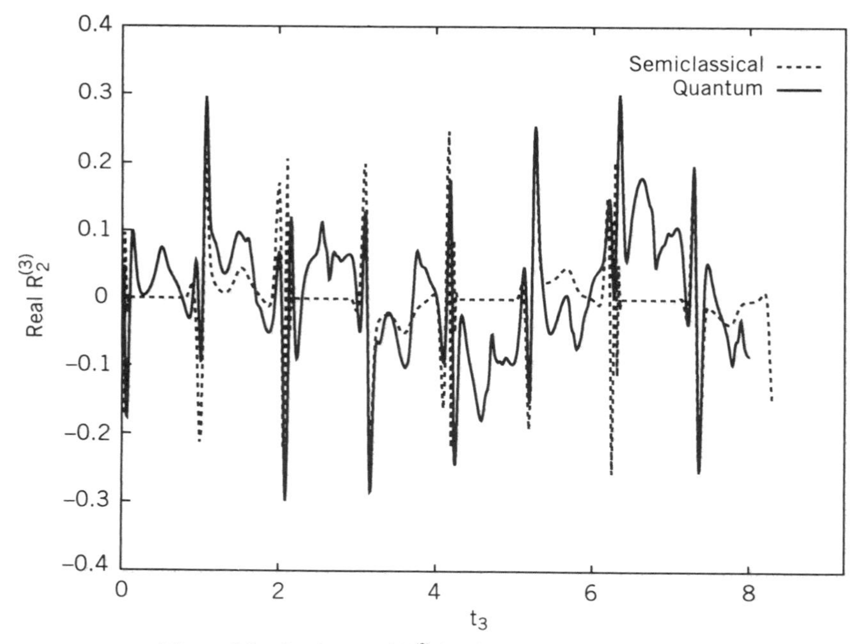

Figure 26. Real part of $R_2^{(3)}$ for fixed t_1, t_2 as a function of t_3.

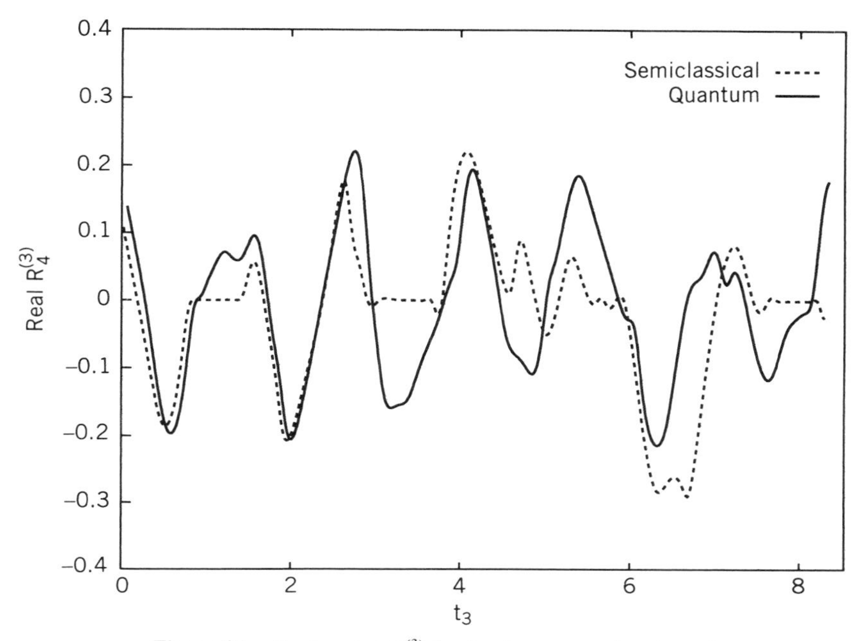

Figure 27. Real part of $R_4^{(3)}$ for fixed t_1, t_2 as a function of t_3.

for expanding the left and right actions we have neglected contributions to the correlation functions from classical paths that make a long excursion away from the two guiding orbits. Nevertheless this method remains extremely useful for short-time dynamics and a large number of degrees of freedom (notice that numerical integration of classical equations of motion scales benevolently versus number of degrees of freedom).

The semiclassical approximation makes it possible to represent in phase space the classical route followed by the density matrix on each of the Liouville paths, $R_k^{(3)}$ (see Figs. 28–31). The corresponding double sided Feynman diagram for the correlation functions is shown in each portrait plus phase space representations of the "single pair" of orbits. In the upper-left we show the orbit associated with the left side of the density matrix propagation and on the lower right the right-side orbit. By interpolating the coordinate excursions of both phase space plots one can extract the Liouville guiding orbit for the time-evolving density matrix; this is shown in the lower-left plot, which is simply a diagram in configuration space of the path followed by the density matrix in time. In all the four figures we can notice that the Liouville guiding orbits obviously correspond to a quasi-integrable type of motion.

Let us understand a little bit better each of the classical paths in order to see the classical correspondence of the electronic coherence. The labels a,

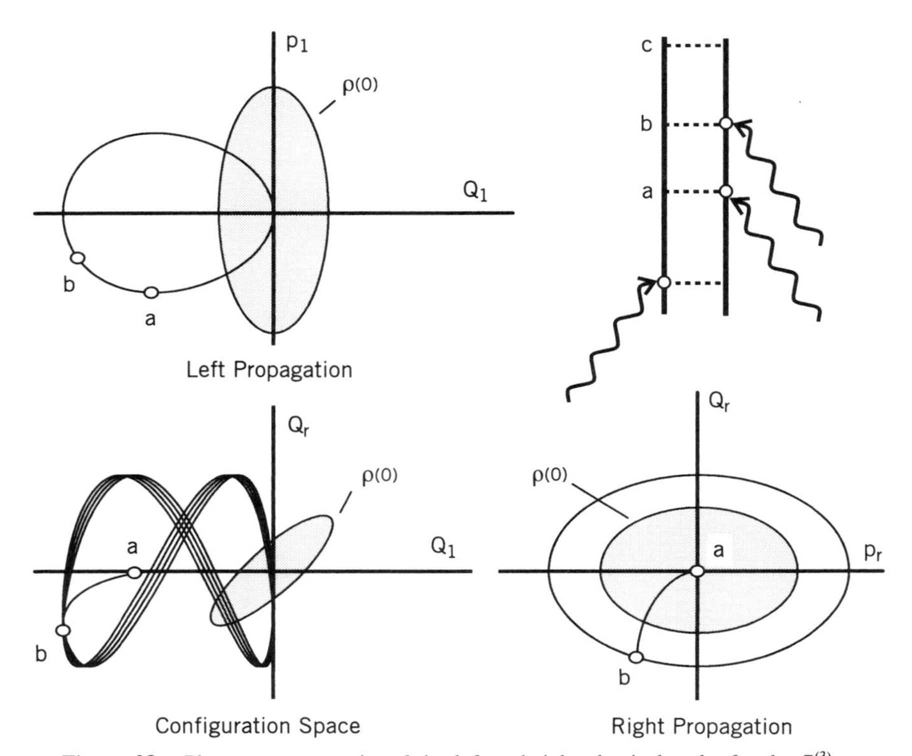

Figure 28. Phase space portraits of the left and right classical paths for the $R_1^{(3)}$ term. The coordinate representation of the guiding orbit for the density matrix is shown on the lower-left diagram.

b, and c have been used to identify the instants in time after the t_1, t_2, and t_3 propagations (see Feynman diagrams). Simultaneously the location of the system at these moments is also indicated on the classical paths. Each path (left and right) is thus formed by three classical orbits: from time 0 to a, a to b, and finally b to c. Not all the fragments necessarily evolve on the same potential surfaces. By looking at Fig. 28 the Feynman diagram indicates that the left path runs on the excited state due to the first pulse excitation by the left, while the right path starts running on the ground state then moves to the Morse potential after the second pulse interaction, and finally, after the third and last pulse, continues on the ground state (similarly for the other Liouville terms in the third-order response). For that reason if we study the part of the path that corresponds to the t_3 time propagation we will see (lower-left diagram) that the orbit described is nothing but the combination of motions in the ground and excited states. Here the true nature of the

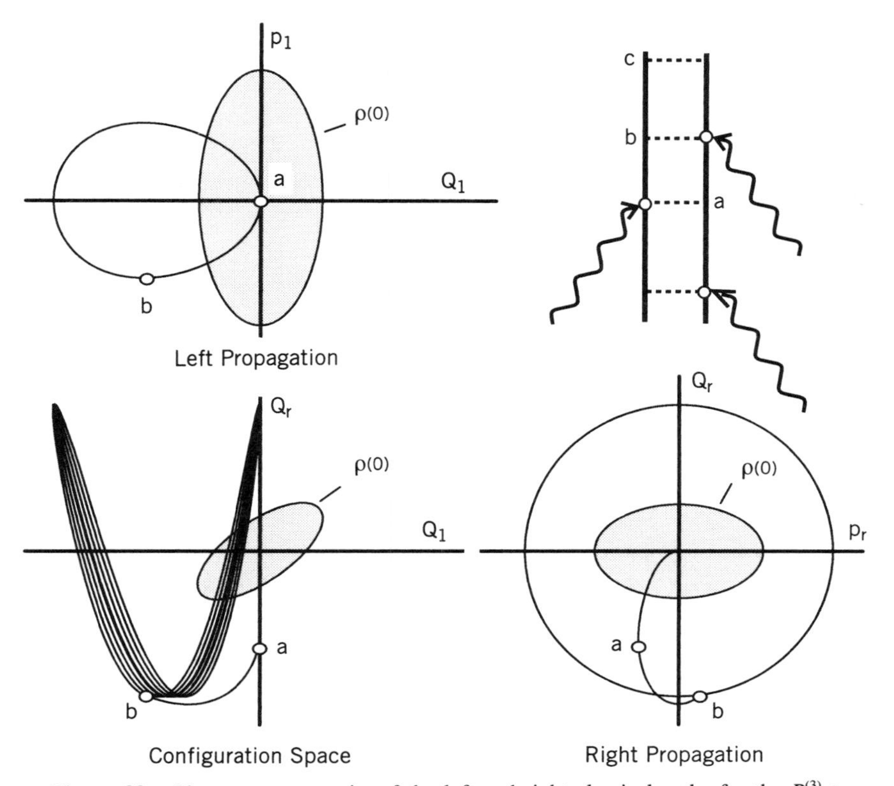

Figure 29. Phase space portraits of the left and right classical paths for the $R_2^{(3)}$ term. The coordinate representation of the guiding orbit for the density matrix is shown on the lower-left diagram.

quantum electronic coherence stimulated during multiphoton experiments starts to emerge.

Although the individual Liouville components of the third-order response can be measured [72], most experiments only probe a limited fraction of the correlation function. Any pump-probe or two-photon experiment explores at least two dimensions of the $R_k^{(3)}$ [73]. For instance, a two-photon echo experiment is described by the function $R_2^{(3)}(t_1, 0, t_3)$ where t_1 is the time difference between the two light pulses and t_3 the time interval from the last pulse to the echo detection. Along the $t_1 = t_3$ direction it is possible to eliminate the inhomogenous broadening that hinders the measurement of spectral lines in a normal linear absorption spectrum.

Figure 32 shows the absolute value of the $R_2^{(3)}$ term for our harmonic plus

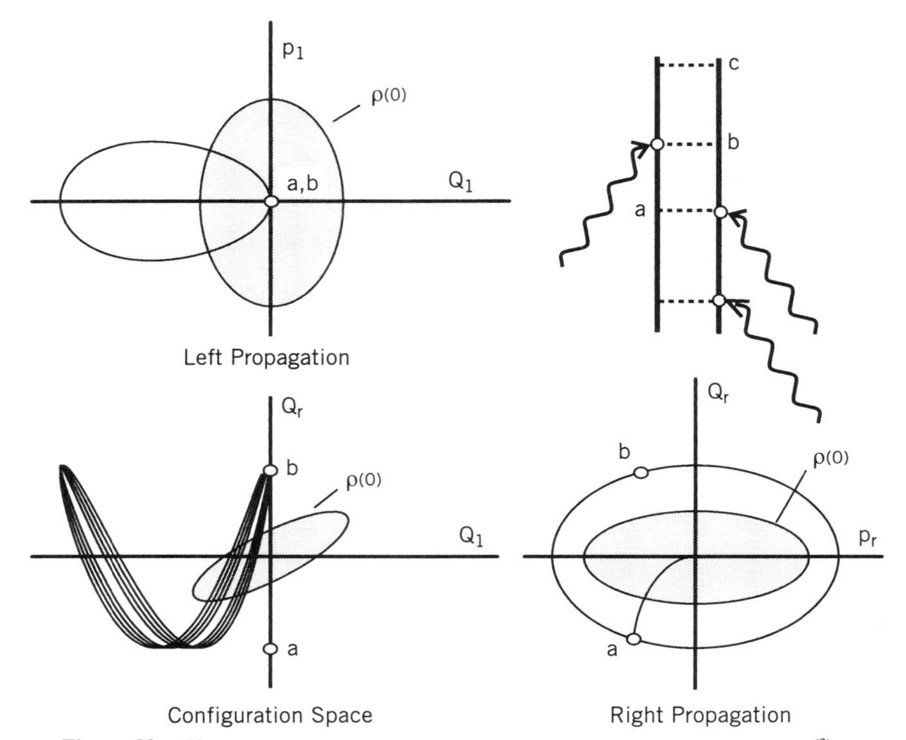

Figure 30. Phase space portraits of the left and right classical paths for the $R_3^{(3)}$ term. The coordinate representation of the guiding orbit for the density matrix is shown on the lower-left diagram.

Morse model as a function of t_1 and t_3. Two clear types of recurrence emerge in the response, labeled A and B. Reading the correlation function along the t_3 or the t_1 axis we observed a periodic sequence of type A recurrences. The time between these features is the period of the Morse orbit that serves as the guiding path in the excited state. Thus designing an experiment that only explores either direction would yield no more information than a normal photoabsorption experiment. However in the $t_1 = t_3$ direction both recurrences occur, i.e., type A and B.

A classical analysis of both types of recurrences will demonstrate the intimate relationship between the generated electronic coherences and classical mechanics. Each recurrence can be identified with certain combinations of classical motions on the Morse and harmonic surfaces. Let us study first type A:

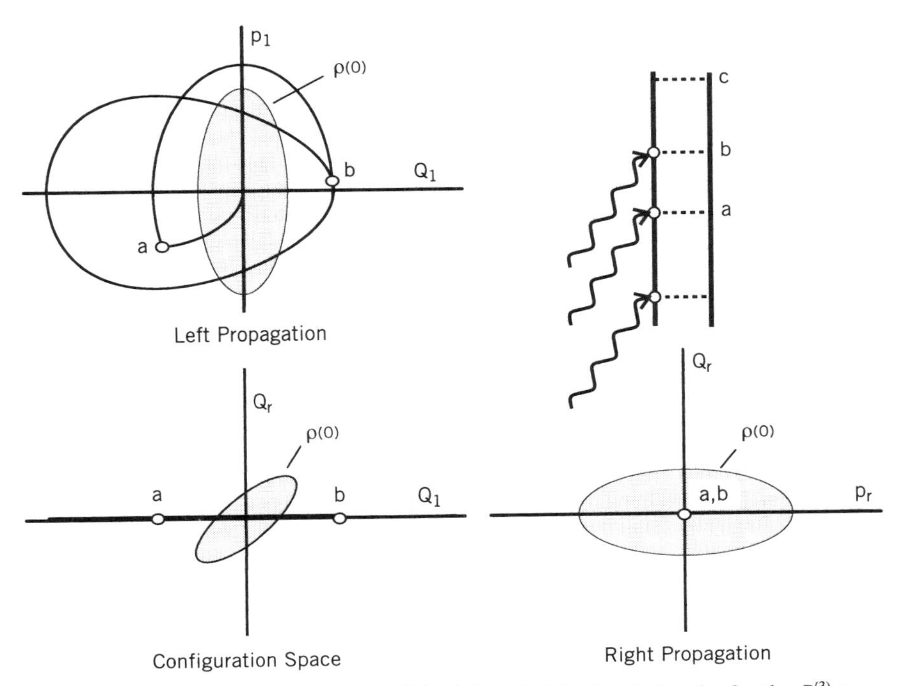

Figure 31. Phase space portraits of the left and right classical paths for the $R_4^{(3)}$ term. The coordinate representation of the guiding orbit for the density matrix is shown on the lower-left diagram.

- From the Feynman diagram in Fig. 32 we see that during the t_1 interval the bra side of the density matrix (right path) has been excited to the upper Morse potential while the ket side (left path) remains in the ground state. The contour plot shows that, for $t_3 = 0$ and as t_1 increases, the correlation function first decays and then rebuilds in a series of type A recurrences. The moment when this happens for the first time coincides with the first period of motion of the right Morse orbit. Even if the t_1 motion continues forever, the left guiding orbit remains stationary at $q_l = 0$, $p_l = 0$. Consequently only when the right orbit goes through $q_l = 0$ with small or zero momentum will there be a recurrence A in the correlation function. The event A happens periodically along the t_3 axis due to the motion of the density matrix along the Morse orbit. A similar scenario takes place along the $t_1 = 0$ axis.
- Other recurrences of type A also occurs along the $t_3 = t_1$ line. Classically this is understood by looking at the phase space portraits of the

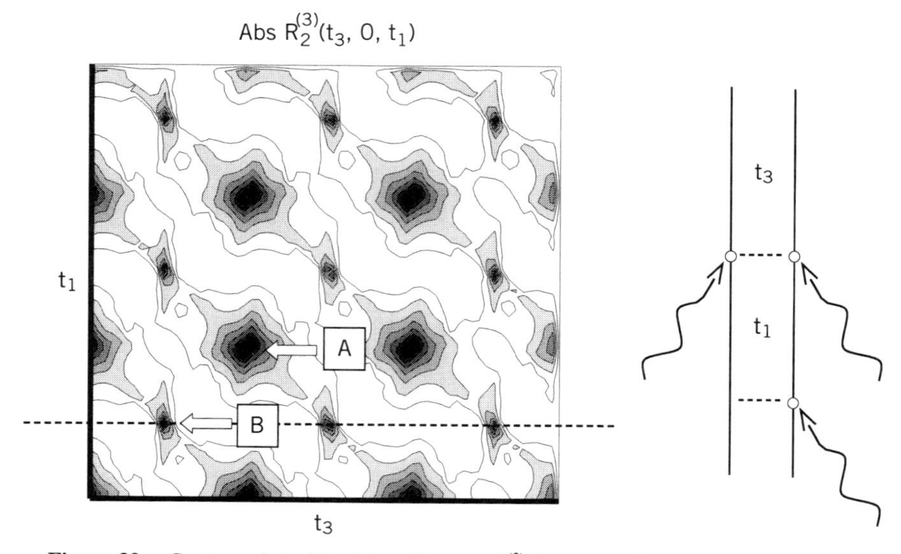

Figure 32. Contour plot of the Liouville term $R_2^{(3)}$ for a two-photon echo measurement.

left and right guiding orbits. Every one of these recurrences is due to a combination of several periods along the Morse orbit by the left- and right-guiding orbits. The Morse orbits complete a full circle in phase space; therefore they always bring the density matrix back to its departure position, and the trace is brought back almost to unity.

The second type of recurrence, *B*, is the most interesting. Figure 33 shows the left and right orbits for a Liouville classical path that takes the system directly into the recurrence labeled *B*. During the first interval t_1, the dynamics moves along the t_3 axis, and we observe the initial decay of the correlation function. After a half period on the Morse orbit of the right guiding path the density matrix has moved along the q_r axis (see bottom-left diagram) away from the diagonal $q_r = q_l$. This diagonal is very important because the trace of the time-evolved density matrix picks up amplitude along that line. When the density matrix crosses $q_l = q_r$ it produces a buildup of the correlation function. During the t_3 period of the propagation, the right orbit describes a full period of motion in the harmonic potential, while the left orbit moves in the Morse excited state for half a period. As a result there are two intersections of the (q_l, q_r) path with the $q_l = q_r$ diagonal. The first one fails to create a recurrence in the correlation function because the

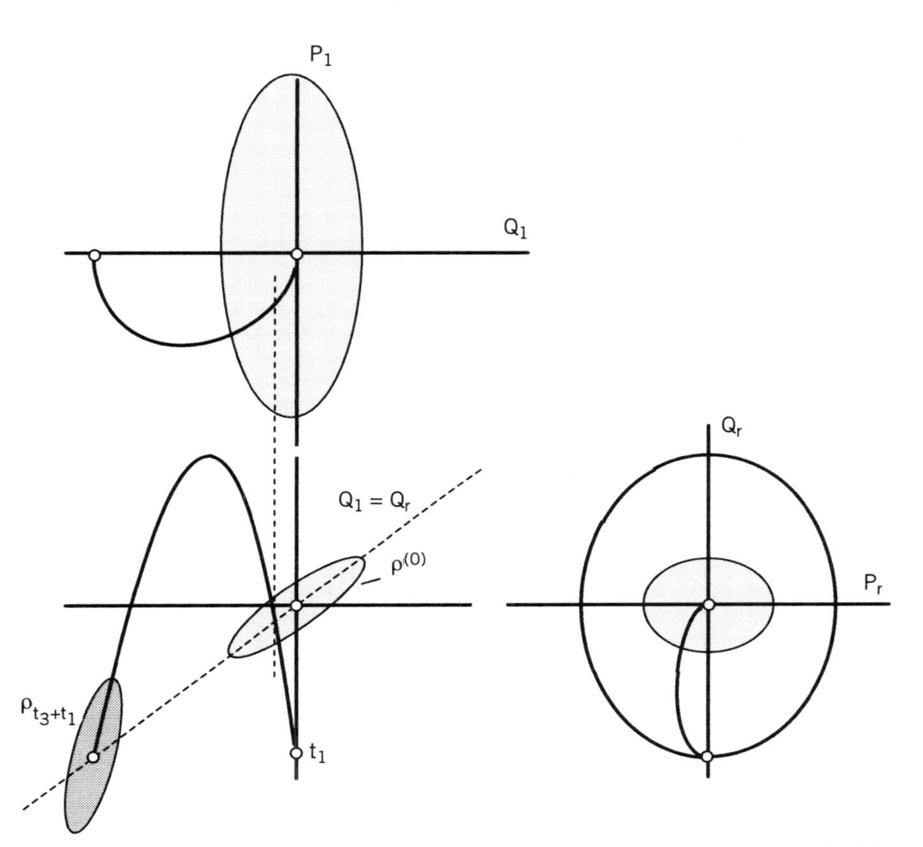

Figure 33. Phase space diagram of type *B* recurrence. Upper left: phase space for left orbit. Lower right: phase space for right orbit. Lower left: configuration space.

average momentum of the distribution is extremely high in the right orbit. The second intersection, however, happens after one full period of the harmonic oscillator in the right orbit. In the second intersection the average momentum of the density matrix is zero (same as the initial value). Only then is there a full revival of the trace. If the dynamics continue for longer t_3, i.e., along the horizontal dashed line in Fig. 32, we see another couple of identical recurrences of type *B*, which are again due to a combination of one full harmonic and half a Morse period.

Classical mechanics identifies recurrences of type *A* as purely excited state dynamics while recurrence of type *B* involve a combination of both ground and electronic state orbits.

F. Multiple Orbit Expansion: Results for H_2, K_2, Na_2

The generalized Gutzwiller propagator can also be implemented using a ''cellular dynamics'' scheme. In the last section we saw an application of the ''single-pair'' expansion of the dynamics in Liouville space. The two-orbit approximation has the same limitations as the old Gaussan wavepacket method; short-time dynamics only and sensitivity to strong nonlinear terms in the potential. A multiple-pair expansion of the same Liouville equations has been developed [65], but we will not describe it here. Instead we will switch arenas and now implement the generalized gutzwiller propagator in the wave function formalism.

Closing this chapter we will focus our attention on describing the response of a diatomic molecule with three electronic states in the energy region of anterest. H_2, K_2, and Na_2 happen to be very interesting examples and each offers an archetypical behavior.

Let us first briefly remember the basic structure of the Hamiltonian and the perturbative solutions

$$H = \sum_{i,k} h_i|i(q)\rangle\ \langle i(q)| + \lambda\mu_{ik}|i(q)\rangle\ \langle k(q)| \tag{6.42}$$

The h_i's are again the nuclear Hamiltonians for a particular electronic state $|i(q)\rangle$. This state is parametrically depending on the nuclear coordinate.

The full wave function for Hamiltonian Eq. (6.42) is a sum of the corresponding vibrational wave functions for each electronic level $|k\rangle$

$$|\Psi(t)\rangle = \sum_k |\Psi_k(t)\rangle|k\rangle \tag{6.43}$$

After expansion of the wavefunctions for each electronic level in the parameter λ

$$|\Psi(t)\rangle = \sum_{k,n} \lambda^n|\Psi_k^{(n)}(t)\rangle|k\rangle \tag{6.44}$$

and a few algebraic manipulations, we can cast the tems in the expansion as

$$|\Psi_k^{(0)}(t)\rangle = e^{-ih_k t/\hbar}|\Psi_k^{(0)}(0)\rangle \tag{6.45a}$$

$$|\Psi_k^{(n)}(t)\rangle = \frac{1}{i\hbar}\sum_l dt_l e^{-ih_k(t-l)/\hbar}\ \mu_{kl}|\Psi_l^{(n-1)}(t_l)\rangle \tag{6.45b}$$

The second iterative equation immediately leads to a generalized Gutzwiller propagator in the $\hbar \to 0$ limit. Any typical nonlinear response of the system (again within the Condon approximation) leads to a correlation function of the form

$$C(t_i, \ldots, t_1) = \langle \Psi(0)|U_i(t_i) \ldots U_1(t_1)|\Psi(0)\rangle \tag{6.46}$$

Everything so far is analogous to the previous treatment in the density matrix formalism. The initial state of the system will be approximated by a Gaussian state (perfectly valid for the molecules we mentioned before)

$$\langle q|\Psi(0)\rangle = \left(\frac{\alpha}{\pi}\right)^{1/4} \exp\left[-\frac{\alpha}{2}(q - q_e)^2\right] \tag{6.47}$$

From now on we will follow the same steps as in the cellular dynamics method on one potential surface (Section V). In the mixed representation

$$\langle p|\exp(-ih_it_i/\hbar) \ldots \exp(-ih_1t_1/\hbar)|q'\rangle$$
$$= (2\pi i\hbar)^{-1/2} \left|\frac{\partial^2 S_T}{\partial p\, \partial q'}\right|^{1/2} \exp\left[\frac{iS_T(q_t, q')}{\hbar} - \frac{i}{\hbar}pq_t - \frac{i\pi}{2}\phi\right] \tag{6.48}$$

Now p and q' are independent variables, and the final position q_t is a function of the initial position and final momentum, $q_t(p, q')$. The phase ϕ takes care of the shifts due to the caustics and the discontinuities in the classical manifolds. An initial value representation for the coordinate propagator can be obtained by Fourier transformation

$$\langle q|\exp(-ih_it_i/\hbar) \ldots \exp(-ih_1t_1/\hbar)|q'\rangle$$
$$= \int dp_t\, \langle q|p_t\rangle\, \langle p_t|\exp(-ih_it_i/\hbar) \ldots \exp(-ih_1t_1/\hbar)|q'\rangle \tag{6.49a}$$
$$\approx (4\pi^2 i\hbar^2)^{-1/2} \int dp' |\mathfrak{M}_{11}|^{1/2} \exp\left[\frac{iS_T(p', q')}{\hbar} + \frac{i}{\hbar}p_t(q - q_t) - \frac{i\pi}{2}\phi\right] \tag{6.49b}$$

The stability matrix $\mathfrak{M}$ refers to the *total* stability matrix along the classical path that moves through the different potential surfaces. Calculation of the phase shift is also similar to the single Born surface case

$$\phi = \mu + \tfrac{1}{2}\,\mathrm{Sgn}\,(-\mathfrak{M}_{11}\mathfrak{M}_{21}) \tag{6.50}$$

With a proper representation for the propagator we can now apply it to the initial state and obtain the wave function at time $T = (t_i, \ldots t_1)$,

$$\langle q|\exp(-ih_it_i/\hbar) \ldots \exp(-ih_1t_1/\hbar)|\Psi(0)\rangle$$
$$= (4\pi^2 i\hbar^2)^{-1/2} \left(\frac{\alpha}{\pi}\right)^4 \int dp' \, dq'$$
$$\cdot \exp\left[\frac{iS}{\hbar} - \frac{ip_t}{\hbar}(q_t - q) - \frac{i\pi}{2}\phi - \frac{\alpha}{2}(q' - q_e)^2\right] \quad (6.51)$$

The previous expression can be evaluated numerically by inserting "cells," as we did in the cellular dynamics method, and expanding the actions locally. For the diatomic molecules in which we are interested, a single cell expansion in initial coordinate is not sufficient because the initial states are not very narrow with respect to the nonlinearities along the coordinate axis. For this reason we must also expand the coordinate dependence of the integral in "cells." This is accomplished by inserting unity in Eq. (6.51) as

$$2\sqrt{\frac{1}{\pi}} \sum_j \exp[-\beta_1(p' - p_j)] \approx 1 \quad (6.52a)$$

$$2\sqrt{\frac{1}{\pi}} \sum_k \exp[-\beta_2(q' - q_k)] \approx 1 \quad (6.52b)$$

Then the propagated wave function becomes

$$\eta \sum_{j,k} \int dp' \, dq' |\mathfrak{M}_{11}|^{1/2} \exp \Theta \quad (6.53)$$

with

$$\eta = \frac{4}{\pi}(4\pi^2 i\hbar^2)^{-1/2}\left(\frac{\alpha}{\pi}\right)^{1/4}$$

$$\Theta = \frac{iS}{\hbar} - \frac{ip_t}{\hbar}(q_t - q)\frac{i\pi}{2}\phi - \frac{\alpha}{2}(q' - q_e)^2 - \beta_1(p' - p_j)^2$$
$$- \beta_2(q' - q_k)^2 \quad (6.54)$$

Each integrand in the sum is a strongly peaked Gaussian, in which widths $\beta_1 = 4/\Delta p^2$ and $\beta_2 = 4/\Delta q^2$ are dependent on the size of the grids in initial position and momentum.

Now actions and trajectories must be expanded using the linearized analysis about the cell centers (p_j, q_k). At the end, each integrand becomes a Gaussian. Finally we have (see Table II for explicit expresions for the coefficients)

$$\langle q|U_i(t_i) \ldots U_1(t_1)|\Psi(0)\rangle \approx \eta \sum_{k,j} \frac{|\mathcal{M}_{11}^{1/2}|^{1/2}}{(\det A)^{1/2}} \exp\left(Z_0 + \frac{1}{4}\mathbf{B}\cdot A^{-1}\cdot\mathbf{B}\right) \tag{6.55}$$

To test formula Eq. (6.55) we will first consider two simple diatomic molecules: H_2 and K_2. Both systems have three lower electronic states that are a popular object of study using femtosecond lasers ($X\ ^1\Sigma_g^+$, $B\ ^1\Sigma_u^+$, and $C\ ^1\Pi_u$) [74]. Potential surfaces for both systems are drawn in Figs. 34 and 35. The most recent calculations of these potential curves have been fitted to Morse functions [75] (See Tables III and IV for the parameters.)

Both systems start on the ground state $\Psi(0)$ of the $X\ ^1\Sigma_g^+$ surface (h_1) and are stimulated by two laser pulses arriving with a certain time delay. The first pulse brings some amplitude of the initial state to the B state (h_2), which evolves in time until the arrival of the second laser pulse. Finally a small amount of probability amplitude makes it to the upper C state (h_3) where it

TABLE II

Coefficients for Multiorbit Expansion of the Generalized Gutzwiller Propagator[a]

$$B_1 = -\frac{i}{\hbar}\mathcal{M}_{11}(\bar{q}_t - q)$$

$$B_2 = -\frac{i}{\hbar}\mathcal{M}_{12}(\bar{q}_t - q) - \alpha(q_k - q_e) - \frac{i}{\hbar}p_j$$

$$A_{11} = \frac{i}{2\hbar}\mathcal{M}_{12}\mathcal{M}_{22} + \beta_2 + \frac{\alpha}{2}$$

$$A_{22} = \frac{i}{2\hbar}\mathcal{M}_{12}\mathcal{M}_{22} + \beta_2 + \frac{\alpha}{2}$$

$$A_{12} = \frac{i}{2\hbar}\mathcal{M}_{11}\mathcal{M}_{22}$$

[a]The "bars" indicate variables associated with the guiding cell trajectory.

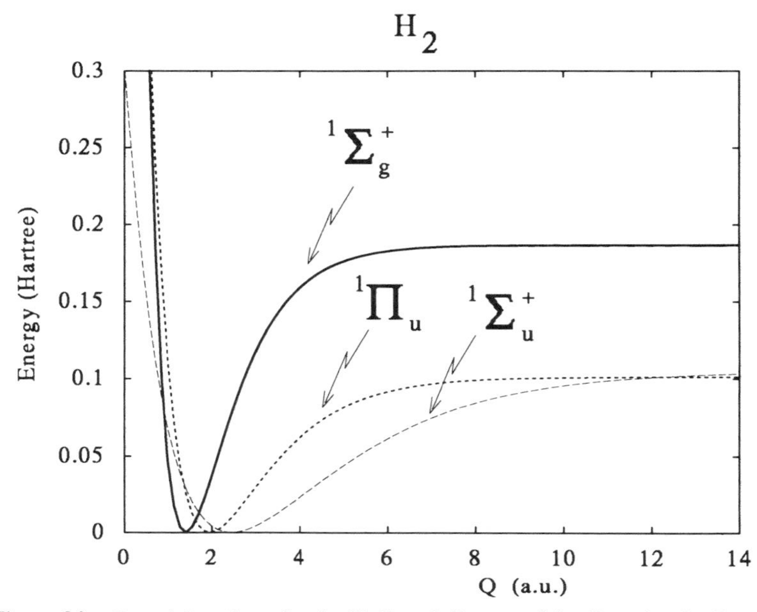

Figure 34. Potential surfaces for the X, B, and C states of the H_2 molecule. Energy and distance in atomic units.

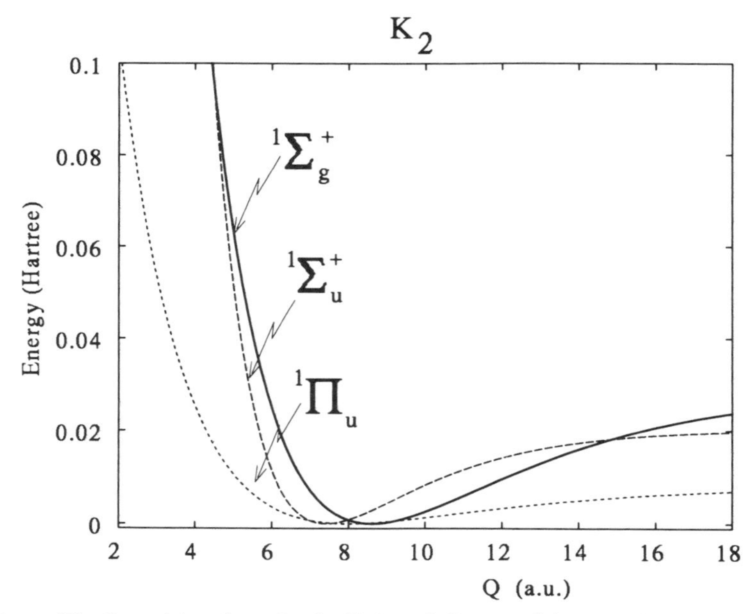

Figure 35. Potential surfaces for the X, B, and C states of the K_2 molecule. Energy and distance in atomic units.

TABLE III
Potential Parameters for H_2

State	T_e (cm^{-1})	ω_e (cm^{-1})	$\omega_e x_e$ (cm^{-1})	α_e	q_e (Å)	D_e (cm^{-1})
$X\ ^1\Sigma_g^+$	0	4395.2	117.9	2.993	0.7416	40962.22
$B\ ^1\Sigma_u^+$	91689.9	1356.9	19.932	1.1933	1.2927	23093.24
$C\ ^1\Pi_u$	100043.0	2442.72	67.03	1.626	1.0331	22254.51

TABLE IV
Potential Parameters for K_2

State	T_e (cm^{-1})	ω_e (cm^{-1})	α_e	q_e (Å)	D_e (cm^{-1})
$X\ ^1\Sigma_g^+$	0	92.399	0.394	3.924	4452.14
$B\ ^1\Sigma_u^+$	11107.9	70.55	0.253	4.55	6331.40
$C\ ^1\Pi_u$	15377	74.891	0.253	4.236	1798.60

moves forward and back on the well of the potential until the time of measurement.

We will not go through all the detail of calculating the full detected signal in the experiment described. For the purpose of testing the semiclassical Eq. (6.55) let us calculate two important correlation functions:

$$\langle \Psi(0) | U_2(t) | \Psi(0) \rangle \tag{6.56}$$

$$\langle \Psi(0) | U_3(t - t_2) U_2(t_2) | \Psi(0) \rangle \tag{6.57}$$

In Fig. 36 we show the quantum and semiclassical results for H_2 (t_2 = 100 femtoseconds). There is a trivial oscillatory term of the form $e^{i\Delta Et}$, where ΔE is the energy difference between the B and C states that have been removed to make the pictures more clear. Both correlation functions, linear and second order, are shown on the same plot. The upper drawing corresponds to the actual "cellular dynamics" of the generalized Gutzwiller propagator. The agreement is quite satisfactory. It is not so, however if one applies a single-orbit expansion (lower drawing). The GWD method yields a very poor estimate of the linear response and fails totally for the second-order correlation function. The results for K_2 using the CD method are equally satisfactory. In this case t_2 = 2000 fs. This demonstrates: (1) The CD method can describe the quantum interference due to the nonlinear dynamics; (2) GWD cannot be used for nonlinear optical calculations except for simple harmonic surfaces.

It is interesting to compare the second-order correlation functions for H_2 and K_2 (Fig. 36 vs. Fig. 37). In the first case, the number of frequency

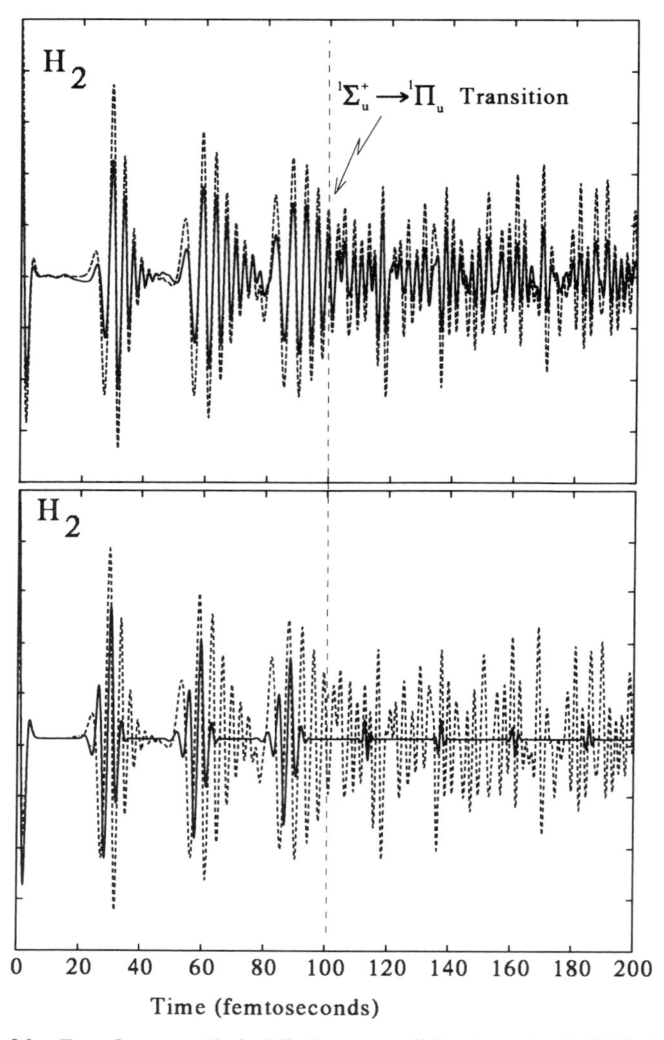

Figure 36. Top: Quantum (dashed line) versus cellular dynamics (solid line) correlation functions. Bottom: Quantum (dashed line) versus Gaussian wavepacket (solid line) correlation functions.

components in the signal increases in the $X \rightarrow B$ transition; meanwhile the opposite occurs for the K_2 molecule. This can be explained if we look at the potential surfaces for the B and C states in each case. The first derivative of the potential gives us an idea of the restoring force exerted by the electron cloud on the nuclei. In the H_2 case, this restoring force decreases when

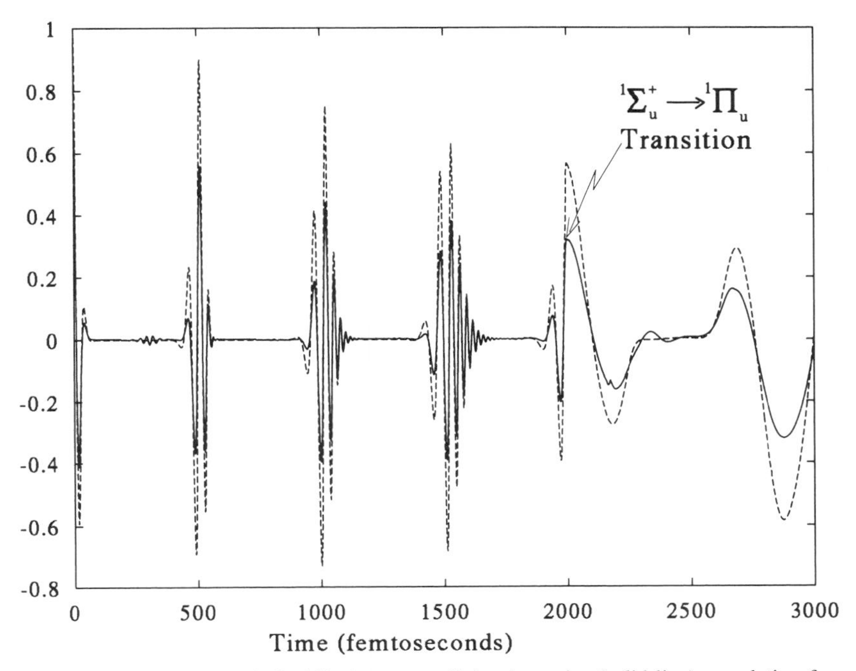

Figure 37. Quantum (dashed line) versus cellular dynamics (solid line) correlation functions. The first 2000 fs correspond to the linear response and the remaining 1000 fs to the second-order response.

switching from the *B* to the *C* state. Consequently the dynamics becomes unstable, and the density of states increases. On the other hand, in the K_2 case, the electronic restoring force increases, making the vibrational force constant harder and slowing down the correlation rate.

Phase space again helps us visualize the change of stability in the dynamics during the pulse excitations. In Fig. 38 two drawings show the location of a set of classical manifolds used for the cellular dynamics calculation of K_2. The upper diagram shows an instantaneous snapshot of phase space after the dynamics on the *B* state; the lower drawing shows the same after the *C* state propagation. It is important to notice here that the basin of phase space spanned by the propagated wave function has shrunk drastically. This is a consequence of the stabilization of the dynamics. On the other hand, looking at the results for H_2 in Fig. 39 (only a single x-manifold is shown), one observes that after the second propagation the classical manifold has stretched and folded quite rapidly in only 100 fs of dynamics. Actual measurement of the stability elements along the manifold indicates a quasifractal dependence

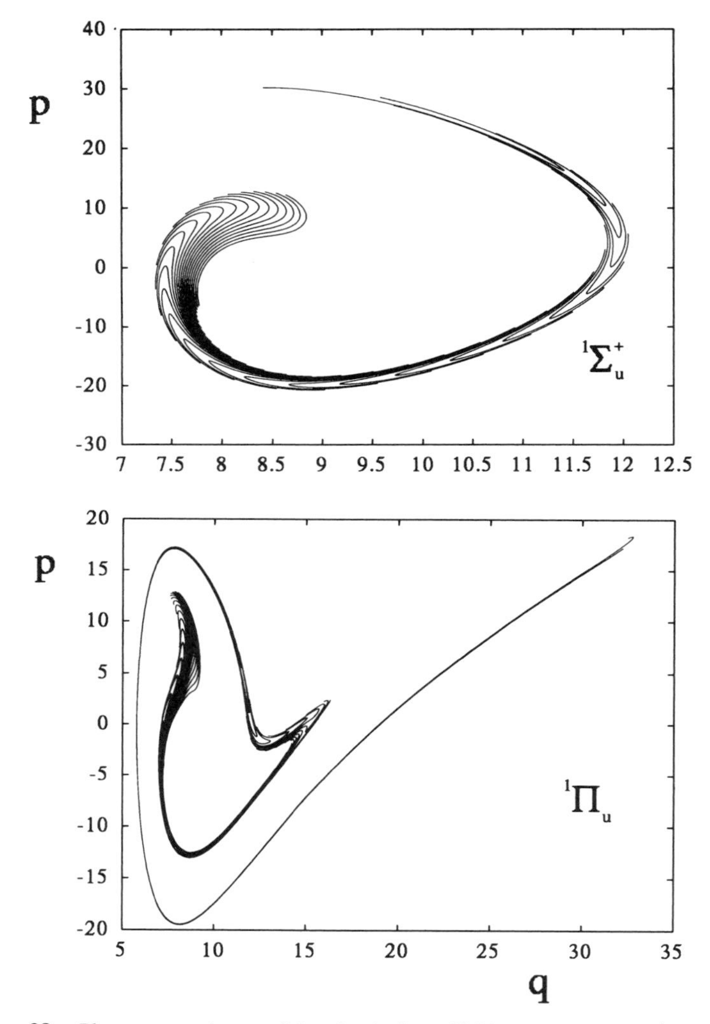

Figure 38. Phase space picture of the classical manifolds associated with $|\Psi(t)\rangle$ 2000 fs after the first propagation (top) and 1000 fs after the second propagation (bottom).

[70]. The transition to the upper C state has caused the molecule to become extraordinarely elastic and unstable.

In summary the semiclassical expression Eq. (6.55) is tested in two quite different scenarios and gives the best asymptotic limit we are capable of. The small differences in the intensities of the recurrences are due to the breakdown of the $\hbar$ expansion in certain regions of the classical manifolds

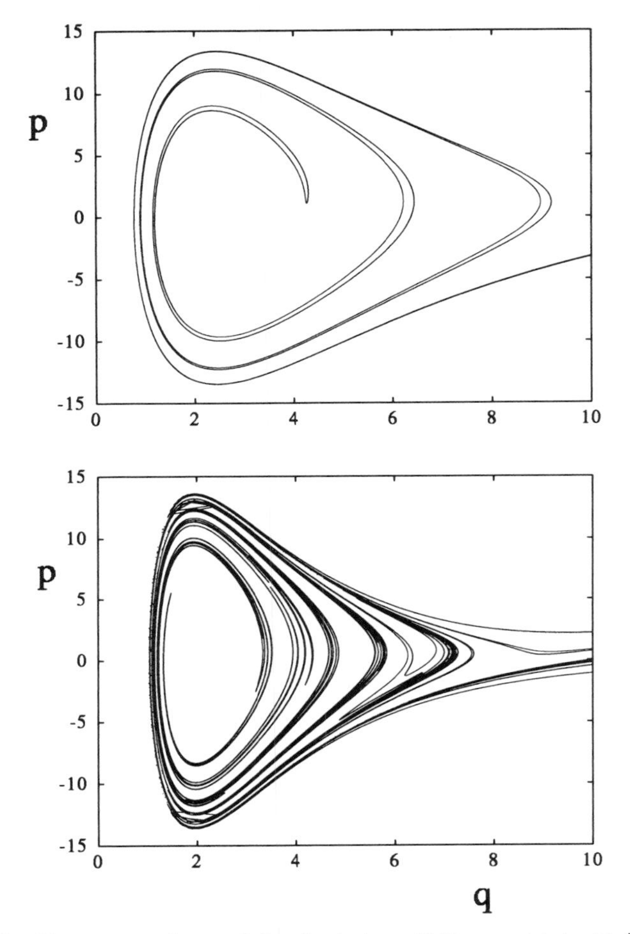

Figure 39. Phase space picture of the classical manifolds associated with $|\Psi(t) > 100$ fs after the first propagation (top) and 100 fs after the second propagation (bottom).

(caustics). However, we always found a very good agreement in the frequency components and phases of the correlation functions. This data determines the nonlinear susceptibilities of the molecule.

The semiclassical formula can be implemented for very long times as well. The times for the previous calculations were selected as a function of the phase space portraits, in order to illustrate the concept of stabilization of the molecular motion by photons and its interpretation in phase space. The Na_2 molecule serves as an example of typical results for longer time dynamics. This molecule has been exhaustively studied. In Fig. 40 we show

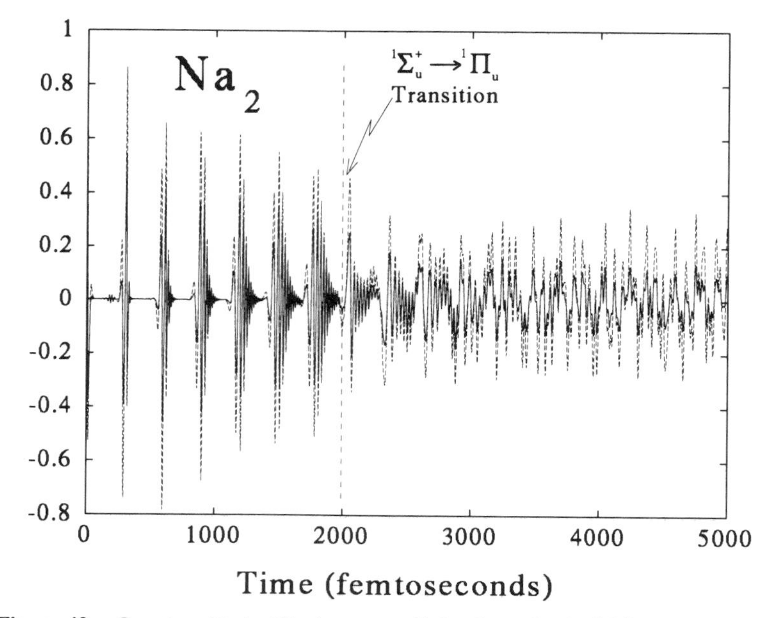

Figure 40. Quantum (dashed line) versus cellular dynamics (solid line) correlation functions. The first 2000 fs correspond to the linear response and the remaining 3000 fs to the second-order response.

the first 3 ps of dynamics on the second-order correlation function. The agreement between frequency components and phases is still perfect. The lack of some of the applitude in the semiclassical correlation function is associated to the normal limit of our asymptotic approximation.

VII. DISSIPATIVE SEMICLASSICAL MECHANICS

In this section we will review the use of a simplified semiclassical scheme to propagate the reduced density matrix and calculate survival probabilities for nonharmonic quantum systems in contact with a dissipative environment. Initiated primarily by the work of Caldeira and Leggett [76, 77], considerable efforts have been devoted, over the last decade, to understanding the effects of the interaction of a quantum system with a heat bath. The theoretical tool used by these and many other authors is the path-integral method, generalized by Feynman and Vernon over 30 years ago [78] to account for dissipative effects. It was shown by Grossman [45] that by applying a simple semiclassical approximation to the generalized path integral that appears in the Feynman–Vernon influence functional formalism one is able to use the methodology reviewed in this paper for dissipative systems also.

The path-integral formalism provides us with the propagator of a quantum system of interest (test system), which is coupled to a set of harmonic oscillators (heat bath) after tracing over the bath coordinates. The environment is then represented by the influence functional, which depends on the temperature and spectral density of the oscillator bath. The influence functional manifests itself in the generalized action that appears in the exponent of the path integral. In principle, the total action has to extremized for a semiclassical approximation of the propagator. For reasons of simplicity, however, the calculation will be restricted to using classical trajectories that extremize the real part of the generalized action only. The imaginary part is then incorporated in zeroth order. A sum over all possible classical paths and the inclusion of the Maslov index are crucial for the applicability of the semiclassical theory to nonlinear test systems. The numerical example of a Morse oscillator interacting with a high temperature heat bath will be given.

A. Reduced Description and Influence Functional

The model of a particle coupled to harmonic environmental modes has proven extremely useful to describe such diverse phenomena as vibrational relaxation [79, 80], condensed phase reaction rates [81], macroscopic quantum tunneling [82], and the transport of hydrogen isotopes or muons between interstitial sites in solids [83]. In all of these applications, the underlying microscopic model consists of a test system that is bilinearly coupled to a set of harmonic oscillators. The total Hamiltonian [84–86] is given by

$$H = \frac{p^2}{2M} + V(q) + \sum_{n=1}^{\infty} \left[\frac{p_n^2}{2m_n} + \frac{1}{2} m_n \omega_n^2 \left(x_n - q \frac{\chi_n}{m_n \omega_n^2} \right)^2 \right] \tag{7.1}$$

i.e., as a sum of the Hamiltonian of the test system (a particle of mass M moving in a potential $V(q)$ with the (generalized) coordinate q and a sum of infinitely many bath oscillator terms together with an interaction term. The coupling constants are denoted by $\{\chi_n\}$.

The information about the harmonic bath is not what one is interested in. In order not to treat the bath coordinates explicitly one therefore traces over them. To this end, let us now summarize the results of the calculation of the influence functional, which was first derived by Feynman and Vernon [78] in their seminal work on dissipation. The time evolution of the density matrix of a quantum system governed by the total Hamiltonian in Eq. (7.1) is formally given by [compare Eq. (6.2)]

$$\rho(t) = \exp\left(-\frac{iHt}{\hbar}\right) \rho(0) \exp\left(\frac{iHt}{\hbar}\right) \tag{7.2}$$

We will consider the case where the density matrix can be factored at the initial time into a system part and a bath part,

$$\rho(0) = \rho_{\mathrm{red}}(0)\rho_{\mathrm{B}}(0) \tag{7.3}$$

In the coordinate representation for the reduced density matrix

$$\rho_{\mathrm{red}}(q_f, q_f', t) \equiv \int d\underline{x}_f \rho(q_f, \underline{x}_f, q_f', \underline{x}_f, t) \tag{7.4}$$

the bath coordinates can be integrated out analytically, making use of the harmonic nature of the bath and the coupling being bilinear. One then finds for the time evolution

$$\rho_{\mathrm{red}}(q_f, q_f', t) = \int dq_i \int dq_i' \; J(q_f, q_f', t; q_i, q_i', 0)\rho_{\mathrm{red}}(q_i, q_i', 0) \tag{7.5}$$

where the propagator is given in the form of a double path integral[2]

$$\begin{aligned} J(q_f, q_f', t; q_i, q_i', 0) \\ \equiv \int_{q(0)=q_i}^{q(t)=q_f} d[q] \int_{q'(0)=q_i'}^{q'(t)=q_f'} d[q'] \\ \cdot \exp\left\{\frac{i}{\hbar}(S_S[q] - S_S[q'])\right\} F[q, q'] \end{aligned} \tag{7.6}$$

with the action S_S of the test system. Equation 7.6 can be interpreted as a path integral for the coordinate q, moving forward in time, and another one for the mirror coordinate q', moving backwards in time, with an influence functional $F[q, q']$ coupling these two path integrals. It is given by[3]

$$F[q, q'] \equiv \exp\left[-\frac{1}{\hbar}\Phi(q, q')\right] \tag{7.7a}$$

$$\begin{aligned} \Phi[q, q'] \equiv & \int_0^t dt'[q(t') - q'(t')] \int_0^{t'} dt'' \, A^{\mathrm{r}}(t' - t'') \, [q(t'') - q'(t'')] \\ & + \frac{i}{2}\int_0^t dt'[q(t') - q'(t')] \frac{d}{dt'} \int_0^{t'} dt'' B(t' - t'') \\ & \cdot [q(t'') + q'(t'')] \end{aligned} \tag{7.7b}$$

[2]$\int d[q]$ is a short hand notation for an infinite dimensional integral.

[3]Note that the argument of a functional is a function and the functional dependence is denoted by [ ].

where the terms defined by ($\beta = 1/kT$)

$$A^{\mathrm{r}}(t) \equiv \int \frac{d\omega}{\pi} I(\omega) \coth\left(\frac{\beta\hbar\omega}{2}\right) \cos(\omega t) \tag{7.8a}$$

$$B(t) \equiv \int \frac{d\omega}{\pi} \frac{2I(\omega)}{\omega} \cos(\omega t) \tag{7.8b}$$

$$I(\omega) \equiv \frac{\pi}{2} \sum_{n=1}^{N} \frac{\chi_n^2}{m_n\omega_n} \delta(\omega - \omega_n) \tag{7.8c}$$

denote the real part of the influence kernel, the damping kernel, and the spectral density of the harmonic bath, respectively (see e.g. [87]). It is only if the kernels defined by Eqs. (7.8a) and (7.8b) are proportional to delta functions that the theory would be memory free (Markovian).

B. A Simple Semiclassical Propagator

In order to apply the semiclassical ideas reviewed here, the case of an Ohmic heat bath at high temperatures will furthermore be considered. Then one will be allowed to make a semiclassical approximation to the dissipative propagator for the reduced density matrix that can be treated numerically by using cellular dynamics. To do so, the generalized action and Lagrangian in the double path integral of Eq. (7.6) are defined by

$$\Sigma[\underline{q}] \equiv S_S[q] - S_S[q'] + i\Phi[q, q'] = \int_0^t dt' \Lambda(\underline{q}, \dot{\underline{q}}, t') \tag{7.9a}$$

$$\begin{aligned} \Lambda(\underline{q}, \dot{\underline{q}}, t') \equiv\; & \frac{M}{2} \dot{q}^2(t') - \frac{M}{2} \dot{q}'^2(t') - V(q, t') + V(q', t') \\ & - \frac{1}{2} [q(t') - q'(t')] \left\{ \frac{d}{dt'} \int_0^{t'} dt'' B(t' - t'') \, [q(t'') + q'(t'')] \right. \\ & \left. - i \int_0^t dt'' \, A^{\mathrm{r}}(t' - t'') \, [q(t'') - q'(t'')] \right\} \end{aligned} \tag{7.9b}$$

with $\underline{q} \equiv (q, q')$ here denoting the general path variable (meaning that it does not need to extremize Σ). These expressions, however, are too general yet to be amenable to the semiclassical scheme because of the kernels describing memory effects. Therefore the density of states of the oscillator bath will be specified as Ohmic

$$I(\omega) = M\gamma\omega\Theta(\omega_c - \omega) \tag{7.10}$$

with a cutoff. It seems worthwhile to mention that by introducing a continuous density of states one has actually introduced dissipation, meaning that energy flowing from the test system into the bath will not return to the test system over reasonable physical time scales. In the following it will be assumed that the cutoff frequency of the reservoir is much larger than a typical system frequency, so that the damping kernel can be approximated by

$$B(t) \approx 2M\gamma\delta(t) \tag{7.11}$$

In order to free the theory also from memory effects in the temperature-dependent fluctuation term, Eq. (7.8a), it will furthermore be restricted to high temperatures, $\hbar\beta\omega_c << 1$. Then the real part of the influence kernel can be approximated by [88, 89]

$$A^{r}(t) \approx \frac{2M\gamma}{\beta\hbar}\,\delta(t) - \frac{M\gamma\beta\hbar}{6}\,\ddot{\delta}(t) \tag{7.12}$$

After making these two approximations it seems appropriate to emphasize that it is only by retaining both terms in Eq. (7.12) that the resulting dissipative master equation belongs to the Lindblad class [90].

It is still unusual, however, that the Lagrangian in Eq. (7.9) is complex. To circumvent the problems introduced by complex trajectories, a simple semiclassical approximation, well-known in the context of scattering from a complex potential [91], was applied in [45]. This procedure amounts to approximating the propagator of the reduced density matrix by expanding it around paths that extremize the *real part* of the complex action. After the insertion of Eq. (7.11) and (7.12) into Eq. (7.9a), and in the high temperature limit, the full action is

$$\begin{aligned}
\Sigma[\underline{q}] &\equiv \Sigma^{r}[\underline{q}] + i\Sigma^{i}[\underline{q}] \\
&= \int_0^t \left[\frac{M}{2}\dot{q}^2(t') - \frac{M}{2}\dot{q}'^2(t') - V(q, t') + V(q', t')\right]dt' \\
&\quad - M\frac{\gamma}{2}\int_0^t [q(t') - q'(t')]\,[\dot{q}(t') + \dot{q}'(t')]dt' \\
&\quad + \frac{iM\gamma}{\hbar\beta}\int_0^t [q(t') - q'(t')]^2\,dt' + \frac{iM\gamma\beta\hbar}{12}\int_0^t [\dot{q}(t') - \dot{q}'(t')]^2 dt'
\end{aligned} \tag{7.13}$$

where the procedure outlined in [76, 89] was used to deal with the delta functions with zero argument. For a simple semiclassical approximation to Eq. (7.13) one keeps terms up to second order in the deviations for the expansion of the real part, and terms up to zeroth order for the expansion of the imaginary part of the action. This procedure is known to be exact for the harmonic oscillator [87] and leads to

$$J^{\text{ssc}}(\underline{q}_f, t; \underline{q}_i, 0) \equiv \sum_{\underline{q}} \exp\left\{\frac{i}{\hbar}\Sigma^{\text{r}}[\underline{q}] - \frac{1}{\hbar}\Sigma^{i}[\underline{q}]\right\} \cdot \int_{\underline{\eta}(0)=\underline{0}}^{\underline{\eta}(t)=0} d[\underline{\eta}] \exp\left\{\frac{i}{2\hbar}\delta^2\Sigma^{\text{r}}[\underline{q}]\right\} \tag{7.14}$$

for the propagator, where $\underline{q}$ now extremizes Σ^{r}, that is the Newtonian equations of motion which have to be fulfilled by $\underline{q}$ are given by

$$M\ddot{q} + M\gamma\dot{q}' + \partial_q V(q) = 0 \tag{7.15}$$

$$M\ddot{q}' + M\gamma\dot{q} + \partial_{p'} V(q') = 0 \tag{7.16}$$

It is instructive to note that Eqs. (7.15) and (7.16) are coupled through the velocity-dependent terms. The remaining path integral in Eq. (7.14) may finally be expressed in the form of a Van Vleck prefactor [92]. The simple semiclassical propagator for dissipative quantum dynamics,

$$J^{\text{ssc}}(\underline{q}_f, t; \underline{q}_i, 0) = (2\pi i\hbar)^{-1} \sum_{\underline{q}} \left|\det\left(\frac{\partial^2\Sigma^{\text{r}}}{\partial\underline{q}_f\,\partial\underline{q}_i^{\text{T}}}\right)\right|^{1/2} \cdot \exp\left\{\frac{\text{i}}{\hbar}\Sigma^{\text{r}}[\underline{q}] - \frac{1}{\hbar}\Sigma^{\text{i}}[\underline{q}] - i\pi\frac{\nu}{2}\right\} \tag{7.17}$$

with the action defined in Eq. (7.13) and the Maslov index calculated according to [93] is the central result of [45]. We want to remind the reader that the matrix of second derivates of the Hamiltonian (corresponding to the generalized Langrangian in the Markovian case) with respect to the momentum has one positive and one negative eigenvalue; therefore the initial index is -1 for every trajectory. A propagator like Eq. (7.17) can now easily be implemented, using the cellular dynamics method. The next subsection gives a numerical example using a Morse oscillator coupled to a heat bath.

C. Dissipative Morse Oscillator

In this section we will show the numerical comparison of our semiclassical result with the one calculated from a quantum mechanical master equation, corresponding to the Ohmic high temperature case [88], as well as to the one derived using the purely classical Kramers equation [94]. The test system used in the following is a Morse oscillator

$$V(r) = D[1 - \exp(-\kappa r)]^2 \tag{7.18}$$

representing a diatomic molecule with reduced mass M and equilibrium separation q_0. $D \equiv E_D/(\hbar\Omega)$ is the dimensionless dissociation energy, where $\Omega \equiv \hbar/(Mq_0^2)$, while $r \equiv (q - q_0)/q_0$. For our numerical example we are using the same parameters, $D = 30$, $\kappa = 0.08$ for the anharmonic oscillator as in [30]. The Morse oscillator is investigated under the influence of a harmonic bath with dimensionless damping strength, $\gamma_{\mathrm{dl}} = \gamma/\Omega$ and dimensionless temperature, $T_{\mathrm{dl}} = 1/(\hbar\beta\Omega)$.

The simple semiclassical propagator for the reduced density matrix was applied to the pure initial state

$$\rho_{\mathrm{red}}(r_i, r_i', 0) = \left(\frac{\sigma}{\pi}\right)^{1/2} \exp\left[-\frac{\sigma}{2}(r_i - r_0)^2 - \frac{\sigma}{2}(r_i' - r_0)^2\right] \tag{7.19}$$

corresponding to a Gaussian wave function that is centered at r_0. The dissipative propagator was implemented by using the cellular dynamics method. In the density matrix formalism we already have to deal with a two-dimensional configuration space for a one-dimensional test system. A matrix version of cellular dynamics was therefore outlined in the appendix of [45]. In doing cellular dynamics we must furthermore keep in mind that the expansion of the path has to be inserted into the real part of the exponent of Eq. (7.14). With the propagation carried through the quantity

$$P(t) \equiv \mathrm{tr}[\rho_{\mathrm{red}}(t)\rho_{\mathrm{red}}(0)] = \int dr_f \int dr_f' \, \rho_{\mathrm{red}}(r_f, r_f', t)\rho_{\mathrm{red}}(r_f', r_f, 0) \tag{7.20}$$

was calculated. This quantity is called survival probability or "probability to stay." For the initial state we have chosen a width parameter of $\sigma = 12$ and an initial displacement of $r_0 = 2$, far to the soft side of the Morse well.

Figure 41 is included to display the nondissipative case. We see that semiclassics reproduces the fast quantum oscillations of $P(t)$ and the shift of the recurrences to longer times, as compared to the classical survival probability. The discrepancy in the height of the recurrences is due to the

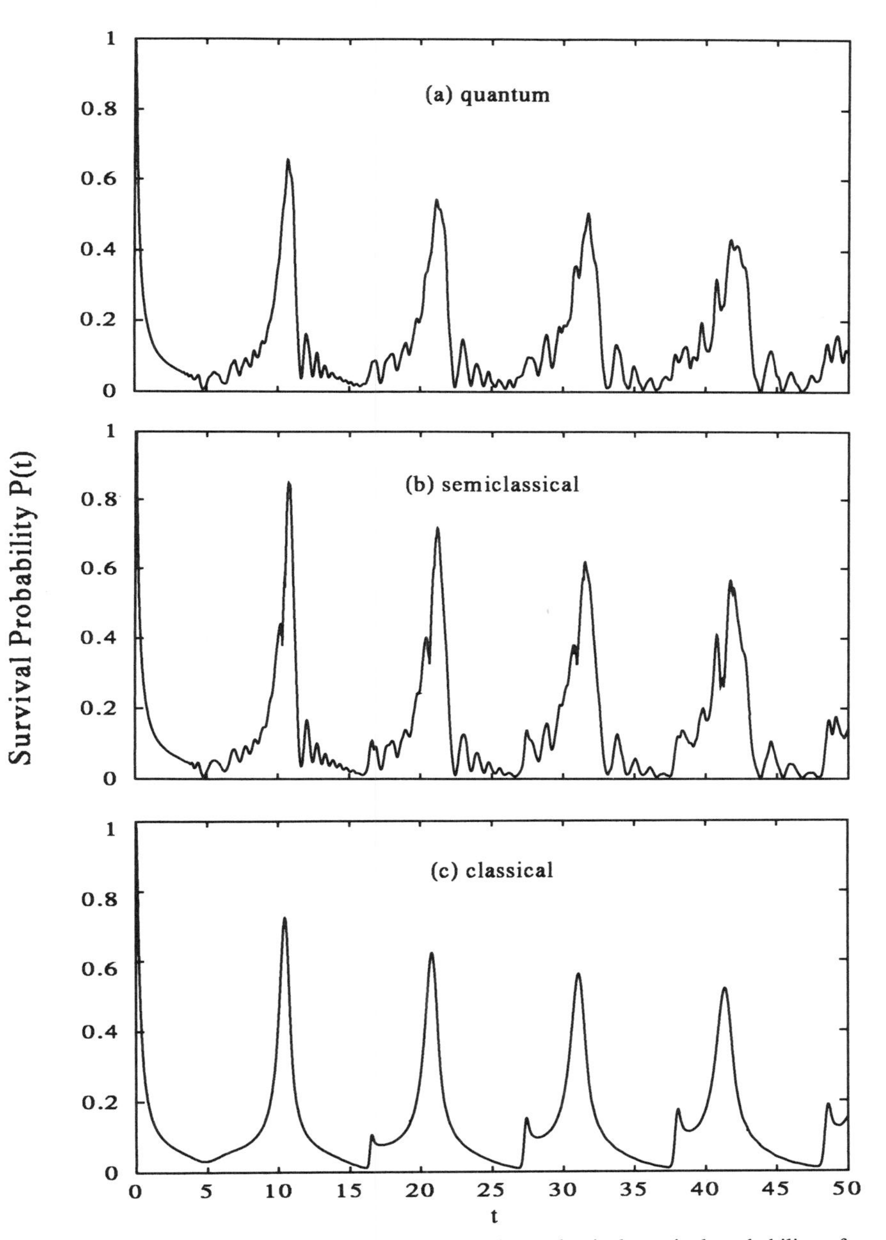

Figure 41. (*a*) Quantum, (*b*) semiclassical, and (*c*) classical survival probability of a Gaussian density matrix in a one-dimensional Morse oscillator without dissipation (the potential parameters and initial density matrix parameters are given in the text).

caustics that the system goes through in the coordinate representation. In a mixed representation this failure of semiclassics would occur at different times. In Fig. 42 dissipation with $\gamma_{\mathrm{dl}} = 0.0001$ and $T_{\mathrm{dl}} = 10$ was included. By comparing the quantum with the classical result one sees that the high-frequency quantum oscillations are getting damped out. This is in accordance with the expectation that the system should behave more classically under the influence of a high-temperature bath. The simple semiclassical approximation does a good job of representing the reduction of the height of the recurrences, as well as the diminishing amplitudes of the high-frequency oscillations, which still distinguish the quantum from the classical result.

The basic physics of the dissipative process can be understood by looking at the classical trajectories starting from the center of the initial density matrix with zero momentum. The effect of the damping kernel on its classical evolution [see the Newtonian Eqs. (7.15) and (7.16)] is to damp out the amplitude of the motion along the diagonal, which leads to a reduction in the height of the recurrences. The real part of the influence kernel does not affect the classical motion in our simple approximation, but it has a temperature-dependent influence on the width of the final wavepacket. The negative exponent thereby reduces the height of the recurrences. Generally, the picture is more difficult because, in a cellular dynamics treatment, trajectories with different conditions have to be considered. The temperature-dependent term, however, tends to reduce the survival probability for all different contributions through its decaying exponent.

One further remark seems to be in order. By using only the single trajectory described above in the calculation of the survival probability, one would effectively be generalizing Gaussian wavepacket dynamics from Section IV to the dissipative case. We have done some promising numerical studies along these lines, which show that the method works well for near-harmonic systems or very short times. For harmonic test systems this procedure would be exact (as would be the simple approximation). Nevertheless, it turns out that cellular dynamics, which is able to propagate correctly for far longer times than GWD, is already a very fast computational method, as compared to grid-based methods, for the solution of the corresponding quantum master equation. This is especially true if only few cells are needed in coordinate space.

Furthermore we want to point out that the inclusion of time-dependent external forces poses no essential problem in our formalism. With a suitable generalization, the systems of Section VI can then be treated also under the influence of a dissipative environment.

While the parameter space for which the simple semiclassical approximation is valid is limited such that the product of temperature times damping strength is not too high, a full semiclassical approximation, using the ex-

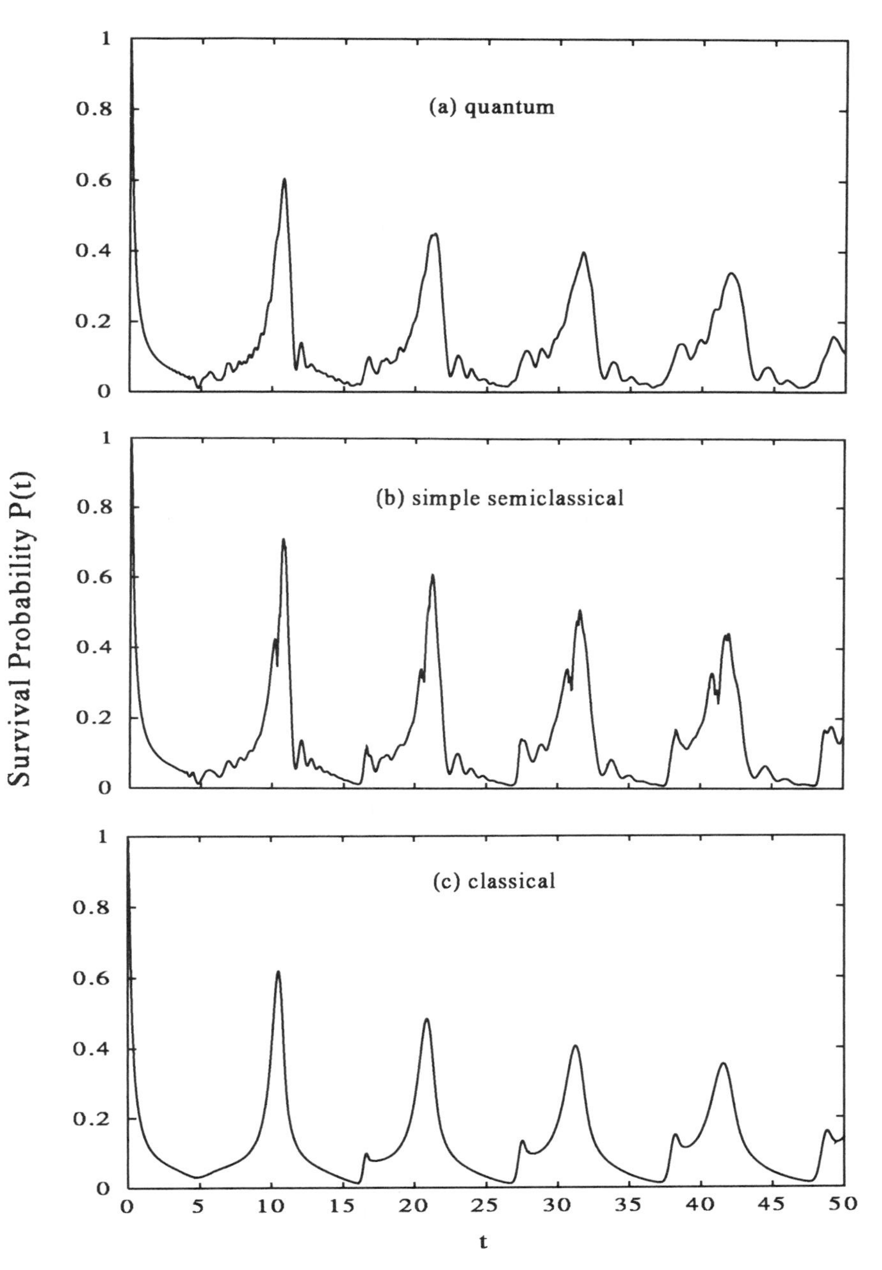

Figure 42. (*a*) Quantum, (*b*) semiclassical, and (*c*) classical survival probability in the dissipative case; damping strength $\gamma_{\mathrm{dl}} = 0.0001$, temperature $T_{\mathrm{dl}} = 10$.

tremization of the complex action might enhance the applicability of the method presented. One nice feature of the simple semiclassical method will then be lost, however. Memory effects resulting from low temperatures (see the kernel (Eq. (7.8a)) can be dealt with easily in the simple approximation: just storing the trajectories and calculating the appropriate memory integrals is enough to deal with those effects. A solution of integral-differential equations would not be necessary in the approch presented here.

VIII. CLASSICALLY CHAOTIC SYSTEMS

We have learnt so far that the VVG propagator is an extremely powerful and general tool. It goes beyond the standard WKB approximation, which suffers from the serious limitation of being dynamics dependent. In contrast the VVG formula is independent of the type of the underlying dynamical structure. So far, we have seen that the semiclassical propagation can be applied successfully not only to classically integrable but also to mixed systems, the most challenging and general kind of system. There are however very important and fundamental questions to be answered concerning the validity of the semiclassical propagator and, more importantly, of the meaning of the "correspondence principle."

As is well known, the solutions to classical equations of motion offer a more complex spectrum of solutions than quantum mechanics. The classical time evolution of a system can be exponentially sensitive to a small variation of the initial conditions, quantum mechanically; however, the same strong dependence on the initial state of the system is not found due to the linear nature of the Schrödinger Equation. One intriging question one would like to answer is how the classical complexity that develops in classically chaotic systems somehow disappears or is washed out by the uncertainty principle. But prior to that, one would like to know if that classical detail is even meaningful quantum mechanically?

Investigating an asymptotic theory like the semiclassical one, which is built on the VVG propagator, should allow one to answer these kind of questions. Unfortunately the validity of the VVG formula itself in the limit $t \to \infty$ (or as classical chaos develops structure inside an area of the size of Planck's constant in phase space) is questionable.

It has been thought for a long time that the uncertainty principle gives the limit of validity to asymptotic approximations. As the classical propagation develops more structure within a size of area $\hbar$ in phase space, the VVG propagation "should" break-down because quantum mechanics cannot depend on what is happening on a scale finer than that of a Planck cell. Interestingly enough, there is growing evidence that this naive viewpoint is wrong [15, 95, 96]. Although it is true that it is impossible to retrieve information on a scale finer than $\hbar$ from quantum mechanics, this does not imply that there is no link between the classical and quantum worlds on these scales. Our numerical evidence [15, 95, 96] shows that careful accounting of all that classical detail finally adds up to the quantum amplitude and phase.

The first serious attempt to identify the time-scale barrier for the breakdown of semiclassics was that of Berry and Balazs [97], and Berry et al. [98]. Their argument is based on the time it takes a typical state to access most cells of size $\hbar$ through mixing dynamics. This time varies as $O(\ln \hbar^{-1})$ (also dubed "log-time scale"). As a comparison, let us remember that in order to resolve a typical eigenstate it is necessary to propagate the dynamics up to the Heisenberg time, $O(\hbar)$. The last one is a much longer time scale and therefore one would predict a total collapse of the semiclassical approximation, due to the chaotic dynamics, much earlier than eigenstates can be isolated.

Fortunately O'Connor's studies on the quantum baker's map (a purely chaotic system) indicated that the log-time played no role. The breakdown of semiclassics was representation-dependent: in a well-adapted representation it was possible to observe a linear breakdown, while in a bad representation the breakdown was immediate. Because the structures of the classical manifolds for the baker's map are nongeneric (manifolds end abruptly in sharp cuts) it was somehow felt that these results, although optimistic, were not general. Nevertheless subsequent tests on the quantum stadium billiard, whose manifolds are not chopped during the dynamics, also indicated a better than log-time result for the breakdown of semiclassics [$O(\hbar^{-1/2})$] [99]. These results were explained by using an argument based, not on how much structure develops within a size $\hbar$ cell, but how much of it is usable by the stationary phase approximation [96]. Calculations on another model, the quantum kicked rotor, whose phase space structure resembles the one of mixed conservative Hamiltonians, also suggests a time scale for breakdown better than the log-time ($O(\hbar^{-1/3})$).

None of the calculations mentioned was carried out using the cellular dynamics approach. Due to the exponential expansion of classical manifolds, any cellular approach would have required sometimes millions, of orbits. Besides, there would always be a doubt about whether the number of cells was large enough to converge the results. In order to study the breakdown of semiclassics it was necessary to perform a "root search" on all the classical paths contributing to the dynamics. The baker's map [95] and the kicked rotor [50, 96] calculations were carried out by using asymptotics on path integrals, while Tomsovic–Heller's calculations involved a *heteroclinic sum*, which we will describe in the remainder of the paper.

A. Heteroclinic Sum

The complex structure developing classically in a chaotic system is due to an entangled network of hyperbolic points, defined as the intersection of stable and unstable manifolds of a single fixed point (homoclinic orbits) or of distinct fixed points (heterocolinic orbits).

If we were to select a localized distribution of orbits in phase space and propagate it in a purely chaotic system, we would observe a progression as in Fig. 43. We will assume that the center of the initially circular distribution is a fixed point in phase space. During the early stages of the dynamics, the distribution rapidly stretches along the unstable direction of the fixed point. For conservative systems, it is impossible for the distribution to continue expanding *ad infinitum*. Nonlinear terms in the dynamics are then responsible for the folding of the stretched distribution. Its branches return to the neighborhood of the fixed point following the stable manifold of the fixed point. The whole procedure of stretching and folding is repeated successively, often generating a complex layer of the manifold across the initial distribution. If one were to apply the GWD method to the initial state, the result would be a long and narrow javelin at the fixed point in the direction of the unstable manifold. The overlap of the linearly propagated state with the initial one would be wrong because it dismisses the contributions of the segments of the evolving state that eventually came back.

Let us consider the calculation of the overlap between two classical distributions of orbits $\rho_b(t)$ and ρ_a. The first one evolves in time according to classical equations of motion. Formally the integral $\langle \rho_a | \rho_b(t) \rangle$ is equivalent to $\langle \rho_a(t/2) | \rho_b(t/2) \rangle$, where one distribution has evolved backwards for a time $t/2$ and the other one forward for the same time.

Figure 44 shows a real calculation for the kicked rotor map [50]. From left to right we can see the phase space drawings of the initial disks, the propagated distributions and a detail of one intersecting region. Both distributions overlap in a finite number of sets, each one indentifies a small section of the initial disk. The overlap integral can be easily evaluated by summing over this set of initial conditions. The behavior of the propagated density of orbits in each of the small intersecting regions can be obtained from the linearization of the dynamics. Tomsovic et al. [15] pointed out that the most appropriate orbit for such a linearization is the heteroclinic orbit underlying each of the dark regions in Fig. 44.

If, instead of propagating the density matrix, we use wave functions and take into account the phase information in the local linear expansions, then we end up with a summation that is very similar to cellular dynamics. In the CD method, orbits were distributed on a uniform grid, while in the heteroclinic sum, the orbits (heteroclinic) must first be "root searched" by some numerical method. This reduces the number of orbits to be run for the propagation of the wave function to the most compact and smallest set of

Figure 43. Hyperbolic structure in phase space. The dark figure indicates the evolving distribution of orbits; time progresses from the upper-left to the lower-right corners. The light dark indicates the initial distribution of orbits and the medium gray javelin shows the local linearized approximation.

orbits possible. The price we pay is having a method that can only be used for energy scaling and low-dimensional systems (like the stadium billiard and other enclosures), and that is inherently nonvectorizable. The local linearization of the propagator and the final cell overlap follows the discussions in previous sections; therefore we will omit any further details.

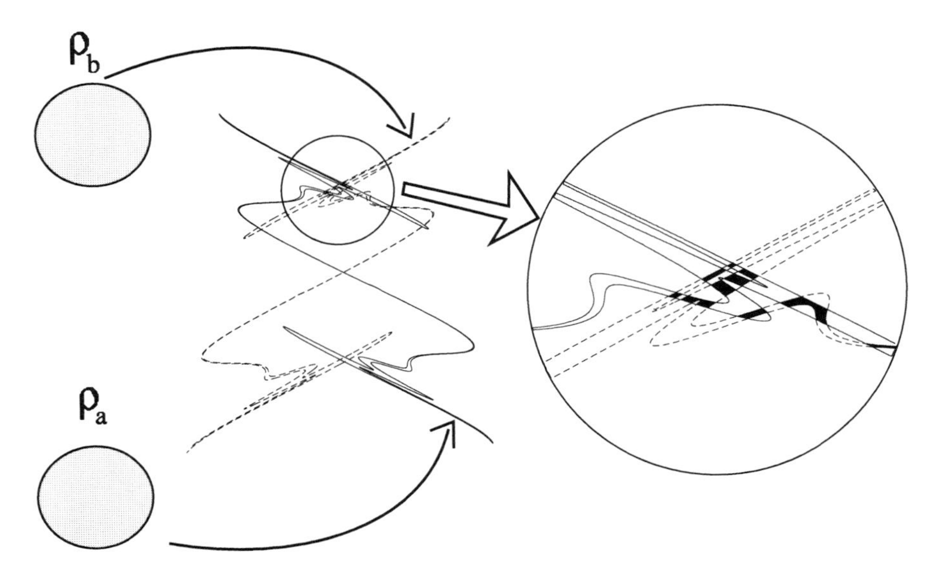

Figure 44. Phase space portrait of the heteroclinic summation technique. *Left*: Initial distributions. *Middle*: Forward and backwards propagated distributions. *Right*: Enlargement of the intersecting regions showing heteroclinic structure.

B. Stadium Billiard

The stadium billiard consists of a free particle specularly reflecting off hard walls whose shape is pictured in Fig. 45. The value μ is the ratio of the length of the straight edge to the semicircular diameter. For $\mu > 0$ the stadium is fully chaotic [100]. In the present calculations (Tomsovic et al. [15]) the mass, the radius of the cap, and μ are unity.

As we mentioned earlier the stadium billiard has energy scaling properties. This means that studying the behavior at a fixed energy is sufficient to understand the whole dynamics. This makes the heteroclinic (homoclinic,

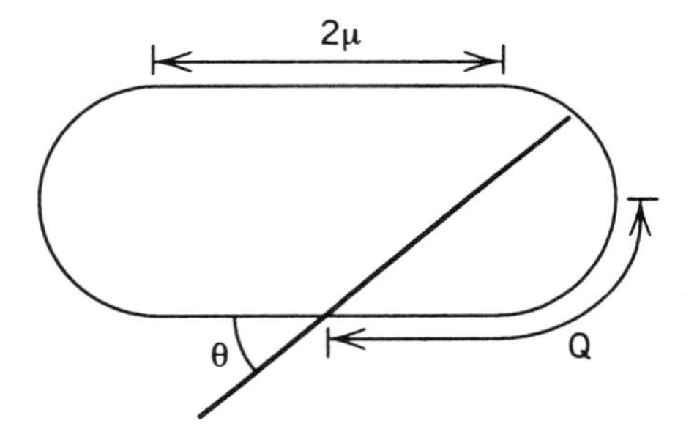

Figure 45. Birkhoff coordinates, normally used for the description of bouncing orbits in the stadium-shaped billiard. The position coordinate Q measures the distance along the perimeter from the middle of the right semicircle. The momentum P is the $\cos\theta$, for energy equals $\frac{1}{2}$.

for an autocorrelation function) sum possible; one only needs to do a root search at a single energy and can scale the solutions to any arbitrary energy. Therefore every heteroclinic orbit we find for the wave propagation corresponds to a family of orbits, all related by the scaling relationship.

Figure 46 shows five snapshots of the time evolution of an initial Gaussian wavepacket launched towards the right cap. Shown are the contours of the real part of the wave function. The progression clearly shows the highly unstable nature of the dynamics; the initially localized wavepacket rapidly delocalizes the covers all of the enclosure available. By $t = 2$, the time of the shortest unstable periodic orbit, the wave function has passed the so-called "Ehrenfest time," i.e., it is no longer recognizable as having originated from a localized state. Beyond this point the obvious classical–quantum correspondence has passed.

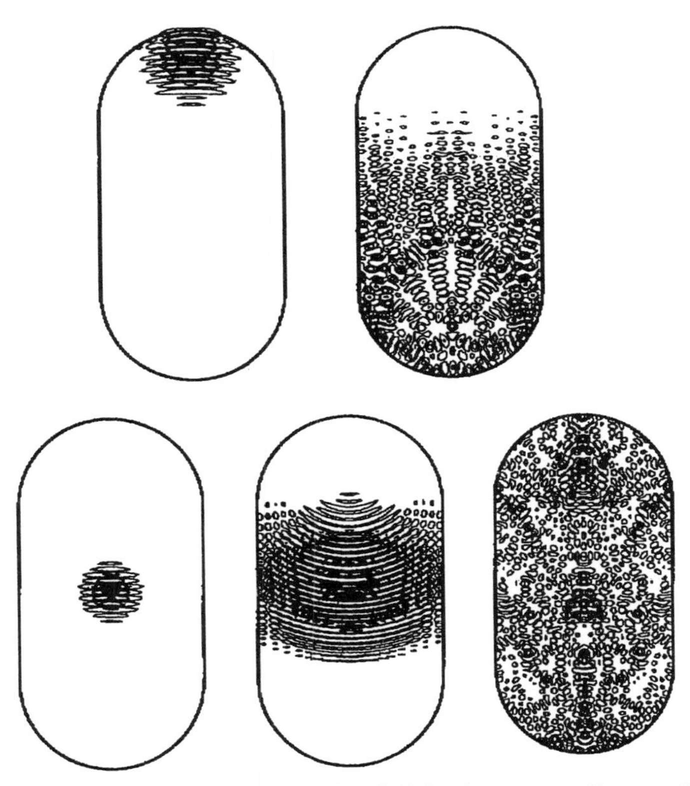

Figure 46. The quantum propagation of an initial coherent state. Four equally spaced contour lines of the real part of the wave function are plotted in successive time snapshots at $t = 0, \frac{1}{2}, 1, 2, 6$ (moving from the upper left to the lower right).

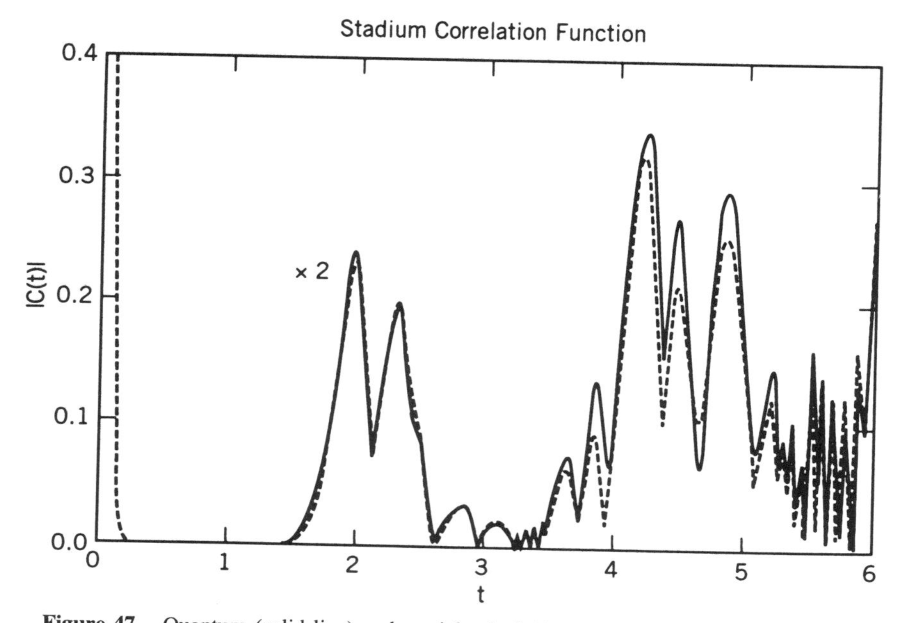

Figure 47. Quantum (solid line) and semiclassical (dashed line) correlation functions.

Calculations of the autocorrelation function $\langle \Psi(0)|\Psi(t)\rangle$, both quantum mechanically and semiclassically are shown in Fig. 47. The dashed line represents the quantum solution, while the solid line is the semiclassical. The agreement between both results is extraordinary, even past the log-time barrier, Figure 48 shows the corresponding semiclassical and quantum spectrum; although, for numerical reasons, the dynamics could not be carried out until the Heisenberg time, it is still possible to see that several eigenstates were resolved by the semiclassical homoclinic sum. According to Tomsovic et al. about a dozen families of orbits contribute to the correlation function by $t \approx 2, 3$. This number increases to several hundred by $t \approx 4$ to 5 and to more than 30,000 families of orbits by $t \approx 6$, i.e., 30,000 orbits crossing the initial disk at each energy! Despite this enormous development of structure within a phase space cell of size $\hbar$ the meticulous accounting of classical contributions adds up to the correct quantum amplitude and phase.

Studying these results we learn an important lesson about the classical–quantum connection: there is a very subtle interference of classical paths within $\hbar$ cells through which information about the classical system is lost; however the final quantum amplitude still depends uniquely on that detail.

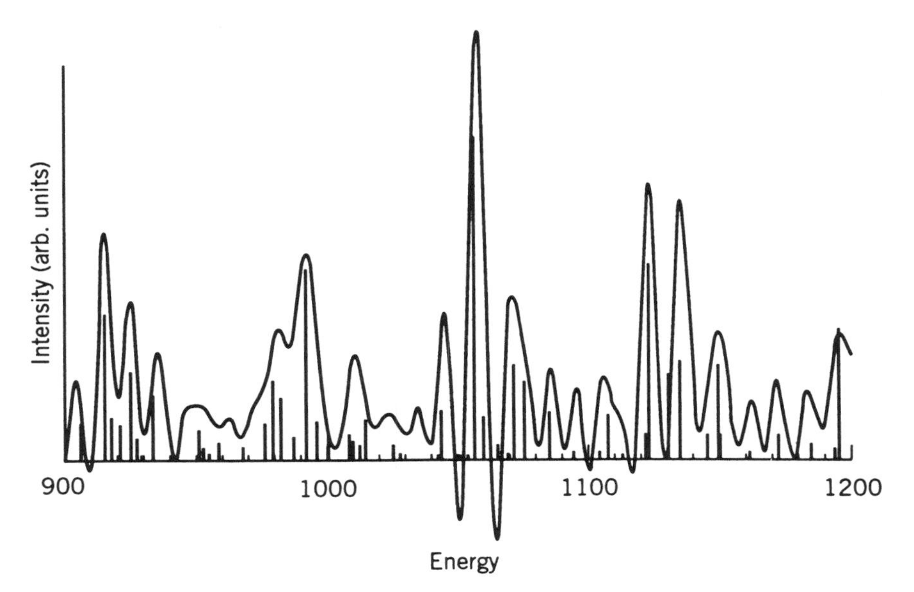

Figure 48. A fraction of the spectrum showing the quantum (vertical lines) spectrum and the semiclassically resolved eigenstates (solid envelop). Since the dynamics was not carried out up to the Heisenberg time not all eigenvalues were isolated.

IX. SUNDRY TOPICS IN TIME-DEPENDENT SEMICLASSICS

The last two topics we will treat in this review are very exciting. First, we want to introduce the reader to the brand new subject of Monte Carlo implementation of the semiclassical Green function. Second, there has been considerable interest over the last few years in the question of whether a semiclassical propagation of a wavepacket over a barrier using only real classical trajectories can reproduce the quantum mechanical tunneling effects. The time-dependent method presented in this review is ideally suited to shed more light on this question, and recent progress will be reported.

A. Monte Carlo Implementation of the Semiclassical Propagator

Not long ago Kinugawa [101] reported a Monte Carlo implementation of the semiclassical Green function. The subject is still very much under development, and some of what we are going to describe here might have developed dramatically by the time of publication of this review. Nevertheless, we want to direct the reader's attention here to an exciting emerging area of research.

In the cellular dynamics method, whether for a single potential or for several coupled surfaces, the basic steps for the representation of the Green function are setting up a constant grid in initial phase space and running the associated classical orbits. The rest of the effort consists of a local expansion and Gaussian quadrature. But what if, instead of using a constant grid, one samples initial phase space stochastically? Does the propagation converge? Are quantum phases reproduced correctly? If the answers to these questions were positive one might ask how many sample points are necessary to converge the propagation? Is it possible to apply this stochastic scheme to systems with many degrees of freedom? These questions were first brought up by R. Coalson [102]. Unfortunately his initial efforts to apply Monte Carlo techniques to the integration of initial value representations like those in Section V.A suffered from large error bars and poor convergence properties as a function of the number of sampling points.

Kinugawa's recent work [101], however, demonstrates that the partial answers to our previous questions are positive! It is possible to integrate the highly oscillatory Van Vleck propagator using Monte Carlo methods, and the agreement of both amplitude and quantum phases with the full quantum result is as satisfactory as the cellular dynamics. In his first account, not only correlation functions but also complete wave functions for a nonlinear one-dimensional case were calculated. Kinugawa used the initial value representation introduced by Heller [30]

$$\begin{aligned}\Psi(q,t) \approx (2\pi i\hbar)^{-1/2} &\int_{-\infty}^{+\infty}\int_{-\infty}^{+\infty} dq'\,dp'\,[M_{21}]^{1/2}\,\delta(q-q_t(q',p'))\\ &\cdot \exp\left(\frac{iS}{\hbar}-\frac{i\pi\mu}{2}\right)\Psi(q',0)\end{aligned} \tag{9.1}$$

where S is the action accumulated along the classical orbit q_t.

Similarly any overlap of the propagated wave function with a localized state can be written as

$$\begin{aligned}\langle\Phi|\Psi(t)\rangle \approx (2\pi i\hbar)^{-1/2} &\int_{-\infty}^{+\infty}\int_{-\infty}^{+\infty} dq'\,dp'\,|M_{21}|^{1/2}\\ &\cdot \Phi^*(q_t)\Psi(q',0)\exp\left(\frac{iS}{\hbar}-\frac{i\pi\mu}{2}\right)\end{aligned} \tag{9.2}$$

The basic idea behind the Monte Carlo integration of Eq. (9.2) is considering the variables of integration q', p' as stochastic variables and the integrand of Eq. (9.2) as its distribution. Taking a large number, N, of samples from the distribution, it is possible to approximate the overlap as a discrete sum

$$\langle \Phi | \Psi(t) \rangle \approx \frac{C}{N} (2\pi i\hbar)^{-1/2} \sum_{q',p'} |M_{21}|^{1/2} \Phi^*(q_t) \Psi(q', 0) \exp\left(\frac{iS}{\hbar} - \frac{i\pi\mu}{2}\right) \tag{9.3}$$

From the form of the initial state it is clear that the distribution function for the variable q' is Gaussian. Meanwhile p' must be selected from a uniform distribution in $(-\infty, +\infty)$. Practically speaking, this is unnecessary because the initial state occupies only a localized region of phase space; particles which have momentum too far from the average initial momentum will contribute neglegibly to the general integral. In Kinugawa's work [101] a smooth cut-off function was used.

As an example, Fig. 49, illustrates a comparison between the semiclassical Monte Carlo and the exact spectra for a hypothetical photoabsorption from the ground state of a harmonic into an upper Morse potential. The intensities and eigenvalues are described quite accurately, both in the high and low end of the spectra. The number of sampling points required for the convergence of the calculation is still to high ($\sim 10^5$) though. The main reason for this is that, as opposed to the cellular dynamics method, each of the classical trajectories in Eq. (9.2) samples an infinitesimally small portion of the integrand. Cellular dynamics on the other hand linearizes the dynamics over a large portion of phase space. A way to improve the convergence rate of the present calculations is, therefore, the combination of stochastic sampling with the linearization idea [103].

B. Tunneling with Classical Trajectories?

The semiclassical treatment of scattering processes stood at the very beginning of the whole field of semiclassical mechanics in the time domain [21]. Nevertheless, up to now the inclusion of the quantum mechanical tunneling effect into semiclassical calculations has been a great conceptual and numerical challenge. Some early progress has been made for tunneling in two-dimensional reactive scattering situations, using trajectories evolving in complex time in Miller's celebrated classical S-matrix formulation [104]. Very recently, Miller and Kesahvamurty [37] raised the intriguing issue of how well the semiclassical propagator can account for tunneling, using only classically allowed trajectories. The authors checked that question for the exactly solvable Eckart barrier [105]. A recent study in the time domain addressing the same question will be reported here [38]. It turns out that the time domain formulation, based on the representation of the S-matrix as the Fourier transform of a correlation function (see Section II), provides the appropriate

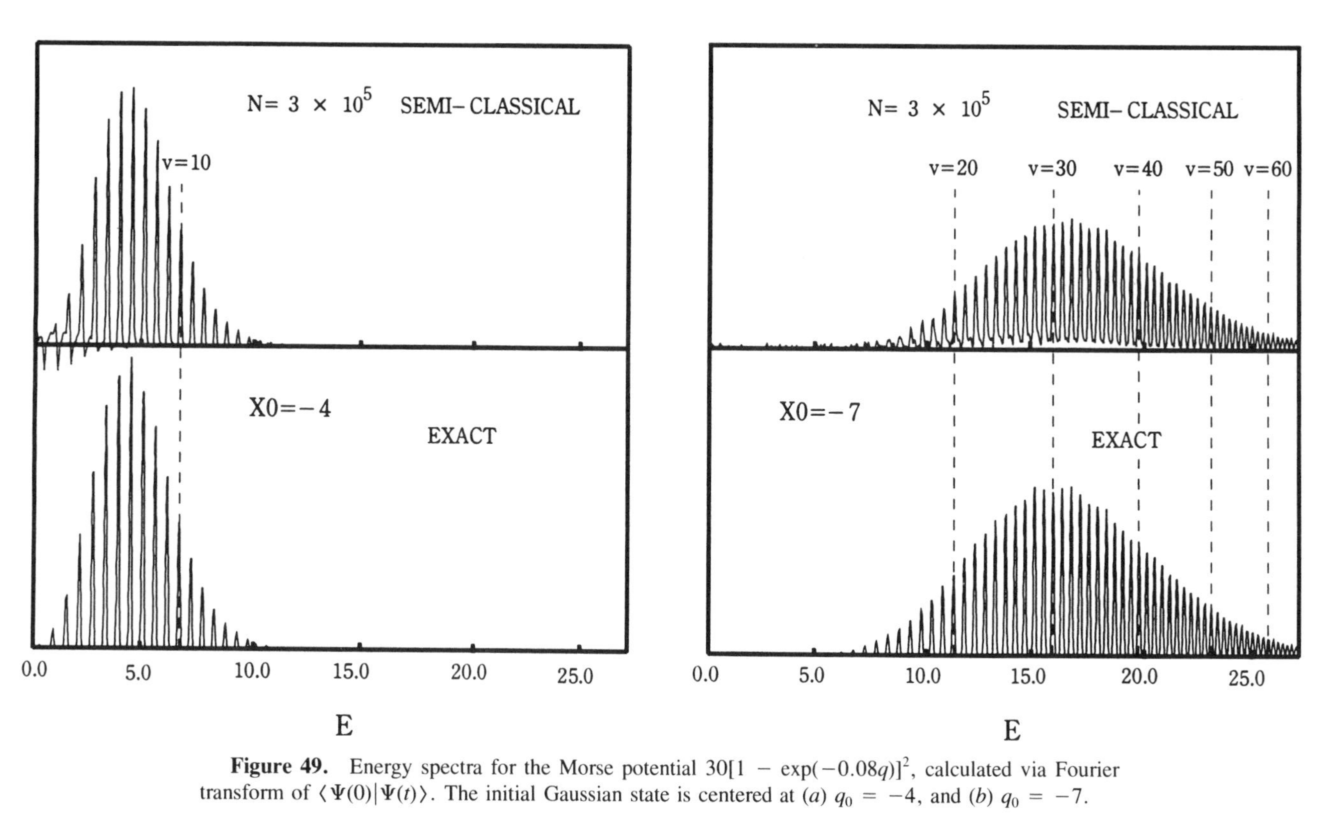

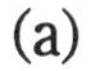

Figure 49. Energy spectra for the Morse potential $30[1 - \exp(-0.08q)]^2$, calculated via Fourier transform of $\langle \Psi(0)|\Psi(t)\rangle$. The initial Gaussian state is centered at (*a*) $q_0 = -4$, and (*b*) $q_0 = -7$.

conceptual means for unravelling the shortcomings of the semiclassical method and pointing out ways for improvements thereof.

In [38] the correlation function formalism of scattering has been applied to a one-dimensional test system given as a symmetric Eckart barrier [105], for which

$$H = T + V(q) = \frac{p^2}{2m} + V_0 \cosh^{-2}(\lambda q) \tag{9.4}$$

The parameters were so chosen that the Hamiltonian represents a simple one-dimensional model of the H_2 + H reaction studied previously [37]. The parameters then are $V_0 = 0.425$ eV, $m = 1060\ m_e$, and $\lambda = 1.3624\ a_0^{-1}$.

In this potential, the propagation of Gaussian wavepackets of the form

$$\phi_a(q) = \left(\frac{2\alpha}{\pi}\right)^{1/4} \exp[-\alpha(q - q_\alpha)^2 + ip_\alpha(q - q_\alpha)] \tag{9.5}$$

with dimensionless width parameter α, dimensionless center coordinate q_α (position in units of a_0), and dimensionless p_α (momentum in units of $\hbar/a_0$), centered far out in the reactant region was studied. The final wave functions Φ_β were of analogous form but centered far out in the product region.

In principle to perform the propagation semiclassically, all possible classical boundary value problems, with $q(0) = q_i$, $q(t) = q_f$, have to be solved for all times t, and the corresponding actions, stability matrix elements, and Maslov indexes have to be calculated. By choosing Gaussian initial and final wave functions, however, the domain over which one needs to perform this task is restricted. Furthermore the correlation function turns out to be important only over a finite time interval. This, along with the fact that much of the root search for the problem studied can be done analytically, allowed the authors to take this route for the study of the fundamental question asked in [38]. In a multidimensional system, however, where the root search becomes prohibitive, a method like cellular dynamics seems to be ideally suited for the numerical calculation of the semiclassical correlation function.

In continuing with the root search method, it was noted that the classical equations of motion can be derived exactly analytically for the Hamiltonian Eq. (9.4). Integrating the law of conservation of energy for the Eckart barrier yields

$$\cosh\left(\sqrt{\frac{2E_{\rm cl}}{m}}\lambda t\right)\sinh(\lambda q_i) + \sinh\left(\sqrt{\frac{2E_{\rm cl}}{m}}\lambda t\right)\sqrt{\cosh^2(\lambda q_i) - \frac{V_0}{E_{\rm cl}}} - \sinh(\lambda q_f) = 0 \tag{9.6}$$

This is an implicit equation for the energy E_{cl}, which is the only undetermined quantity in the classical problem, since the initial and final position and the time t are dictated by the argument of the propagator. In principle one must solve this highly nonlinear equation in E_{cl} for all possible roots. Following Keshavamurthy and Miller [37], a restriction was made to real solutions of Eq. (9.6), of which there is only one for a given time. In calculating the semiclassical overlap, only trajectories that go over the barrier with an energy $E_{cl} > V_0$ were used. This is because the initial (final) wave function is sharply peaked in the reactant (product) region. The root search for these energies was done numerically. After having calculated the correlation function, the Fourier transform was performed numerically and the S-matrix was determined as prescribed in Eq. (2.5a). Besides the usage of the VVG propagator, the only approximation in the semiclassical calculation of [38] is the restriction to real trajectories. All other parts of the calculation of the transmission probability, $P(E) = |S_{\beta\alpha}|^2$, were either numerically or analytically exact. The Fourier transform of the correlation function was approximated using a discrete Fourier series; this implies introducing a time cutoff for the correlation function, $C(t)$. However, this correlation function is well localized in time and the cutoff was moved to regions where $C(t)$ vanishes; therefore a finite discrete Fourier series is a good approximation.

Figure 50 shows a comparison of the semiclassically calculated correlation function, $c(t)$, with the corresponding quantum result. For short times one finds very good agreement between both results. At longer times, however, the semiclassical result starts to lag behind the quantum one. It is important to note that because of the normalization procedure prescribed in Eq. (5a), even small changes in the time signal, $c(t)$, can lead to pronounced differences in $P(E)$. Figure 51 shows the transmission probability calculated from the two correlation functions of Fig. 51 and from another pair (semiclassical and quantum) with a different set of center parameters. It is stressed that the quantum results for the tunneling probability are independent of the choice of q_α, q_β. This is as it should be, because no approximation was made in performing the propagation. Unfortunately the semiclassical result depends on this choice. To illustrate this fact $P(E = 0.26 \text{ eV})$ is listed for five different values of q_α in Table V. Over the range of q_α shown, the results differ roughtly by a factor of 2. In total, one finds that the semiclassical $P(E)$ decreases (increases) as a function of q_α for E smaller (larger) than approximately V_0. For increasing q_α, the results are therefore slowly tending toward the classical result, which is a step function at V_0. The semiclassical correlation function in Fig. 50 as compared to the quantum result is revealing. There is a small error seen at long times. Since the Fourier transform from time to energy (below the barrier top) filters this signal at a low Fourier frequency, the long-time, lower-frequency part of the correlation

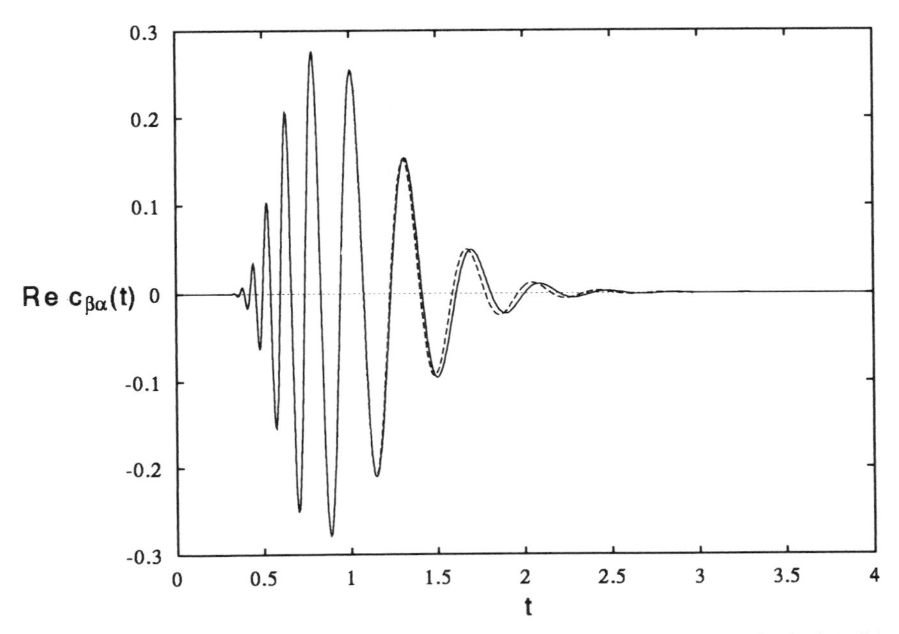

Figure 50. Real part of the semiclassical (dashed line) and quantum mechanical (solid line) time correlation function that enters the calculation of the *S*-matrix for the Eckart barrier. The dimensionless wavepacket parameters are: $\alpha = \beta = 10$; $q_\alpha = -q_\beta = -3$; $p_\alpha = p_\beta = 6.5$.

function is most important. Its errors signal the error in the derived tunneling probabilities.

The Eckart barrier with the parameters studied is rather sharply peaked and a harmonic approximation around the barrier top would break down rapidly. In the limit of small λ, the barrier is better approximated by a quadratic over a longer range of coordinates and the results of the present "over the barrier" method converge to exact quantum result. It has to be noted, however, that even for very anharmonic potential parameters the semiclassical result does not need uniformization to show a smooth transition from energies below to energies above the barrier, which is unlike the standard WKB case. The qualitative correspondence of the semiclassical result with the quantum one was also noted in [37]. These facts, and the slow dependence of the results on the initial state parameter, make the semiclassical method practically interesting as it stands, but we would like to do better and get rid of the nonconvergence, which was first exposed in reference [38]. In order to do so, it was mentioned that there are other "roots" to Eq. (9.6), and it is possible that inclusion of these analytic continuations

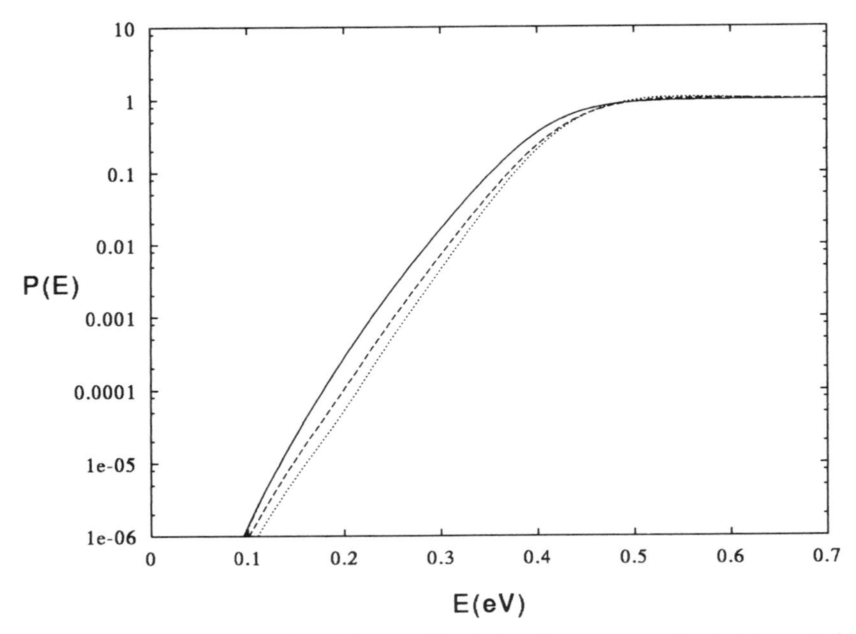

Figure 51. Logarithmic plot of the semiclassical and quantum mechanical transmission probabilities for the Eckart barrier for two different sets of wavepacket parameters. The first set is derived from the time-domain results presented and is depicted with dashed (semiclassical) and solid (quantum) lines. The second set of parameters is identical to the first except for the center parameters, which were chosen as $q_\alpha = -q_\beta = -6$. The corresponding transmission probabilities are given as dotted (semiclassical) and dashed-dotted (quantum) lines. The two quantum results are nearly identical and coincide to within line thickness.

of the usual classical mechanics might be incorporated to improve the results, especially for sharp barriers. For example, by restriction to a symmetric trajectory (with real $q_f = -q_i$) one finds that all negative energies with $E_{\mathrm{cl}} = m(n\pi)^2/2\lambda^2 ta2$, where $n = 1, 3, 5, 7, \ldots$, solve the root search for every q_i in real time t. For these nonclassical trajectories (with imaginary initial momentum) all the components of the semiclassical propagator are independent of q_i except the prefactor, which is proportional to

TABLE V
Semiclassical Transmission Probability for the Eckart Barrier at Energy $E = 0.26$ eV on the Center Coordinate, q_α, of the Initial Wavepacket

q_α	-3	-4	-5	-6	-7
$P(E = 0.26\ \text{eV})$	1.43×10^{-3}	1.16×10^{-3}	9.60×10^{-4}	8.21×10^{-4}	7.20×10^{-4}

$\sqrt{E_{el}/(1 - V(q_i)}$; therefore, a correction based on these trajectories will become more important for initial states centered further away from the barrier top. Thus, they are potent candidates to remove the nonconvergence.

X. CONCLUSION

In this review, we described the methodology and recent progress made in the semiclassical mechanics of a variety of nontrivial systems of general interest. First, the basic object of semiclassical theory, the Van Vleck–Gutzwiller propagator was reviewed, and its original inception as well as the derivation from the path integral were sketched. The connection between the propagator and elements of the monodromy matrix, important in studying the stability properties of classical systems, has been reviewed. In this context some remarks on the correct calculation of the Maslov index and its alteration under the change of representations have been made. It is only by including these indexes that the semiclassical propagator can be used beyond caustics, which are omnipresent in classical mechanical systems.

At the core of our numerical methodology stands the newly developed cellular dynamics method. This was set forth along with the error function corrections needed for the cure of cutoff effects introduced through the use of a finite initial manifold. Cellular dynamics was contrasted with the single trajectory GWD methodology. Very long propagation times can be achieved using the new method, whereas the application of GWD is restricted to short times or systems with maximally quadratic Hamiltonians. For mixed phase space systems we showed that cellular dynamics makes it possible to achieve astonishing coincidence with fully quantum mechanical results. In particular, the semiclassically calculated spectrum of the Barbanis potential demonstrates the capability of the CD to tackle mixed dynamical systems with strongly chaotic regions in phase space. Examination of the benzophenone spectrum using classical orbits allowed the assignment of dynamical features in the hierarchical tree of its spectrum. We mentioned also several other approaches to the semiclassical calculation of correlation functions. The heteroclinic sum by Tomsovic and Heller [15] turned out to be very useful for purely chaotic systems. Even a full-fledged root search could be done (semianalytically) for the barrier penetration problem in one dimension. Neither of the two methods, however, has the same promising prospects for the application to true multidimensional problems as does cellular dynamics.

We have furthermore introduced the generalized Gutzwiller propagator, which allows the semiclassical treatment of nonlinear spectroscopies. For a long time, spectroscopic applications of semiclassics were limited to single potential propagations. This is no longer the case, thanks to the combination of the CD method and the standard ordering of the nonlinear interactions with the fields.

The application of cellular dynamics was furthermore extended to dissipative systems. The newly developed simple semiclassical Van Vleck–Gutzwiller propagator for a system interacting with a bath of harmonic oscillators was used for the propagation of the reduced density matrix of a Morse oscillator. For a high-temperature heat bath with Ohmic spectral density, the comparison with fully quantum mechanical and purely classical results shows a good reproduction of the remaining quantum mechanical features in the simple semiclassical result. There is great potential for semiclassics in dissipative systems because in a density matrix formulation the dimension of the test system is already twice the number of ordinary dimensions and the computational advantages of the semiclassical method might make calculations feasible. Furthermore, by including the imaginary part of the action in the extremization procedure a wider area of (γ, T) parameter space could probably be investigated. One would need to deal with complex trajectories, however; the complexity being introduced through the temperature-dependent imaginary part of the potential. This is not an easy task and several fundamental problems would have to be dealt with, e.g., the possible crossing of Stokes lines in the complex plane. The growing industry of Monte Carlo-like procedures in dissipative quantum theory that deal with wave functions instead of density matrices [106] might provide interesting applications for the semiclassical theory presented here. The problem with these stochastic methods lies in taking on ensemble average over many different Monte Carlo wave function realizations. These time consuming steps could be shortened considerably by employing semiclassical propagation techniques.

With a more fundamental perspective we have reviewed some recent work concerning the validity of the Van Vleck formula for purely chaotic systems. We now know that semiclassical propagation of wave functions and resolution of eigenvalues can be carried out to time scales not previously expected. The new results give us a better insight not only into the correspondence between classical and quantum mechanics but also into the uncertainty principle. Similarly, driven by the fundamental question of whether tunneling can be accounted for semiclassically by using real trajectories, we have reviewed a very recent study. It seems that convergence of the results can only be achieved by incorporating complex trajectories. The complexity would now arise due to the neccessity of circumventing the barrier in complex phase space.

Finally, we described some recent results on the Monte Carlo implementation of the cellular dynamics method. The convergence of the semiclassical propagation in this case is not as good as in the CD method, but expect future advances in the field to close the gap between both methods and offer a fantastic tool to treat a large number of degrees of freedom. Both methods are easily adaptable for parallel computation and offer us perhaps the only chance to simulate large molecular systems while including the most important quantum effects. Let us remember that as the mass of the molecule increases one is effectly moving into the realm of semiclassical approximations ($\hbar \rightarrow 0$). This is the area towards which we expect to see most future efforts being devoted.

ACKNOWLEDGMENTS

This work was supported by the Japanese Society for the Promotion of Science, by a Grant-in-Aid from the Ministery of Education, Science and Culture of Japan, the Alexander von Humboldt Foundation (Bonn, Germany) and the Deutsche Forschungsgemeinschaft under Sonderforschungsbereich 276.

We thank our colleagues Mr. T. Kinugawa and Professor S. Tomsovic for providing us with the figures releated to their work and valuable comments on some of their forthcoming publications. Special thanks go to Professors S. Rice and E. J. Heller for getting us started on this project. We are also indebted to Professors Heller and Tomsovic as well as Drs. Pat O'Connor and Jan-Michael Rost for many invaluable discussions on semiclassical methodology.

One of us (M. A. S.) wishes to thank Professor K. Fukui and Dr. Tasaki for their support while at the Institute for Fundamental Chemistry.

REFERENCES

1. M. C. Gutzwiller, *Chaos in Classical and Quantum Mechanics*, Interdisciplineary Applied Mathematics, Vol. 1, Springer-Verlag, New York (1990).
2. M. V. Berry and K. B. Mount, *Rep. Prog. Phys.* **35,** 315 (1972); M. V. Berry, in T. H. Seligman and H. Nishioka, Eds., *Quantum Chaos and Statistical Nuclear Physics*, Lecture Notes in Physics **263,** Springer-Verlag, Berlin (1986).
3. R. G. Littlejohn, *Phys. Rep.* **138,** 193 (1986).
4. J. B. Delos, *Adv. Chem. Phys.* **65,** 161 (1986).
5. E. J. Heller, *Accts. Chem. Res.* **14,** 368 (1981).
6. W. H. Miller, *Adv. Chem. Phys.* **25,** 69 (1974).
7. E. J. Heller, in: M. J. Giannoni, A. Voros and J. Zinn-Justin Eds., *Chaos and Quantum Physics*, Les Houches, Session LII, 1989, Elsevier, 1991.
8. A. Einstein, *Verth. Dtsch. Phys. Ges.* **19,** 82 (1917); J. B. Keller, *Ann. Phy.* (NY) **4,** 180 (1958); J. B. Keller and S. C. Rubinow, *Ann. Phys.* (NY) **9,** 24 (1960).
9. V. P. Maslov, *Théorie des Perturbations et Méthodes Asymptotiques*, Dunod, Paris (1972); V. P. Maslov and M. V. Fedoriuk Eds., *Semi-classical Approximations in Quantum Mechanics*, Reidel, Boston 1981.
10. M. Gutzwiller *J. Math. Phys.* **11,** 1791 (1970); **12,** 343 (1970).
11. Comp. Phys. Comm. **63** (1991) (thematic issue on time-dependent methods for quantum dynamics, ed. by K. C. Kulander).
12. R. P. Feynman, *Rev. Mod. Phys.* **20,** 367 (1948).
13. C. Morette, *Phys. Rev.* **81,** 848 (1951).

14. J. H. Van Vleck, *Proc. Nat. Acad. Sci.*, **14,** 178 (1928).
15. S. Tomsovic, E. J. Heller, *Phys. Rev. E* **47,** 282 (1993).
16. E. J. Heller, S. Tomsovic, and M. A. Sepúlveda, *Chaos* **2,** 105 (1992).
17. M. A. Sepúlveda, and E. J. Heller, *J. Chem. Phys.* **101,** 8004 (1994).
18. M. A. Sepúlveda, and E. J. Heller, *J. Chem. Phys.* **101,** 8016 (1994).
19. K. G. Kay, *J. Chem. Phys.* **101,** 2250 (1994); *ibid.* **100,** 4432 (1994).
20. M. V. Berry and M. Tabor, *Proc. Roy. Soc. London A* **349,** 101 (1976).
21. E. J. Heller, *J. Chem. Phys.* **62,** 1544 (1975).
22. E. J. Heller, *J. Chem. Phys.* **91,** 27 (1981).
23. M. F. Herman and E. Kluk, *Chem. Phys.* **91,** 27 (1984); E. Kluk, M. F. Herman, and M. Davis, *J. Chem. Phys.* **84,** 326 (1986).
24. D. Huber and E. J. Heller, *J. Chem. Phys.* **87,** 5302 (1987).
25. R. Schinke, *Photodissociation Dynamics*, Cambridge University Press, New York (1993).
26. N. E. Henriksen and E. J. Heller, *J. Chem. Phys.* **91,** 4700 (1989).
27. Soo-Y. Lee, E. J. Heller, *J. Chem. Phys.* **71,** 4777 (1979); E. J. Heller, R. L. Sundberg, and D. J. Tannor *J. Phys. Chem.* **86,** 1822 (1982) and references therein.
28. G. Drolshagen and E. J. Heller, *J. Chem. Phys.* **79,** 2072 (1983); *Surface Science* **139,** 260 (1984); *J. Chem. Phys.* **82,** 226 (1985).
29. R. T. Skodje and D. G. Truhlar, *J. Chem. Phys.* **80,** 3123 (1984).
30. E. J. Heller, *J. Chem. Phys.* **94,** 2723 (1991).
31. O. Zobay, Diplomas Thesis, University of Freiburg, 1992; for a periodic orbit type calculation see: O. Zobay and G. Alber, *J. Phys. B* **26,** L539 (1993).
32. R. Sadeghi and R. T. Skodje, *J. Chem. Phys.* **99,** 5126 (1993).
33. A. Isele, C. Meier, V. Engel, N. Fahrer, Ch. Schlier, *J. Chem. Phys.* **101,** 5919 (1994).
34. D. J. Tannor and D. E. Weeks, *J. Chem. Phys.* **98,** 3884 (1993).
35. E. J. Heller, *J. Chem. Phys.* **62,** 1544 (1975).
36. W. H. Miller, S. D. Schwartz, and J. W. Tromp, *J. Chem. Phys.* **79,** 4889 (1983).
37. S. Keshavamurthy and W. H. Miller, *Chem. Phys. Lett.* **218,** 189 (1994).
38. F. Grossmann and E. J. Heller, *Chem. Phys. Lett.* **241,** 45 (1995).
39. G. Wentzel, *Zeits. Physik*, **38,** 518 (1926).
40. L. Brillouin, *J. Physique* **7,** 353 (1926).
41. H. A. Kramers, *Zeits. Physik* **39,** 828 (1926).
42. P. A. M. Dirac, *Proc. Roy. Soc. London* **113A,** 621 (1927).
43. P. Jordan, *Zeits Physik* **40,** 809 (1927); *ibid* **44,** 1 (1927).
44. A. J. Lichtenberg, and M. A. Lieberman, *Regular and Chaotic Dynamics*, Springer-Verlag N.Y. 2nd Edition (1992).
45. F. Grossman, *J. Chem. Phys.* **xx,** yyyy (1995).
46. M. C. Gutzwiller, *J. Math. Phys.* **8,** 1979 (1967).
47. M. C. Gutzwiller, *J. Math. Phys.* **10,** 1004 (1969).
48. M. Morse, *Am. Math. Soc. Colloquium Publ.* **18**, (1934).
49. R. G. Littlejohn, *J. Stat. Phys.* **68,** 7 (1992).
50. M. A. Sepúlveda, Ph.D. Thesis, University of Washington, Seattle (1993).
51. E. J. Heller, *J. Chem. Phys.* **68,** 2066 (1978).

52. J. D. Kress and A. E. DePristo, *J. Chem. Phys.* **89,** 2866 (1988).
53. E. J. Heller, J. R. Reimers, and G. Drolshagen, *Phys. Rev. A* **36,** 2613 (1987).
54. M. Messina and K. Wilson, *Chem. Phys. Lett.* **41,** 502 (1995).
55. J. H. Frederick, E. J. Heller, J. L. Ozment, and D. W. Pratt, *J. Chem. Phys.* **88,** 2169 (1988).
56. E. J. Heller, E. B. Stechel, and M. J. Davis, *J. Chem. Phys.* **73,** 4720 (1980).
57. M. J. Davis, *J. Chem. Phys.* **98,** 2614 (1993).
58. E. Shalev, J. Klafter, D. F. Plusquellic, D. W. Pratt, *Physica A* **191,** 186 (1992).
59. K. W. Holtzclaw and D. W. Pratt, *J. Chem. Phys.* **84,** 4713 (1986).
60. S. Kamei, T. Sato, N. Mikami, and M. Ito, *J. Chem. Phys.* **90,** 5615 (1986).
61. D. W. Noid, M. L. Koszykowski, and R. A. Marcus, *J. Chem. Phys.* **78** 4018 (1983); T. Uzer, D. W. Noid, and R. A. Marcus *J. Chem. Phys.* **79,** 4412 (1983); D. Farrelly and T. Uzer, *J. Chem. Phys.* **85,** 308 (1986).
62. J. C. Tully, Nonadiabatic Processes in Molecular Collision, in *Dynamics of Molecular Collisions, Part B*, W. H. Miller, Ed., Plenum, New York 1974.
63. M. S. Child, *Molecular Collision Theory*, Academic Press, London (1974); B. C. Eu *Semiclassical Theories of Molecular Scattering*, Springer-Verlag, Berlin (1984).
64. L. D. Landau, *Phys. Z. Sow.* **2,** 46 (1932); C. Zener, *Proc. R. Soc. Ser.* **AB7,** 696 (1932); E. C. G. Stueckelberg, *Helv. Phys. Acta* **5,** 369 (1932); O. Atabeck, R. Lefebvre, and M. Jacon *J. Chem. Phys.* **81,** 3874 (1984).
65. M. A. Sepúlveda and S. Mukamel, *J. Chem. Phys.* **102,** 9327 (1995).
66. M. A. Sepúlveda, Submitted.
67. S. Mukamel, *Principles of Nonlinear Optical Spectroscopy*, Oxford University Press, Oxford, 1995.
68. R. W. Boyd, Ed., *Nonlinear Optics*, Academic Press; New York 1992.
69. N. Bloembergen, *Nonlinear Optics*, Benjamin, New York, 1965.
70. M. A. Sepúlveda, S. Tasaki, in preparation.
71. M. Berman and R. Kosloff, *Comp. Phys. Comm.* **63,** 1 (1991); R. Kosloff, *Annu. Rev. Phys. Chem.* **45,** 145 (1994).
72. N. F. Scherer, R. J. Carlson, A. Matro, M. Du, A. J. Ruggiero, V. Romero-Rochin, J. A. Cina, G. R. Flemming, and S. Rice, *J. Chem. Phys.* **95,** 1487 (1991); N. F. Scherer, A. Matro, L. D. Ziegler, M. Du, R. J. Carlson, J. A. Cina, and G. R. Flemming, *J. Chem. Phys.* **96,** 4180 (1992).
73. M. Gruebele and A. H. Zewail, *J. Chem. Phys.* **98,** 883 (1992); Y. Yan and S. Mukamel *J. Chem. Phys.* **94,** 179 (1990); Y. Yan and S. Mukamel *J. Chem. Phys.* **41,** 6485 (1990).
74. T. Baumert, V. Engel, C. Meier, G. Gerber *Chem. Phys. Lett.* **200,** 488 (1992); V. Engel and H. Metiu *J. Chem. Phys.* **100,** 5448 (1994); C. Meier and V. Engel, *Chem. Phys. Lett.* **212,** 691 (1993).
75. Ephraim Hyabaev and Uzi Kaldor *J. Chem. Phys.* **98,** 7126 (1993); S. Magnier, Ph. Millié, O. Dulieu and F. Masnou-Seeuws, *J. Chem. Phys.* **98,** 7113 (1993).
76. A. O. Caldeira and A. J. Leggett, *Physica A* **121,** 587 (1983); *ibid.* **130** 374(E), (1985).
77. A. O. Caldeira and A. J. Leggett, *Ann. Phys.* **149,** 374 (1983); *ibid.* **153** 445(E), (1984).
78. R. P. Feynman and F. L. Vernon, *Ann. Phys.* **24,** 118 (1963).
79. S. H. Lin and H. Eyring, *Ann. Rev. Phys. Chem.* **25,** 39 (1974).

80. P. Riseborough, P. Hängi, J. Weizz, *Phys. Rev. A* **31,** 471 (1985).
81. P. Hänggi, P. Talkner, and M. Borkovec, *Rev. Mod. Phys.* **62,** 251 (1990) and references therein.
82. A. J. Leggett et al. *Rev. Mod. Phys.* **59,** 1 (1987).
83. H. Grabert and H. Wipf, in *Festkörperprobleme—Advances in Solid State Physics*, Vol. 30, Vieweg, Braunschweig, 1990, p. 1.
84. I. R. Senitzky, *Phys. Rev.* **119,** 670 (1960); **124,** 642 (1961).
85. G. W. Ford, M. Kac, and P. Mazur, *J. Math. Phys.* **6,** 504 (1965).
86. R. Zwanzig, *J. Stat. Phys.* **9,** 215 (1973).
87. H. Grabert, P. Schramm, and G.-L. Ingold, *Phys. Rep.* **168,** 115 (1988).
88. L. Diosi, *Europhys. Lett.* **22,** 1 (1993).
89. L. Diosi, *Physica A* **199,** 517 (1993).
90. G. Lindblad, *Commun. Math. Phys.* **48,** 119 (1976).
91. T. Koeling and R. A. Malfliet, *Phys. Rep.* **22,** 181 (1975).
92. S. Levit and U. Smilansky, *Ann. Phys.* **103,** 198 (1977); *Proc. Amer. Math. Soc.* **65,** 299 (1977).
93. S. Levit, K. Möhring, U. Smilansky, and T. Dreyfus, *Ann. Phys.* **114,** 223 (1978).
94. H. A. Kramers, *Physica* **7,** 284 (1940).
95. P. W. O'Connor, S. Tomsovic, and E. J. Heller, *Physica D* **55,** 340 (1992); P. W. O'Connor and S. Tomsovic *Ann. Phys. (N.Y.)* **207,** 218 (1991).
96. M. A. Sepúlveda, S. Tomsovic, and E. J. Heller, *Phys. Rev. Lett.* **69,** 402 (1992).
97. M. V. Berry, N. L. Balazs, *J. Phys. A* **12,** 625 (1979).
98. M. V. Berry, N. L. Balazs, M. Tabor, A. Voros *Ann. Phys. (N.Y.)* **122,** 26 (1979); M. V. Berry, *Ann. Phys. (N.Y.)* **357,** 183 (1983).
99. S. Tomsovic and E. J. Heller *Phys. Rev. Lett.* **67,** 664 (1991).
100. L. A. Bunimovich, *Funct. Anal. Appl.* **8,** 254 (1974); *Commun. Math. Phys.* **65,** 295 (1979).
101. T. Kinugawa, *Chem. Phys. Lett.* **235,** 395 (1995).
102. R. Coalson, private communication (1992).
103. T. Kinugawa, in progress.
104. T. F. George and W. H. Miller, *J. Chem. Phys.* **57,** 2458 (1972).
105. C. Eckart, *Phys. Rev.* **35,** 1303 (1930).
106. K. Molmer, Y. Castin, and J. Dalibard, *J. Opt. Soc. Am. B* **10,** 524, (1993).

AUTHOR INDEX

Numbers in parentheses are reference numbers and indicate that the author's work is referred to although his name is not mentioned in the text. Numbers in *italic* show the pages on which the complete references are listed.

SUBJECT INDEX